Integrating Evolution and Development

Integrating Evolution and Development

From Theory to Practice

edited by Roger Sansom and Robert N. Brandon

WITHDRAWN

A Bradford Book
The MIT Press
Cambridge, Massachusetts
London, England

© 2007 Massachusetts Institute of Technology

All rights reserved. No part of this book may be reproduced in any form by any electronic or mechanical means (including photocopying, recording, or information storage and retrieval) without permission in writing from the publisher.

MIT Press books may be purchased at special quantity discounts for business or sales promotional use. For information, please e-mail special_sales@mitpress.mit.edu or write to Special Sales Department, The MIT Press, 55 Hayward Street, Cambridge, MA 02142.

This book was set in Stone Serif and Stone Sans on 3B2 by Asco Typesetters, Hong Kong, and was printed and bound in the United States of America.

Library of Congress Cataloging-in-Publication Data

Integrating evolution and development : from theory to practice / by Roger Sansom and Robert N. Brandon, editors.
  p. cm.
"A Bradford book."
Includes bibliographical references and index.
ISBN-13: 978-0-262-19560-7 (hardcover : alk. paper)
ISBN-13: 978-0-262-69353-0 (pbk. : alk. paper)
1. Evolution (Biology) 2. Developmental biology. I. Sansom, Roger. II. Brandon, Robert N.
QH366.2I52 2007
576.8'2—dc22   2006030814

10 9 8 7 6 5 4 3 2 1

Contents

Preface vii
Acknowledgments xiii

1 Embryos, Cells, Genes, and Organisms: Reflections on the History of Evolutionary Developmental Biology 1
Manfred D. Laubichler and Jane Maienschein

2 The Organismic Systems Approach: Streamlining the Naturalistic Agenda 25
Werner Callebaut, Gerd B. Müller, and Stuart A. Newman

3 Complex Traits: Genetics, Development, and Evolution 93
H. Frederik Nijhout

4 Functional and Developmental Constraints on Life-Cycle Evolution: An Attempt on the Architecture of Constraints 113
Gerhard Schlosser

5 Legacies of Adaptive Development 173
Roger Sansom

6 Evo-Devo Meets the Mind: Toward a Developmental Evolutionary Psychology 195
Paul E. Griffiths

7 Reproducing Entrenchments to Scaffold Culture: The Central Role of Development in Cultural Evolution 227
William C. Wimsatt and James R. Griesemer

Index 325

Preface

Development and *evolution* both, fundamentally, mean change, and both terms have long been applied to change in life. Over the nineteenth and twentieth centuries, these terms came to refer to quite different processes. Development is the process that an individual organism goes through over the course of its life, and evolution is the process that a population goes through as its members reproduce and die. August Weismann successfully argued for the conceptual separation of the germ line, which can evolve, from the soma, which can develop.

Advances in genetics produced the modern synthesis, which combined genetics and the theory of evolution by seeing evolution as a change in the genomes of a population over time. An early champion of this separation was G. C. Williams, and its current popular form is Dawkins's selfish gene. Dawkins (1976) argued that the gene is *the* unit of natural selection. On this view, natural selection looked through the organism right to the genome. Thus, the process of development was rendered epiphenominal to the process of evolution. The development of an organism with various traits was just a mechanism to differentiate the fitness of genes. Developmental biology continued to be studied, but, other than assuming various general ideas about adaptation, that study was largely segregated from the study of evolution. The conceptual separation of development and biology was widely seen as an important step in the rapid advance of biology in the twentieth century, because it allowed evolution to be studied without getting bogged down in the messy details of development.

This segregation has always had its dissenters, however. To see genes as the unit of natural selection misconstrues their role in evolution. Gould (2001, 203) echoed an argument from Wimsatt (1980) when he said, "Units of selection must be actors within the guts of the mechanism, not items in the calculus of results." Gould went on to do other work that is beyond the view of the modern synthesis. He coined a new intellectual sin, "adaptationism," which involves seeing all traits as adaptations, rather than as the results of developmental constraints, and his view of punctuated equilibrium challenged the steady-as-she-goes, gene-by-gene sorting implied by the modern synthesis. Ironically, it was advances in genetics that lead to a greater interest the mechanisms of developmental genetics and their evolution, which brought development back into the fold. The remarkable conservation of developmental genes, such as Hox genes, cried out for an explanation that could only be given by considering both evolution and development.

There is now growing interest in the developmental synthesis (also known as evo-devo). Old ideas, such as bauplan, are being reviewed in a new light, and relatively new ideas, such as canalization, modularity, and evolvability, all essentially involving both evolution and development, are increasingly being incorporated into theoretical and empirical work. Evolutionary biologists are investigating developmental constraints and discovering how evolutionary transitions came about. A nice example of such work is Brylski and Hall's (1988a, 1988b) study on the evolution of external furry cheek pouches in geomyoid rodents. Pocket gophers and kangaroo rats store food in external cheek pouches. Developmental data showed that these pouches evolved from internal cheek pouches, which are not as adaptive because they are smaller and lose moisture to the food. Both types of cheek pouches develop from the bruccal epithelium by epithelial evagination. Brylski and Hall discovered that the change to external pouches was due to a small change in location and magnitude of epithelial evagination at the corner of the mouth to include the lip epithelium. The corner of the mouth then became the opening to the external pouch as the lips and snout grew. That small change in the developmental mechanism produced a significant coordinated change in adult morphology, thereby contributing to the direction taken by evolution. In particular the developmental mechanism determined that the first external pouch was lined with fur (see Robert 2002 for further discussion of this and other examples of the developmental mechanisms of evolutionary novelty).

Widespread interest in the developmental synthesis is a relatively new phenomenon. It remains unclear just how much of a revision of evolutionary theory it requires (see Sterelny 2000 and Robert 2002 for opposing views). We hope that this book will act as a focus for this growing project. It is a relatively young project, and like so many young things, it is still unclear what it will be when it grows up. The more modest result of the new developmental synthesis is that developmental theory will supplement evolutionary theory. That is, theoretical and empirical work on development will answer questions that have troubled evolutionary theorist or soon will. Almost certainly, the more modest project will be successful in some manner; work on development is bound to contribute to our understanding of evolution, because, after all, evolution is a process of the evolution of things that develop.

The more ambitious and more significant result of the developmental synthesis would be a fundamental theoretical rethinking of evolution itself. The developmental systems approach of Susan Oyama, Paul Griffiths, and Russell Gray (2001) is an example of this, although it seeks to integrate more than development into its reinterpretation of evolution. James Griesemer (2000) offers a purer developmental synthesis that hinges on what is distinctive to evolution and development (i.e., reproducers). The cases for these more extravagant developmental syntheses are still being made, and the jury is still out. All of the chapters in this book argue for the significance of evo-devo; some arguments are direct, but mostly the work here contributes to the synthesis itself. The success of these chapters would be a part of the success of the developmental synthesis.

In chapter 1, Manfred D. Laubichler and Jane Maienschein offer some historical vignettes to show how the study of biological generation separated into the study of development and biology in the late twentieth century; that there is indeed a growing interest in their reintegration; but this faces the difficulty that work in evo-devo is itself experiencing centrifugal tendencies. The most obvious is that some integrate development and evolution by using information about evolution to learn about developmental mechanisms, while others use information about developmental mechanisms to learn about evolution. This could be the result of the previous division between developmental and evolutionary biology as each camp continues to be biased in the questions that they want to answer. Laubichler and Maienschein suggest that recent history shows there is the will and even possibly some funding to bridge these two cultures and truly balance an interdisciplinary field. However, they warn that the history of evo-devo (as described above) may ultimately be judged as a naive myth, unless a unifying set of theoretical principles for evo-devo are established. A new genuine synthesis may remain elusive, due to a lack of experimental success and theoretical structure.

Werner Callebaut, Gerd B. Müller, and Stuart A. Newman's organismic systems approach to biology—described in chapter 2—offers one set of unifying principles. Their view is founded on emphasizing causation over correlation. They see development as the causal mechanism for the process of evolution. This turns evolution on its head. Rather than evolution producing organisms that develop, "Development has resulted in populations of organisms that evolve." They too investigate the potential for evo-devo to change evolutionary theory, and like Laubichler and Maienschein, investigate the forces that integrate and disintegrate science in general and evo-devo in particular.

The modern synthesis ignored how genotype determines phenotype, which was left to be studied by developmental biologists. In chapter 3, H. Frederik Nijhout offers a mathematical model for representing the genotype-phenotype relationship in an *n*-dimensional hyperspace. The model is based on plausible developmental assumptions. Combinations of trait values determine phenotypes, allowing all possible phenotypes to be represented. By considering changes in just one gene on one aspect of a phenotype, we can see how that gene influences that phenotype. By considering variations in other genes, too, we can see how the way the first gene influences that phenotype can change. That is, other genes determine the developmental program, which may be understood as providing constraints. One novel result of this is a distinction between evolution that occurs within the constraints of a given set of developmental mechanisms, versus evolution that results from changes in developmental mechanisms. This concrete representation of constraints coming from a developmental biologist provides one promising way that abstract concepts of evo-devo may be empirically studied and quantified in order to produce specific predictions.

As chapter 4 suggests, Gerhard Schlosser builds his integrated view of development and evolution around a broader notion of constraints than is usually considered in evolutionary biology. For him, constraints arise from the necessity to maintain a stable/functional organization after variation. Changing one trait will tend to require changing some other traits, but not all others. Significantly, this includes not only generative dependencies typically thought of as determining constraints, but functional dependencies that are necessary for organism viability too. Mutually constraining factors bundle together to form the units of evolution. Because these units of evolution can correspond to modules of development or behavior, results from physiology can play an important role in their discovery.

In chapter 5, Roger Sansom argues for the general adaptive value of gradual mutation and that this can only be selected at a multigenerational level. Therefore, there is another unit of selection—a legacy. Because generative entrenchment is a generic feature of complex organisms, he suggests that this selection will encourage developmental modules that are functionally integrated. The nontriviality of this thesis requires that the identities of functions are not determined by physiology. Instead, Sansom looks to ecology for an answer.

Paul E. Griffiths largely assumes that evolutionary developmental biology has been productive for the study of evolution and, in chapter 6, applies its lessons to psychology. Recent work in evolutionary psychology has assumed the modularity of the brain. However, Griffiths argues that this work is suspect because it has failed to take account of homology as an organizing principle as well as the ecology of psychological development. The evolutionary developmental ecological psychology Griffiths endorses is not unlike the classical ethology that was eclipsed by sociobiology in the 1960s. This completes an intellectual circle, because it was this sociobiology that inspired Gould's attack on adaptationism, which was an important step toward the current interest in the developmental synthesis.

In chapter 7, William C. Wimsatt and James R. Griesemer attempt to get a handle on identifying the units in cultural evolution by applying the notion of development. They make use of the notion of scaffolding in cultural evolution—the idea that permanent or recurring social and material structures are important to inheritance in culture. Incorporating one of the earliest ideas in evo-devo, Wimsatt's "generative entrenchment" and Griesemer's more recent "material transfer," they begin the work of characterizing the dynamic interplay between channels of inheritance to identify the units of cultural change. Incorporating development is of particular importance to understanding cultural evolution, because many enthusiasts have become enthralled by Dawkins's idea of a cultural meme (a replicating unit analogous to a gene in biological evolution). The search for memes is a search that has been blind to what would count as development in culture and the insight that incorporating development might bring. However, the complexity of cultural evolution results in memetics having less to offer than gene selectionism and, as the developmental synthesis does in general, Wimsatt and Griesemer attempt to investigate the complexities of cultural evolution, rather than abstracting them away.

References

Brylski, P. and B. K. Hall. 1988a. Ontogeny of a macroevolutionary phenotype: The external cheek pouches of geomyoid rodents. *Evolution* 43: 391–395.

Brylski, P. and B. K. Hall. 1988b. Epithelial behaviors and threshold effects in the development and evolution of internal and external cheek pouches in rodents. *Zeitschrift für Zoologische Systematik und Evolutionforschung* 26: 144–154.

Dawkins, R. 1976. *The Selfish Gene*. Oxford: Oxford University Press.

Gould, S. J. 2001. The evolutionary definition of selective agency, validation of the theory of hierarchical selection, and fallacy of the selfish gene. In R. S. Singh, C. B. Krimbas, D. B. Paul, and J. Beatty, eds., *Thinking about Evolution*, Vol. 2: *Historical, Philosophical, and Political Perspectives*, 208–234. Cambridge: Cambridge University Press.

Griesemer, J. R. 2000. Reproduction and the reduction of genetics. In P. Beurton, R. Falk, and H.-J. Rheinberger, eds., *The Concept of the Gene in Development and Evolution: Historical and Epistemological Perspectives*, 240–285. Cambridge: Cambridge University Press.

Oyama, S., P. E. Griffiths, and R. D. Gray, eds. 2001. *Cycles of Contingency: Developmental Systems and Evolution*. Cambridge, MA: MIT Press.

Robert, J. S. 2002. How developmental is evolutionary developmental biology? *Biology & Philosophy* 17: 591–611.

Sterelny, K. 2000. Development, evolution, and adaptation. *Philosophy of Science* (proceedings) 67: S369–S387.

Wimsatt, W. 1980. Reductionistic research strategies and their biases in the units of selection controversy. In T. Nickles, ed. *Scientific Discovery, vol. 2: Case Studies*, 213–259. Dordrecht: D. Reidel.

Acknowledgments

Primarily, we thank the contributors for their earnest and in one or two cases monumental work. We also thank Michael Ruse for his early and continued support of the project and Tom Stone and his reviewers as well as Sandra Minkkinen, Elizabeth Judd, Chryseis Fox, and others at The MIT Press for their efficiency in helping to bring it to a successful conclusion.

1 Embryos, Cells, Genes, and Organisms: Reflections on the History of Evolutionary Developmental Biology

Manfred D. Laubichler and Jane Maienschein

Evolution and development are the two biological processes most associated with the idea of organic change. Indeed, the very notion of evolution originally referred to the unfolding of a preformed structure within the developing embryo and only later acquired its current meaning as the transformation of species through time. It seems, therefore, only logical to assume that the biological disciplines that study these two different phenomena—embryology, later transformed into developmental biology, and evolutionary biology, especially phylogenetics—would be closely related.

From Aristotle until the late nineteenth century, history in the context of the life sciences was always understood as life history. As such, history always stretched across generations. What we today identify as three distinct processes of development, inheritance, and evolution (each investigated by several separate research programs), were previously all part of an inclusive theory of generation. This fact, often overlooked in recent discussions of the prehistory of evolutionary developmental biology, is important, because the conceptual topology and epistemological structure of these earlier discussions is quite different from today's attempts to resynthesize evolution and development (see also Laubichler 2007). This earlier concept of generation conceptualized as organic nature unfolding as one grand historical process is distinctly pre-Weismannian, whereas today's attempts to integrate evolution and development implicitly accept and even reify Weismann's idea of a separation of the soma and the germ line as the respective domains of these divergent research programs (see, e.g., Weismann 1892; Buss 1987). Synthesis in our current context therefore means finding a way to integrate the results from one discipline within the theoretical structure of the other, whether as research into the evolution of developmental mechanisms or as research into the evolutionary consequences of developmental processes. There are a few exceptions to these two paradigmatic cases of integration, but those have not yet become sustained research programs within evolutionary developmental biology.

During the second half of the nineteenth century several attempts were made to hold on to the idea of the unity of generation in light of the growing specialization of research within the life sciences. To be clear, the idea of generation itself had undergone several transformations since its canonical formulation in the eighteenth century, but its main focus on a continuous historical connection through the life cycle of organisms remained intact. Ernst Haeckel's program of evolutionary morphology and phylogenetics with its focus on the biogenetic law provided one such attempt, as did August Weismann's theoretical and Wilhelm Roux's experimental systems. (Neither Weismann nor Roux was as radical as their followers believed they were; both are transitional figures mostly concerned with establishing sustainable research programs within the conceptual topology of generation.) However, at the turn of the twentieth century, with the sustained success of *Entwicklungsmechanik* and other experimental approaches to the study of development and inheritance, the situation began to change. While most late nineteenth-century scientists did not consider the evolutionary process to be truly separate from development, the focus of the next generation was different. Rather than phylogeny and generation together, organisms and cells and their respective properties such as regulation and differentiation provided the frame of reference for new experiments, observations, and theories. This trend, seeking to account for development on the level of its supposed determinants (cells, genes, or molecules, but also organism-level phenomena, such as fields and gradients), continued throughout the twentieth century (see, e.g., Allen 1975; Gilbert 1994; Mayr 1982; Mocek 1998). A similar pattern can be seen in evolutionary biology where, within the emerging disciplines of population and quantitative genetics, evolution was reconceptualized as the change in the frequencies of certain alleles within populations (Provine 1971). In the context of these models the focus of evolutionary biology shifted from an earlier emphasis on explaining phenotypic change to the study of genetic variation within populations. This view of evolution produced operational models and theories, but completely ignored the crucial question of how a genotype produces a phenotype (e.g., Sarkar 1998).

It has been argued that the success of the modern synthesis was based on the exclusion of the messy phenomenon of development and the correlated claim that denies a difference between micro- and macroevolutionary processes (see, e.g., discussions in Mayr and Provine 1980). The tables were turned when, in the early 1970s, several authors argued that there is something important to be gained by bridging the gap between developmental and evolutionary biology. Initially, these proposals, such as punctuated equilibrium (Eldredge and Gould 1972), developmental constraints (Maynard-Smith et al. 1985), or burden (Riedl 1975), remained minority opinions, but after remarkable new results in developmental genetics showed the widespread conservation of "developmental genes," such "new syntheses" of macro- and microevolution and of evolution and development soon gained momentum (see, e.g., Hall 1992, 1998). This is, at least, the growing myth about the origin of modern evo-devo.

However, despite recent enthusiasm for this "new synthesis" it is not at all clear whether there is enough agreement among the various versions that supposedly fall within this camp to justify such a label of synthesis. Different authors entered the field with different perspectives and from different intellectual traditions. Thus, Brian Hall's recent question "evo-devo or devo-evo?" is more than just an exercise in semantics (Hall 2000). We still find very few universally agreed on concepts or even research questions in evo-devo (e.g., Wagner, Chiu, and Laublicher 2000). Understanding the origins of the different conceptions of evo-devo might thus be a necessary step on the way to a deep synthesis.

Here, we seek to shed some light on the epistemological and theoretical assumptions that lie behind attempts to conceptualize development and evolution and to ask "what is new with evo-devo?", "what are the conceptual resources of different versions of evo-devo?", and "to what extent is evo-devo a continuation of earlier traditions?" Our chapter is decidedly *not* intended as a history of evo-devo, or even as a history of the changing relations between evolution and development. Such a study would require much more space than we have here (for beginnings of a history of evo-devo, see Amundson 2005; Laubichler 2005; Laubichler and Maienschein 2007). Rather, we illustrate through a few short historical vignettes a specific hypothesis related to the conceptual and epistemological shifts that determined the ways researchers have thought about the relationship between evolution and development.

In short, our hypothesis is that there was a crucial conceptual and epistemological break associated with the establishment of several independent and self-sustaining experimental research programs devoted to specific aspects of evolution, development, and inheritance at the turn of the twentieth century. For centuries these phenomena were conceptualized within the single theoretical framework of generation that implied the unity of development, inheritance, and later also of evolution. However, late nineteenth-century adherents of this conceptual framework did not succeed in establishing a sustainable experimental research program, nor could they accommodate all of the new experimental results that emerged within the lines of research made possible by the many technological as well as organizational innovations during that period. As a consequence, the unity of generation disintegrated with the rise of the growing specialization of the experimental disciplines within biology. A small band of theoretical and experimental biologists tried to hold on to the conceptual unity of generation as well as to create a new conceptual structure for biology, but they remained a minority and did not succeed in establishing a conceptual alternative powerful enough to counteract the centrifugal tendencies within experimental biology. As a consequence, the conceptual topology once represented by the idea of generation was transformed into several separate domains represented by the concepts of inheritance, development, and evolution.

For example, embryologists in medical schools focused on "proximate" details of the developing individual human, while the "ultimate" distant evolutionary history seemed of little immediate importance. This lack of attention to evolution persists far outside the world where it is medically explicable, and for a much wider range of related reasons developmental biologists have largely ignored evolution as unimportant to the immediate research at hand. There has been little explicit opposition, but neither has there been a consistent sense of sameness of purpose or a compulsion to bring embryology closer to evolution. From the other side, as both historians and biologists have often noted, embryology was largely not included in the so-called evolutionary synthesis of the 1950s—though whether it was actively left out or just failed to see the point of joining remains an open question. Ron Amundson, for instance, has published his own take on this history, one based on the assumption of an active exclusion of developmental biology by what he calls "synthesis historiography" (Amundson 2005; see also Laubichler 2005 for a critical reading of Amundson).

These separatist tendencies have changed in the last couple of decades, of course, and it is not because researchers have managed to fit embryology belatedly into the now-established synthesis or because development has somehow been tied into the "central dogma" of genetics. Rather, there are new ways of thinking about how to bring the fields together, and new reasons to do so, leading to the search for a new and different synthesis. Hence the perceived need for a lively new name for the integrated field dedicated to stimulating research (and funding), seemingly fulfilled by "evo devo."

However, as our historical reconstruction of the shift in the conceptual and epistemological structure from generation to development, inheritance, and evolution indicates, accomplishing a true synthesis of "evo" and "devo" will actually be quite difficult. This is largely true because, with a few exceptions, most of the current discussion remains within the conceptual topology that separates development, inheritance, and evolution. Furthermore, "evo-devo" or "devo-evo" is already experiencing the same centrifugal tendencies that have led to the earlier separation into different disciplines, and largely for the same reasons of experimental success and the lack of a unifying theoretical structure. Our examples suggest that unless a new conceptual topology is established, within which development, inheritance, and evolution represent different elements of one historical process (as was the case in the earlier conception of generation), a new synthesis of evo-devo might remain elusive.

The Phenomenology of *Entwicklung*

We have stated above that throughout most of the nineteenth century, historical processes in nature were conceptualized as generation. This unified view of generation, or *Entwicklung,* had far-reaching epistemological consequences, especially with regard to the relationship between historical description and mechanical causality. Even though studies of generation had always also referred to mechanical causes (or other forms of the Aristotelian *causa efficiens*), the primary focus of these studies had been historical. *Entwicklungsgeschichte* was foremost a phenomenology of Entwicklung. However, within this framework of Entwicklungsgeschichte the older conception of generation, which focused on the iterative processes of development and inheritance, could be extended to include an evolutionary dimension. In this way it can be argued that the conception of embryology as Entwicklungsgeschichte enabled the formulation of the theory of evolution (see also Richards 1992). The foremost representatives of this trend in the second half of the nineteenth century are Darwin and Haeckel, whereas Weismann and Roux represent transitional figures, who tried to integrate new experimental approaches and results within this conceptual structure of generation and Entwicklungsgeschichte.

Darwin on Development and Generation

Darwin brought development into the foreground of natural history in the first edition of his *Origin*. There he declared his enthusiasm for embryology as providing perhaps the most compelling evidence for evolution by common descent, "second in importance to none in natural history" (Darwin [1859] 1964, 450). He asked,

> How, then, can we explain these several facts in embryology,—namely the very general, but not universal difference in structure between the embryo and the adult;—of parts in the same individual embryo, which ultimately became very unlike and serve for diverse purposes, being at this early period of growth alike;—of embryos of different species within the same class;—of embryos of different species within the same class, generally, but not universally, resembling each other;—of the structure of the embryo not being closely related to its conditions of existence, except when the embryo becomes at any period of life active and has to provide for itself;—of the embryo apparently having sometimes a higher organization than the mature animal, into which it is developed.

The answer lay with evolution, for "I believe that all these facts can be explained as follows, on the view of descent with modification" (Darwin [1859] 1964, 442–443).

Darwin scholars have provided much historical evidence regarding what Darwin knew, when he knew it, how he knew it, and what he concluded, when, and why. Darwin was clearly influenced by German embryological studies, and reinforced by Karl Ernst von Baer's "laws" that embryos remain largely similar for similar types of organisms and only diverge later according to type. Historians have pointed out the irony that empirical reports of what Darwin offered as his best evidence came in large part from those who opposed the idea of evolution. Yet this fits Darwin's pattern of taking what is available (such as William Paley's 1802 argument from design) and brilliantly using it to demonstrate the fit with his theory of evolution as common descent through natural selection. In Darwin's methodological reasoning, if the evidence can be explained by evolutionary theory, it lends confirmation to that theory. Therefore, Darwin had more a devo-evo focus, concerned with taking embryology to inform evolution (or more properly embryo-evo, since what became developmental biology after World War II was called embryology in Darwin's day and "development" often referred to the unfolding that occurs during evolution).

Darwin's focus on evolutionary relationships, especially among embryos, guaranteed that embryology would become a lively subject at the end of the nineteenth century as researchers sought empirical support for evolutionary ideas, or against them. Tracing detailed morphological patterns of development for individual types of organisms provided data, and the apparent ability of embryonic relationships to reveal ancestral and therefore also adult relationships provided work for many embryologists. Darwin had, in effect, issued an invitation to engage in detailed descriptions of embryonic development typical of an individual species. This was not Rudy and Elizabeth Raff's "evolutionary ontogenetics" for the sake of studying development, but rather embryology in aid of constructing evolutionary phylogenies, more devo-into-evo. In other words, Darwin was still arguing within the conceptual framework of generation where embryological data could support claims about descent with modification and the phylogenetic relations between different taxa. While this was not Darwin's emphasis, Ernst Haeckel very quickly provided a most highly visible theoretical structure by which to organize these burgeoning investigations.

Ernst Haeckel and the Biogenetic Law

As Ernst Mayr has explained, Haeckel was practically required reading for intelligent young students early in the twentieth century, and well before. Because of his "monistic materialism," Haeckel was a bit naughty, and public school teachers did not really want their young students discussing such things (Mayr 1999). Yet Haeckel had quickly gained a popularity and credibility that made it impossible to ban him from the classroom. Thus, the clever young student could both annoy the teacher and intrigue other students by quoting Haeckel. Haeckel evidently thus inadvertently helped to start at least one young German man on his way to becoming one of the world's leading evolutionary biologists.

Haeckel built on earlier studies based in a *Naturphilosophie* tradition that stressed the unity of nature. He sought to outline comparisons between the series of changes in the development of individuals (ontogeny) and that of species (phylogeny). Further, he sought to demonstrate the value of comparative ontogeny for revealing otherwise elusive phylogenetic relationships. Haeckel expressed his ideas in different places and in varying forms for both German- and English-speaking audiences, because his major books were quickly translated and published in popular form. Statements of the theory, its corollaries, and implications were often distorted, even by Haeckel himself in some cases. Yet the key principles remained quite clear and consistent, and though familiar to some, Haeckel's views are worth reviewing since they are so often misrepresented.

Most basically, Haeckel saw ontogeny and phylogeny as intimately related, not as separate processes. Indeed, the "fundamental law of organic evolution" was "that Ontogeny is a recapitulation of Phylogeny; or somewhat more explicitly: that the series of forms through which the Individual Organism passes during its progress from the egg cell to its fully developed state, is a brief, compressed reproduction of the long series of forms through which the animal ancestors of that organism (or the ancestral forms of its species) have passed from the earliest periods of so-called organic creation down to the present time" (Haeckel 1876, 6–7). Furthermore, this is a causal relationship in which the phylogenetic changes in one sense cause the ontogenetic series of changes. Therefore, development reveals evolution, or devo takes us into evo. The recapitulation is not perfect, however, but rather ontogeny is the short and rapid recapitulation of phylogeny, "conditioned by physiological functions such as heredity (reproduction) and adaptation (nutrition). The organic individual...repeats during the rapid and short course of its individual development the most important of the form-changes which its ancestors traversed during the long and slow course of their paleontological evolution according to the laws of heredity and adaptation." Deviations and specifics make the patterns, so devo illuminates evo and reveals relationships. Or, to put it in modern terms, devo is seen as reflecting evo (Haeckel 1866, vol. II, 300). All this he offered with special emphasis on the role of changes in the germ layers, which provided a convenient starting point for research and raised questions for the theory. Ultimately, however, the earlier stages prior to germ-layer formation did not show the same visual embryonic parallels that had enthused Haeckel. Haeckel did admit that secondary adaptation can cause divergences from the ancestral pattern, but he saw those as only helping to inform our understanding of the evolutionary process. His biogenetic law, or the law of recapitulation, is the most familiar encapsulation of his views.

To reinforce his lengthy and often repetitious tomes, Haeckel typically provided tables of comparative figures to make his point more persuasive. Haeckel's many long volumes were eagerly received in the United States and elsewhere, as well as in Germany. In fact, they appeared in such large numbers from such popular presses that they are still quite easy to find inexpensively in used bookstores.

While Haeckel was a great authority for claims about embryonic parallels and recapitulation, he later became a repudiated figure regarded as a mere popularizer and an intellectual lightweight, and was accused of deliberate fraud. It is not for us to decide Haeckel's scientific reputation here, nor to chronicle his debates with Carl Gegenbaur, Anton Dohrn, and others, but rather to note that by the early twentieth century Haeckel, more than any other single author besides Darwin himself, focused attention on the relations between embryos and ancestors, between development and Darwinism (Laubichler and Maienschein 2003; Nyhart 1995). The fact that the pages of such leading scientific journals as *Science* and *Nature* still carry notes on Haeckel's contributions (albeit often highly critical) stands as testimony to his impact (see also Richards 1992; Haeckel Haus documents in Jena). By bringing evolution and embryology together in the way he did, however, he also set the stage for repudiation of the particular speculative relationships that the embryological comparisons seemed to suggest. Indirectly, Haeckel's excessive speculation and theorizing helped to stimulate opposition to the goal of phylogenizing, and also led to a rallying to embryology for its own sake separate from evolution. Embryologists increasingly called for exploration of the mechanisms and proximate causes of ontogenies, increasingly pushing evolution into the background. Ironically, this initial interest in development stimulated by interest in evolution helped to drive a sharp wedge between embryology and evolution for most of a century. The connections seemed too weak and strained as biologists called for a stronger, experimentally, and empirically grounded science.

August Weismann and the Gradual Disintegration of "Generation"
Etymology can sometimes lead to interesting insights. It would be a worthwhile undertaking to document all the multiple interpretations of the word *evolution* in the second half of the nineteenth century. This term was still mostly defined in opposition to epigenesis in that it referred to a strictly mechanical theory of development. Development (evolution) was seen as a gradual unfolding of causes (factors) that are already present at the beginning—that is, in the fertilized egg. Epigenesis, on the other hand, implied the gradual emergence of complexity as part of a dynamic process of development. As Weismann, who did more than anybody else to develop this view, stated in the preface to his theory of the germplasm, "So kam ich zuletzt zu der Einsicht, dass es eine epigenetische Entwicklung überhaupt nicht geben kann" ("and thus I finally realized that epigenetic development is impossible"). Weismann, who according to his own admission, had tried to develop several theoretical systems that would include epigenetic processes in development, finally convinced himself that only a strictly deterministic theory of development could account for all the empirical facts and be theoretically satisfactory. The one theoretical problem that Weismann was most concerned with was the causal and material relationships among development (Entwicklungsgeschichte), heredity, and the transmutation of species (Abstammungslehre). The problem, as it presented itself to Weismann, was to find the material cause that would connect all the different elements of generation (including descent with modification).

He starts his discussion of the problem with some remarks about Darwin's theory of pangenesis as well as about Herbert Spencer's notion of "physiological entities," but soon rejects both because of the number of theoretical assumptions that these theories require. Weismann's solution was to focus on the material continuity between the generations (heredity) and separate it from the mechanistic causation of development. This theoretical separation of development from inheritance allowed Weismann to clearly analyze the kind of causation involved in each of these processes and to ask how these chains of causation could be realized materially. His answer was deceptively simple. To account for heredity, Weismann assumed that the germplasm, which contained all hereditary factors, always remains within the germ cells—in other words, that there is a continuity of the germ line. This assumption, for which there was ample empirical evidence, also supported the theoretical separation of development from inheritance. Weismann argued that during development, which represents a differentiation of the zygote into multiple cell types, the material composition of the dividing cells changes; the idioplasm of the differentiating cells is therefore not identical to the germplasm of the gametes. Furthermore, he argued that these changes in the material composition of individual cells are the causes for their differentiation into separate cell types. However, and this was a central part of Weismann's argument, the idioplasm of differentiated cells in the body is completely separate from the germplasm. Weismann did not allow for any form of causal connection that would reach from the differentiated cells of the organism back to the germplasm. This view was the opposite of Darwin's theory of pangenesis (which had already been discredited by Darwin's own cousin, Francis Galton) and also affirmed Weismann's commitment to an evolutionary (unfolding) conception of development.

Conceptually, Weismann's theoretical system introduced a clear distinction between the processes of development and heredity, two aspects of the older concept of generation. However, in his system Weismann still maintained the material unity of generation. The germplasm represents the material connection between generations, and the material changes in the idioplasm provide a mechanical explanation of development as evolution (unfolding). Furthermore, the germplasm contains all the material factors that are needed to build an organism. But the theoretical separation between germplasm and idioplasm also provided the conceptual framework for the emerging experimental research programs in Entwicklungsmechanik and genetics.

The Separation of Entwicklung into Independent Research Programs

By 1900 the conceptual unity of generation had fallen apart. The Haeckel-Gegenbaur program of evolutionary morphology and the biogenetic law could no longer be sustained as a productive research program, largely because it did not solve the fundamental problem of circularity inherent in reconstructing phylogenies based solely on comparative and embryological data (Laubichler 2003; Laubichler and Maienschein 2003; Nyhart 2002). In the meantime, new experimental programs had established themselves as powerful alternatives to earlier descriptive approaches, introducing a conceptual shift from phenomenological Entwicklungsgeschichte to Entwicklungsmechanik and Entwicklungsphysiologie. Consequently, the new focus was predominantly on proximate causes for development (or on the Aristotelian causa materialis and causa efficiens). This was, in a way, inevitable, because the problem of generation was now approached experimentally, and each experimental manipulation defines its own form of causation as correlated changes between measurable parameters. As a consequence, development, inheritance, and evolution were mostly studied as separate experimental problems, soon followed by conceptual developments specific to each of the newly emerging disciplines.

In the context of development the focus was on the causal determinants of differentiation. This required a careful record of cellular differentiation during development and a conceptual reorientation of the question of development from the life history of an organism to the differentiation of cells. Phenomenology was thus still part of Entwicklung, but it was the phenomenology of parts, not wholes, that mattered here. It was studied experimentally through increasingly difficult manipulations, such as selective killing of cells, various forms of constriction experiments, and a whole series of grafting experiments. The conceptual innovations that most characterize this period are the ideas of cell-lineage studies, of tissue cultures, and of the physical-chemical determinants of development, culminating in the idea of the organizer, as well as in the notion of regulation in development.

The study of inheritance took a similar path, focusing mostly on factors of inheritance, although the history of genetics during the first half of the twentieth century is extremely diverse and also includes several research programs that continued to study inheritance and development together as two intricately related biological processes. The most prominent of these alternative approaches were Richard Goldschmidt's program in physiological genetics and Alfred Kühn's related program in developmental genetics (e.g., Geison and Laubichler 2001; Laubichler and Rheinberger 2004). But these locally successful programs were eclipsed by the even greater success of Morgan-style transmission genetics, which established the Drosophila model as the international standard (Kühn and Goldschmidt both used different species of moths as model organisms), the emerging mathematical population genetics, which was soon integrated into the modern synthesis, and, shortly thereafter, by molecular genetics. Besides their experimental work, Kühn and Goldschmidt also made important theoretical contributions that continued to develop the conceptual framework of generation (as well as of epigenetics), but their theories had a similar fate as their model organisms: they were "outbred" by their much simpler and faster reproducing competitors.

An even more ambitious program in experimental biology that explicitly continued within the earlier tradition of the conceptual unity of generation was initiated by Hans Przibram at the Vienna Vivarium. Przibram's program included experimental research into development, regeneration, heredity, and evolution. To that end he and his coworkers developed the most sophisticated techniques to maintain research animals for extended periods and many generations. Research in the Vivarium was explicitly focused on an epigenetic conception of development that included the study of regeneration as well as experiments that investigated the role of the environment in development and evolution. Today, the Vienna Vivarium is mostly associated with the controversy surrounding Paul Kammerer and the final discreditation of neo-Lamarckian theories of inheritance. This is extremely unfortunate, since in many ways, the research program of the Vienna Vivarium is the link between nineteenth-century theories of generation and late twentieth-century attempts to resynthesize evolution and development and lately ecology as well.

But for the reasons sketched above as well as for a variety of others that we could not discuss here, a different set of questions came to dominate the scientific study of development, inheritance, and evolution in the early decades of the twentieth century. Here we will provide two exemplary cases that represent the transition from the earlier focus of Entwicklungsgeschichte and generation to the newly emerging research programs of cell biology, Entwicklungsmechanik, and transmission genetics.

E. B. Wilson

The American biologist E. B. Wilson felt this call to undertake a rigorous study of embryology, in the context of cell theory. Wilson saw evolution and cell theory as the two great foundations for biology, and development as a central part of cell theory (Wilson 1896, 1). In an essay "Some Aspects of Progress in Modern Zoology," this leading cytologist explained the increasing divergence between those interested in evolution and those interested in embryology. While Darwin concentrated attention on evolution and phylogenetic relationships for a while, soon the "post-Darwinians awoke once more to the profound interest that lies in the genetic composition and capacities of living things as they now are. They turned aside from general theories of evolution and their deductive application to special problems of descent in order to take up objective experiments on variation and heredity for their own sake" (Wilson 1915, 6).

This was certainly not because they rejected evolution. Quite the contrary. Evolution became, in effect, a fundamental background condition, against which individual development and behavior were to be understood. Yet the background faded in immediate importance, as the researchers focused on individual structure, function, and their development. Instead of evolutionary relationships, embryologists and geneticists found new areas to explore, and what they saw as the proper exact science of biology quickly moved in those directions. This was devo in the foreground, with evo essentially in waiting as a background assumption. Evo and devo were not yet connected.

Wilson saw embryologists as able to remain on relatively firm ground, with a "rich harvest" of careful, detailed empirical descriptions of the stages of development. In contrast, he feared that the evolutionist phylogenizers often tread on thin metaphysical ice and narrowly miss entering the "habitat of the mystic" in their speculations. Evolution was just too difficult to study rigorously, he felt (Wilson 1915, 8). Embryology, in contrast, is based on chemistry and physics and the close study of cells, and hence more solidly grounded in empirical science.

Complex epistemological preferences dictated this conclusion, shaped by Wilson's own education at Johns Hopkins, and reinforced by his research at the Stazione Zoologica in Naples and the Marine Biological Laboratory in Woods Hole, Massachusetts (Maienschein 1991). He was a leader among biologists, and typical of the new specialists who decades later were called cell and developmental biologists. We can already see embryology diverging by the first decade of the twentieth century from evolution: different questions, different approaches, different methods, different researchers, and different values. Development might be a foundation for biology, and evolution might be a persistent shaping force, but for those who would study biology, these were two separate cornerstones and not integrated profoundly.

For Wilson, the "exact" (and hence most desirable) biological science for his time already lay with embryology and genetics (or heredity and development). For the time being, he pointed out that we did not understand how the inherited material "of the germ-cell can so respond to the play of physical forces upon it as to call forth an adaptive variation." In addition, the distance from the inorganic to the simplest life seems so large that it is difficult to understand how the gap could be bridged. Therefore, Wilson concluded in 1896, "I can only express my conviction that the magnitude of the problem of development, whether ontogenetic or phylogenetic, has been underestimated" (Wilson 1896, 330). Cell theory had made tremendous advances in understanding the basic phenomena of life, but much remained to be done. Study of development based on cells and evolution remained far apart. Devo sat alongside evo. We had not discovered how to bring the two together and retain the proper epistemic commitments of science.

In large part, Wilson remained focused on morphology. By looking at the structure of cells and cell parts, and their changes through the stages of individual development, this premier cytologist did not see how to connect the developing morphological patterns of individuals and those of the species to which they belong and from which they have derived by evolution. It was by going deeper into the cell, and beginning to see the connections of physical-chemical compositions, which took researchers much of the twentieth century, that we moved toward bridging the gap that Wilson saw at the end of the nineteenth century.

T. H. Morgan

Three decades later, another American (a friend and colleague of Wilson's)—Thomas Hunt Morgan—began to suggest one way the increasingly apparent differences might be bridged. By then, he had already carried out the Nobel Prize–winning foundational work in genetics on white-eyed Drosophila. The eighth chapter of his *The Scientific Basis of Evolution* (1932) considered "Embryonic Development and Its Relation to Evolution." As this architect of genetics put it, "One of the important chapters of the Evolution Theory concerns the interpretation of the evidence from embryonic development." All the discussions of recapitulation sparked by Darwin and Haeckel had led to great debate and much new study, Morgan (1932, 171), acknowledged, but they had not made much progress in their attempt to "unravel the remote past" of evolution. Darwin's assertion that "community of embryonic structure reveals community of descent" did not go far toward explaining how that revelation would occur, or its details.

As he pointed out, it does not get us very far to know whether ape adults or ape embryos were the closest human ancestors. For Morgan (1932, e.g., 177), we were going to need to know a lot more than that to make any progress with the important questions of biology.

That knowledge could only come with close study of chemical and physical details of the egg and the embryo. Specifically, genetics held the key for Morgan. It was genes that would tie together individual development and evolution, since genes underlie heredity that ties individuals together across generations and guides development. Genes could lead to changes in early, embryonic, stages of development and thereby have far-reaching effects later on. The risks involved in making early changes would explain why those early stages are so highly conserved, and why the later stages tend to vary more often. Genetics, in short, could explain why we see the patterns we do in embryology and evolution. Yet for Morgan, it clearly remained more important, more legitimate, and more progressive science to build a strong bridge from the evolutionary past to the present developing individual organism, through genetics. He would surely have been happy with others pursuing "evolutionary ontogenetics" as a goal. But then evolution remained entirely in the background of his own research, which focused on heredity primarily and at times on development.

The examples of Morgan and Wilson and many others that we could not mention here demonstrate how productive research programs all too easily take on a life of their own. "Evolutionary ontogenetics" remained a goal for Morgan, but one that gradually disappeared behind mountains of new data and research questions made possible by the establishment of the successful Drosophila model. However, these new data also needed to be interpreted and organized, which, in turn, required the development of a new conceptual structure, that of transmission genetics and the chromosomal theory of the gene. Heredity was now a problem of transmission rules; genes, still identified by their phenotypic effects, were localized on chromosomes; and complications that arose due to development (the genotype-phenotype mapping problem) were soon hidden behind conceptual innovations designed to insulate the core assumptions of transmission genetics from all potential threats to the theory. Concepts such as "penetrance" and "expressivity" allowed researchers to maintain a simple model of genetic determination, while paying lip service to the intricate process of development.

Experimental embryology (soon to be renamed as developmental biology), which had been a major success story from the early 1890s on, hit a roadblock during the mid-1930s, mainly because it had reached the limits of what was technically possible at that time and with the standard-model organisms. Despite intensive research, the chemical nature of Spemann's organizer remained elusive and the connection of development to the newly emerging genetics was even more difficult to accomplish. It took Alfred Kühn, for example, years of almost industrial-scale research and the help of Nobel Prize–winning chemist Adolf Butenandt to uncover the biochemical pathway of an eye-color mutation in the moth Ephestia kühnellia (Rheinberger 2000). No wonder, then, that problems of gene action were soon studied with simpler organisms, such as Neurospora, E. coli, or even phage. But this shift in experimental methodology and technology also initiated another major conceptual change, that of molecular biology. The subsequent history of molecular biology and molecular genetics has been well documented, but it is also important to remember that it represents a further step away from the initial conceptual unity of generation.

A similar story can be told for population genetics. Originally mired in the controversy between Mendelians and biometricians about the nature of variation in natural populations, it soon emerged as the mathematical foundation of the modern synthesis. This was possible because Fisher, Haldane, Wright, and others managed to establish a system of successful mathematical abstractions that reconceptualized the problem of evolution. Under the assumption that specific genetic factors are correlated with specific genotypes—an assumption seemingly supported by the results of transmission genetics—it was possible to establish an operational mathematical theory for the dynamics of gene frequencies within (similarly abstracted) populations. This mathematical theory then provided the theoretical foundation for the modern synthesis in that it connected, through a successful series of abstractions, such as fitness values, the dynamics of alleles (particles) within populations to such phenotypic phenomena as adaptation and speciation.

The initial unity of generation thus disappeared while evolution (now defined by the assumptions of the modern synthesis), genetics (as population genetics, quantitative genetics, and molecular genetics) and developmental biology (soon transformed by molecular biology), established themselves as the cornerstones of twentieth-century biology. To be fair, quite a number of researchers tried to hold on to more integrative questions—the whole movement of theoretical biology in the first decades of the twentieth century is a case in point—but they remained at the margins and often had to fight hard for their reputation. Goldschmidt, for instance, was long vilified and only recently experienced somewhat of a renaissance, largely due to the efforts of Stephen Jay Gould, and C. H. Waddington had to endure unfounded accusations of being a neo-Lamarckian, because his ideas, such as genetic assimilation did not easily fit within the theoretical structure of population genetics. But those "renegades" were the ones who first saw the importance of new results and methods in evolutionary biology and developmental genetics, and in the early 1970s they began to address the old problem of the relations between development and evolution.

The Emergence of an Evo-Devo or Devo-Evo Synthesis?

Embryology may not have made it into the self-declared "evolutionary synthesis" of the 1930s and 1940s, and it has taken a while for researchers to sort out how to study relationships of evolution and embryology in the context of the conceptual topology of late twentieth-century biology. Yet the current evo-devo enthusiasm provides us with an encouraging major shift in thinking that may actually bring about that "new light" that de Beer foresaw. This synthetic field has the potential to bridge the epistemological, methodological, and theoretical gaps separating development and evolution for the past century.

Over the last thirty years many important contributions shaped the gradual emergence of evo-devo. Scientists working within a variety of different traditions of twentieth-century biology began to address the question of the relation between development and evolution. These diverse perspectives have led to several different versions of the evo-devo or devo-evo synthesis that have not yet been fully reconciled. Some of these differences are merely individual idiosyncrasies, but others are symptoms of a fundamental theoretical rift that divides the evo-devo community. Does one incorporate elements of evolutionary biology to better understand development (evo-devo), or does one integrate the results of developmental genetics and developmental biology into evolutionary biology in order to gain a better understanding of phenotypic evolution (devo-evo)? This theoretical rift is largely a consequence of the current conceptual topology of biology after the original unity of generation has been split into the separate concepts of inheritance, development, and evolution. In particular, development and evolution (the two historical processes of biology), have each acquired their own interpretation of dynamical causality, which roughly fits Ernst Mayr's distinction between proximate and ultimate causes. Without fundamentally changing this framework, one field's explanatory structure will always dominate, whereas the other can only contribute supplementary evidence, or, at most, lead to a modification of some of the basic assumptions about the underlying causes of the respective historical dynamic (of development or evolution).

Yet despite this fundamental rift in the theoretical structure, evo-devo is apparently coming of age, with reports in *Science*, a new professional organization, and new journals. Obviously, this new venture concerns evolution and development, but what is the relationship? Is it simply evolutionary ontogenetics, or something more? Or something different?

In 2001 Michael K. Richardson offered a website for an "Evo-Devo Research Group" (which turned up first with popular Internet search engines under "evo-devo" at that time). Richardson explained that "'Evo-devo' is the nickname for a branch of science which aims to understand how developmental mechanisms are modified during evolution. Evo-devo has its origins in the work of scientists such as von Baer and Haeckel. But it emerged in its modern form with the rise of molecular biology." Wallace Arthur (2002) has advocated a similar view. Such accounts sounds like evolutionary ontogenetics, with the emphasis on bringing evolutionary thinking into developmental biology and developmental genetics. They deemphasize the value of development for evolution, which would be part of a truly balanced interdisciplinary field. In contrast, others do emphasize devo-into-evo, which does recognize the value of developmental patterns for phylogenetics.

This mix of views is not surprising for an emerging field. Wade Roush and Elizabeth Pennisi (1997) brought attention to the "Growing Pains" in their article in *Science*, explaining that "Evo-Devo Researchers Straddle Cultures." Typical evolutionary biologists seek to document the course of evolution across species and across time, while developmental biologists look at the developmental changes in the life of individual model organisms. As developmental biologist Greg Wray put it in that article, "Evolutionary biologists have the conceptual background [on evolution], but a lot of the time they don't even understand these data. Developmental biologists have the data, but they are not really up on what to do with it" (Roush and Pennisi 1997, 38). Ideally, at least, evo-devo researchers combine both the data on individuals, with the perspective and data concerning evolutionary patterns. Genetic and molecular information provide links.

Yet such work is not easy since it is difficult to cross deep and well-established disciplinary boundaries, especially when the data remain incomplete on both sides and it is necessary to make assumptions.

Researchers on one side may fear or distrust those on the other, so that those who would straddle both must excel in both to gain credibility. For example, as Günter Wagner points out, evolutionary biology is much more theoretical in focus than molecular biology. The evo-devolutionist therefore must master the molecular details, generate and interpret the developmental data, and also work within the larger theoretical frameworks of evolution. For untenured younger scientists this can be risky.

There are growing pains indeed. As Mark Martindale admitted in 1997, funding at that time remained elusive and limited. "Evo-Devo is what we discuss over Friday beers, but when it comes to paying bills, people are more pragmatic," he reported (Roush 1997, 39). Others manage because as senior scientists they can piggyback this research on their "regular" grants. The situation has reportedly improved since then, and the successes are exciting for those who do succeed. So, only two years later there is growing evidence of that success and of a community of researchers eager to take the risks involved to be part of the efforts to build effective bridges.

The Society for Integrative and Comparative Biology (formerly American Society of Zoologists, now known as SICB, and one of the oldest persisting biological organizations in the United States) announced in its spring 1999 newsletter the formation of a new "Division of Evolutionary Developmental Biology." As the SICB leadership explained, this new field of evo-devo "is attracting growing attention from the life sciences community, academic institutions, funding agencies and major journals" (SICB 1999, 2). As co-organizer with Scott Gilbert, Billie Swalla urged members to attend that first-ever symposium at the annual SICB meeting in Atlanta in January 2000.

That meeting brought together those who embrace evo-devo and those who prefer devo-evo. At times, in the enthusiasms of talking together on the same podium in the course of a day of marvelous papers, the gaps still seemed large, but the will to bridge them is clearly strong and enthusiastic. In the wrap-up session, Rudy Raff called for "new experimental directions" that cooperation between the once-disparate disciplines could bring. And Günter Wagner and his collaborators suggested that the field of evo-devo has moved through what biologist Gunther Stent once called a "romantic" and "enthusiastic" stage to the much more difficult "academic" stage (Wagner, Chiu, and Laubichler 2000). At this more mature point, it becomes necessary to work out what is really at issue and to develop broadly multidisciplinary approaches to evolution, development, and genetics through which the field can make progress. There are methodological, epistemological, conceptual, and very real practical challenges ahead, as is evident in the persistent appeal to different approaches and problems, and in the tendency to see different sorts of things as exciting conclusions. But the enthusiasm for engaging in something like evo-devo that really tries to synthesize the best of evolutionary biology and development, with genetics, is clear and strong. Concrete examples of the way such studies can make a significant difference in the science done are just emerging, but a growing community has the conviction that they are there.

There are ample historical examples that tell us about the commonalities between all successful research programs or scientific disciplines. These include a social organization that supports the field and allows for the recruitment of researchers and students, shared values that establish a sense of community, a core of accepted questions and criteria for the evaluation of contributions, shared technologies, and finally also a shared theoretical structure that adequately reflects the problem at hand. Evo-devo or devo-evo fulfills most of these criteria. The big open questions, however, lie in the conceptual structure that would adequately represent the integration of development and evolution and, related to it, in the evidentiary standards for the field.

The lesson that we can draw from our brief sketch of the history of the problem is that each successful research program developed its own conceptual structure that was adequate for the problem at hand. If the research problem of a unified evo-devo and devo-evo lies in a true integration of the two historical processes of (individual) development and evolution, then a new focus on generation will be required. This means that a conceptual structure that adequately reflects the phenomenology of Entwicklung, including development, reproduction, heredity, and evolution, will have to be developed. It will be based on new forms of mathematical representations centered around decomposition theory and equivalence relations; will focus on concepts such as modularity, robustness, or configurations; and will include a fresh perspective on such old problems as epigenetic effects. These are all sketches, of course, that are based on our understanding of the problem, our review of current theoretical discussion in the field, and our interpretation of the history of the problem.

If, on the other hand, the problem is not so much in the full integration of development and evolution, but rather in two separate approaches of evo-devo and devo-evo, then each new "synthesis" will most likely not really change the conceptual structure of its "parent discipline" in any fundamental way. Consequently, these two versions will not go beyond the current separation of development and evolution in any meaningful way.

Conclusions

So, given the evident enthusiasm for bringing evolution and embryology together in some more truly integrative way, what difference does this make? Will it really make a difference for biology, or matter more broadly? Will it bring us back to thinking in terms of generation again? In fact, we believe that this movement is critically important to the advancement of the understanding of life, and that recovering something of what nineteenth-century biologists recognized as generation can help make biology more whole rather than a sum of disconnected and often discordant parts (such as molecular versus developmental versus physiological versus ecological, and so on). In light of current enthusiasm for what former U.S. National Science Foundation director Rita Colwell calls "biocomplexity," this should lead to new possibilities within science.

The success of evo-devo will crucially depend on whether its practitioners succeed in formulating an adequate conceptual and theoretical structure for the field. Biology at the beginning of the twenty-first century is in a similar position as it was at the end of the nineteenth; it has amassed an overwhelming amount of data compared to very little progress in terms of an integrative theory. In the early twentieth century the centrifugal tendencies dominated; different fields blossomed because they restricted their focus as well as their conceptual structure. An alternative integrative theoretical biology existed but it did not manage to overcome the growing separation of biological disciplines, even though it provided the impetus for many important individual contributions. Today, evo-devo has the potential to develop a conceptual framework that would integrate the ever-increasing mountains of data that characterize the postgenomic era of biology. It can be at the vanguard of what Walter Gilbert (1991, 99) envisioned as the future of theoretical biology, where the "starting point of a biological investigations will be theoretical." Historical awareness can help evo-devo accomplish this goal, if only by suggesting that the answer to the problem of integrating development and evolution might be in rephrasing the question; not development *and* evolution, rather development and evolution *as two sides* of a more unified historical process of generation. Within such a framework many problems, such as the role of epigenetic effects and of the environment, might no longer be as difficult to imagine.

References

Allen, G. E. 1975. *Life Science in the Twentieth Century*. New York: Wiley.

Amundson, R. 2005. *The Changing Role of the Embryo in Evolutionary Thought*. Cambridge: Cambridge University Press.

Arthur, W. 2002. The emerging conceptual framework of evolutionary developmental biology. *Nature* 415(6873): 757–764.

Behe, Michael. 1996. *Darwin's Black Box: The Biochemical Challenge to Evolution*. New York: Free Press.

Buss, Leo. 1987. *The Evolution of Individuality*. Princeton, NJ: Princeton University Press.

Coleman, W. 1971. *Biology in the Nineteenth Century: Problems of Form, Function, and Transformation*. New York: Wiley.

Creath, Richard, and Jane Maienschein, eds. 2000. *Biology and Epistemology*. Cambridge: Cambridge University Press.

Darwin, Charles. 1859. *On the Origin of Species*. London: John Murray; reprinted Cambridge, MA: Harvard University Press, 1964.

De Beer, G. R. 1951. *Embryos and Ancestors*. Oxford: Clarendon Press.

Eldredge, N. and S. J. Gould. 1972. Punctuated equilibria: an alternative to phyletic gradualism. In T. J. M. Schopf. ed., *Models in Paleobiology*, 82–115. San Francisco: Freean.

Gilbert, W. 1991. Towards a paradigm shift in biology. *Nature* 349(6305): 99.

Gilbert, S., ed. 1994. *A Conceptual History of Modern Embryology*. Baltimore: Johns Hopkins University Press.

Gould, Stephen Jay. 1977. *Ontogeny and Phylogeny*. Cambridge, MA: Harvard University Press.

Haeckel, Ernst. 1866. *Generelle Morphologie der Organismen*. Berlin: Reimer.

Haeckel, Ernst. 1876. *Evolution of Man*. Akron, OH: Werner Company.

Hall, B. K. 1992. *Evolutionary Developmental Biology*. London: Chapman and Hall.

Hall, B. K. 1998. *Evolutionary Developmental Biology*. 2nd ed. Dordrecht: Kluwer.

Hall, B. K. 2000. Evo-devo or devo-evo—Does it matter? *Evolution and Development* 2: 177–178.

Laubichler, M. D. 2003. Carl Gegenbaur (1832–1903): Integrating Comparative Anatomy and Embryology. *Journal of Experimental Zoology: Part B Molecular and Developmental Evolution* 300B: 23–31.

Laubichler, M. D. 2005. A constrained view of evo-devo's roots. *Science* 309: 1019–1020.

Laubichler, M. D., 2007. Does history recapitulate itself? Epistemological reflections on the origins of evolutionary developmental biology. In Manfred D. Laubichler and Jane Maienschein, eds., *From Embryology to Evo Devo.* Cambridge, MA: MIT Press.

Laubichler, M. D., and J. Maienschein. 2003. Ontogeny, anatomy, and the problem of homology: Carl Gegenbaur and the American tradition of cell lineage studies. *Theory in Biosciences* 122: 194–203.

Laubichler, M. D., and J. Maienschein, eds. 2007. *From Embryology to Evo Devo.* Cambridge, MA: MIT Press.

Laubichler, M. D., and H.-J. Rheinberger. 2004. Alfred Kühn (1885–1968) and developmental evolution. *Journal of Experimental Zoology Part B: Molecular and Developmental Evolution* 302B: 103–110.

Maienschein, Jane. 1991. *Transforming Traditions in American Biology, 1880–1915.* Baltimore: Johns Hopkins University Press.

Maienschein, Jane. 2000. Competing epistemologies and developmental biology. In Richard Creath and Jane Maienschein, eds., *Biology and Epistemology*, 122–137. Cambridge: Cambridge University Press.

Maynard-Smith, J., R. Burian, S. Kauffman, P. Alberch, J. Campbell, B. Goodwin, R. Lands, D. Raup, and C. Wolpert. 1985. Developmental constraints and evolution. *Quarterly Review of Biology* 60: 265–287.

Mayr, E. 1982. *The Growth of Biological Thought: Diversity, Evolution, and Inheritance.* Cambridge, MA: Belknap Press.

Mayr, Ernst. 1999. Personal discussion during Dibner Institute Seminar in the History of Biology: "Why Haeckel?", Marine Biological Laboratory, Woods Hole, MA.

Mayr, E., and W. B. Provine, eds. 1980. *The Evolutionary Synthesis: Perspectives on the Unification of Biology.* Cambridge, MA: Harvard University Press.

Mocek, R. 1998. *Die werdende Form: Eine Geschichte der kausalen Morphologie.* Marburg: Basilisken Presse.

Morgan, Thomas Hunt. 1932. *The Scientific Basis of Evolution.* New York: Norton.

Nyhart, Lynn. 1995. *Biology Takes Form.* Chicago: University of Chicago Press.

Nyhart, L. K. 2002. Learning from history: Morphology's challenges in Germany circa 1900. *Journal of Morphology* 252: 2–14.

Paley, William. [1802] 1972. *Natural Theology.* Houston: St. Thomas Press.

Provine, W. B. 1971. *The Origins of Theoretical Population Genetics*. Chicago: University of Chicago Press.

Raff, Rudolf A., and Thomas C. Kaufman. 1983. *Embryos, Genes, and Evolution*. New York: Macmillan.

Raff, Rudolf A., and Elizabeth C., Raff, eds. 1987. *Development as an Evolutionary Process*. New York: Alan Liss.

Rheinberger, H. J. 2000. Ephestia: The experimental design of Alfred Kuhn's physiological developmental genetics. *Journal of the History of Biology* 33(3): 535–576.

Richards, Robert J. 1992. *The Meaning of Evolution*. Chicago: University of Chicago Press.

Riedl, R. 1975. *Die Ordnung des Lebendigen*. Hamburg: Parey.

Roush, Wade. 1997. Evo-devo funding: "Still only a trickle." *Science* 277: 39.

Roush, Wade, and Elizabeth Pennisi. 1997. *Growing pains: Evo-devo researchers straddle cultures*. *Science* 277: 38–39.

Sarkar, S. 1998. *Genetics and Reductionism*. Cambridge: Cambridge University Press.

SICB (Society for Integrative and Comparative Biology). Fall 1999. *Newsletter*, 2.

Wagner, G. P., C.-H. Chiu, and M. Laubichler. 2000. Developmental evolution as a mechanistic science: The inference from developmental mechanisms to evolutionary processes. *American Zoologist* 40: 819–831.

Weismann, August. 1892. *Das Keimplasma: Eine Theorie der Vererbung*. Jena: Gustav Fischer.

Wilson, Edmund B. 1896. *The Cell in Development and Inheritance*. New York: Macmillan.

Wilson, Edmund B. 1915. Some aspects of progress in modern zoology. *Science* 41: 1–11.

2 The Organismic Systems Approach: Evo-Devo and the Streamlining of the Naturalistic Agenda

Werner Callebaut, Gerd B. Müller, and Stuart A. Newman

In this chapter we first discuss the emergence of evolutionary developmental biology (evo-devo) as a response to the essential incompleteness of the modern synthesis. In the second section, we provisionally characterize evo-devo in terms of its conceptual framework, methods, and explanatory strategies, and try to introduce some order into the plurality of its theoretical perspectives. Here we also suggest that a philosophical naturalism committed to causal-mechanistic explanation is the philosophical theory that best fits our own organismic systems approach (OSA) to evo-devo and evolution at large. In the third section, we argue that a proper understanding of evo-devo requires a reconceptualization of the relationship between what counts as genetic and what as epigenetic. In the fourth section, which constitutes the heart of the chapter, we sketch some major features of OSA. Our central concerns are the generic, conditional (i.e., unprogrammed) *generation* of primordial organismal form and structure ("origination"), the question of evolutionary *innovation*—how novel elements arise in body plans—and the factors of *organization*, that is, how structural elements and body plans are established. The fifth section deals with some of the major problems that OSA allows us to address: tempo and mode of phenotypic evolution, selection and emergence, integration, and inherency. To round off in the final section, we return to more philosophical issues; in particular, we discuss the kinds of nonreductionistic unification that are at stake in evo-devo in relation to the data- and technique-driven nature of much current research.

As a package, the OSA/naturalistic account allows us to turn the tables on the adaptationist, gene-centric view and take seriously again Walter Garstang's dictum (in 1922) that ontogeny does not recapitulate phylogeny, but rather *creates* phylogeny (Gilbert 2003a, 777). More specifically, in Løvtrup's (1984, 261) words, "Inverting Haeckel's biogenetic law, we may assert that ontogeny is the mechanical cause of phylogeny. And it must be so, for ontogeny is a mechanical process, while phylogeny is a historical phenomenon."

Unfinished Synthesis

Until about 1800, biological phenomena were subsumed under two labels: medicine (comparative anatomy, morphology, and physiology) and natural history.[1] This was a "remarkably perceptive division" (Mayr 1982, 67), for biology can be divided into the study of proximate or short-term causation and the study of ultimate causes. The former are the province of functional biology or the physiological sciences, broadly conceived, which seek answers to "How?"-questions; the latter are the subject of evolutionary biology, which seeks answers to "Why?"-questions (Mayr [1961] 1988, 1982). Even "greedy" reductionists admit the necessity of taking into account *historical contingencies* in evolutionary explanation, and grant that this circumstance makes biology essentially different from (classical) physics.[2] Under the spell of Weismannism, informational accounts of ontogeny such as Mayr's shift attention away from ontogenetic agency.[3] Independently of Mayr, Tinbergen (1963) recognized that every trait requires both proximate and ultimate explanation. Inspired by Lorenz's work in ethology, he also realized that these explanations are quite different depending on whether the explanandum is a single form or a sequence. Tinbergen's "four questions"—now commonly referred to as proximate mechanisms, fitness advantage, ontogeny, and phylogeny—along with his view that they are independent yet complementary components of any complete biological explanation—have become widely accepted not only for animal behavior, but for biology in general.[4]

The interrelation between proximate and ultimate causation is the central problem of any explanatory theory of organismal form (Müller 2005). Hall (2000) characterizes evo-devo, which he views as continuous with the evolutionary embryology of the nineteenth century, as a way of integrating proximate and ultimate causes for the origin of phenotypes. Although embryology—which later came to be called developmental biology—played an important, if not indispensable, role in the inception of evolutionary theory (Richards 1992, 2002), it was conspicuously absent from the modern synthesis. Müller (2005, 89–90) discusses how the disrepute of recapitulationism and of neo-Lamarckian beliefs about environmental influences on embryogenesis contributed to this ostracism. In the past two decades, stimulated by a number of explanatory deficits in the prevailing evolutionary paradigm and by the rise of molecular developmental genetics, developmental considerations have come to the fore again with a vengeance (Laubichler and Maienschein, chapter 1, this volume).

The success story of the modern synthesis is being told and retold every day by both the popular scientific media and the professional Darwin industry. (For more or less critical assessments, see Mayr and Provine 1980; Depew and Weber 1985, 1995; Burian 1988; Futuyma 1988; Gayon 1989; Smocovitis 1996; Weber 1998; Keller 2000a; Gould 2002.) Yet a sentiment has accumulated in the last two decades that extant evolutionary theory, in which the reigning mode of explanation is genetic (Lewontin 2000a) and adaptationist (Gould and Lewontin 1979), represents an "unfinished synthesis" (Eldredge 1985; Reid 1985).[5] The synthetic theory of evolution worked well at the population genetic level it concentrated on, but population genetics is far from the whole story.[6] The modern synthesis was increasingly found to run into difficulties when it came to explaining characteristics of *phenotypic evolution* such as biased variation, rapid changes of form, the occurrence of nonadaptive traits, or the origination of higher-level organization such as homology and body plans (Müller and Newman 1999; Laubichler 2000; Wagner and Laubichler 2004; Müller 2005).

Evo-devo, broadly conceived, is addressing seriously Viktor Hamburger's (1980) complaint that the modern synthesis treated the processes of ontogeny as a "black box." Current calls for a "new evolutionary synthesis" that would genuinely incorporate development (Carroll 2000; Gilbert 2003a, 777–779), or for a straightforward "developmental synthesis" (Amundson 1994), revive Waddington's lifelong but ultimately unsuccessful attempt to forge a "final" unification of embryology and evolution (Wilkins 1997; Keller 2000b, 251; Van Speybroeck 2002).[7] Gene-centered biology comes dangerously close to officially long-rejected preformationism.[8] Impressed by the close mapping between genotype and phenotype in many kinds of modern organisms, it takes an organism's morphological phenotype—the physical, organizational, and behavioral expression of an organism during its lifetime—to be "determined" by its genotype—the heritable repository of information that "instructs" the production of molecules whose interactions, in conjunction with the environment, generate and maintain the phenotype. It thus portrays the development of individual organisms as the unfolding of a "genetic program" alleged to reside in the fertilized egg.[9] (For critical assessments, see, e.g., Varela 1979; Atlan and Koppel 1989; Rose 1998; Moss 2002; Robert 2004.) Yet, as one dissenter puts it, "The organism does not compute itself from its genes. Any computer that did as poor a job of computation as an organism does from its genetic 'program' would be immediately thrown into the trash and its manufacturer would be sued by the purchaser" (Lewontin 2000a, 17). The evolutionary-theoretical version of the notion of genetic program usually remains unchallenged even in glosses on neo-Darwinism that otherwise disagree over such issues as the tempo and mode of phenotypic evolution, the degree to which genetic change can result from selectively neutral mechanisms, and the universality of adaptation in accounting for complex traits (but see Neumann-Held and Rehmann-Sutter 2006).

Life did not necessarily require DNA to get started (see, e.g., Rosen 1990, Goodwin 1994, Maynard Smith and Szathmáry 1998, Fry 2000, and Hazen 2005, for reviews of theories of the emergence of biological order). However, it has long entered a regime in which genes are not only indispensable for both development and evolution, but have become increasingly privileged—though never in an absolute sense (Maynard Smith and Szathmáry 1995, chap. 6; Gilbert 2003b; Newman 2005). We thus agree with Schaffner (1998) that the developmental systems perspective (DSP)[10] goes too far in claiming that other developmental resources are "on a par" with genes (Griffiths and Gray 1994, 283; cf. Oyama 2000 on "parity of reasoning"). As Gilbert (2003b, 349) observes, DSP has generally failed to distinguish between *instructive* and *permissive* influences on epigenetic interactions.[11] The official "interactionist consensus" (Robert 2004) pays lip service to the realization that genes alone are insufficient for developmental agency (see, e.g., Gray 1992 and Moss 2002 for forceful arguments to this effect), but require the genotype of the developing cells to interact with their cellular environment for the production of phenotypes. Yet dominant practice still largely disregards the generative—and hence explanatory—roles the phenotype per se plays in many biological phenomena. This is true in at least two important ways.

First, at the developmental level, there is no basis for assuming that informational models according to which "all the relevant information" resides in the genomic DNA suffice to capture all or even most of the relevant developmental data (Sarkar 2000). In light of this fundamental limitation of the informational account of ontogeny, Keller's (2000b, 245) suggestion that genetics and molecular biology are "committed to the goals of causal-mechanistic explanation" must be qualified. She claims that the "long-sought efficient causality" came to biology "with the identification of genes (coupled with the causal properties attributed to them in prevailing notions of 'gene action'), and with the more definitive identification of DNA as the carrier of genetic information (coupled with the notion of a 'genetic program')." But only on particular contemporary philosophical interpretations of causation, such as the view that a causal process involves the transfer of (a particular token of) information from one state of a system to another (the so-called conserved-quantity view: Dowe 1992, 2000; Salmon 1994; Collier 1999), can the informational account of ontogeny be considered causal.[12] Genuine causal analysis, taking fully into account epigenetic and environmental determinants, remains as indispensable as ever—remember *Entwicklungsmechanik* (Laubichler and Maienschein 2007a)![13]

Second, a similar limitation exists at the evolutionary level, where a full understanding of selection requires that we capture the regulatory phenotypic structures and functions of species. Again the informational picture must remain incomplete: "One can construct mathematical 'laws' of genetics within the information picture, but one cannot assert that they apply, nor understand the limits of their application, nor understand the alternative possibilities where their application breaks down, without recourse to the details of the causal picture" (Hahlweg and Hooker 1989, 85–86).[14] For evolutionary biologists, the emerging focus on developmental mechanisms is "an opportunity to understand the origins of variation not just in the selective milieu but also in the variability of the developmental process, the substrate for morphological change" (von Dassow and Munro 1999, 307). Adding a causal-mechanistic dimension to, say, the historical study of character evolution also allows one to test evolutionary hypotheses more rigorously (Autumn, Ryan, and Wake 2002). Note that the great debates within evolutionary biology focus on the causal mechanisms of evolutionary change rather than on the elucidation of particular phylogenetic relationships (Atkinson 1992). Opposing developmental biology to evolutionary biology, then, in terms of the distinction between causal-mechanistic explanation (answering "How?"-questions) versus functional explanation (answering "Why?"-questions) would be an oversimplification.[15]

"Dawkinsspeak" has by now become so prevalent in our media-driven culture that it may no longer be trivial to point out that, for instance in the case of our own species, there simply *aren't enough genes* to "determine" the phenotype in the simplistic ways envisaged by most behavioral geneticists and evolutionary psychologists, among others. For instance, genes cannot incorporate enough instructions into the structure of the human brain to "program" an appropriate reaction to most—let alone all—conceivable behavioral situations. As Ehrlich (2000, 124) reminds us, there are roughly 100–1,000 trillion connections (synapses) between more than a trillion nerve cells in our brains; that is, "at least 1 *billion* synapses per gene, even if every gene in the genome contributed to creating a synapse." Ehrlich concludes: "Clearly, the characteristics of that neural network can be only partially specified by genetic information; the environment and cultural evolution *must* play a very large, often dominant role in establishing the complex neural networks that modulate human behavior."

It may be too early to say how modest or revolutionary the changes in our thinking about evolution occasioned by evo-devo will be (see Gilbert 2003b). But it is certainly timely to observe the profoundly paradoxical nature of the gene-centrism that characterizes the currently dominant evolutionary paradigm. Its fatalist view of phenotypes as "vehicles" that are the prisoners of their genes (Dawkins's "lumbering robots") reigns supreme at a time when the genomes of plants, animals, and increasingly also humans are dramatically transformed *by human agency* in the name of health, economic profit, and other societal aims, and when findings in molecular biology itself (e.g., Lolle et al. 2005) call into question the very "Central Dogma" that the fatalist view requires. There is much to be said, then, for Tibon-Cornillot's (1992, 17) diagnosis of this paradox as a crisis of the neo-Darwinian paradigm (cf. also Strohman 1997).

To set the scene for our discussion of OSA, we will in the subsequent two sections consider some of the requirements for evo-devo to become a mature scientific field. (For a fuller treatment, we refer to Müller and Newman 2005; Müller 2006.)

Evo-Devo: A Clash of Weltanschauungen?

Reflecting on the renewed interest in the relationship between evolutionary biology and developmental biology that awoke in the 1980s, Atkinson (1992, 93) speculated that their reunion "may result in a new subdiscipline of biology, if there is a set of unique concepts and methods which tie the various research approaches together." Although we think more is involved in the articulation of evo-devo than conceptual and methodological issues, we discuss concepts and methods in turn as a first approximation to evo-devo and add two more issues: its explanatory strategies and what we call "evo-devo packages."

Elements of a Conceptual Framework for Evo-Devo

Anticipating current attempts to differentiate evo-devo from more established research in developmental biology and evolutionary biology (Gilbert, Opitz, and Raff 1996; Hall 2000; Raff 2000; Wagner, Chiu, and Laubichler 2000; Robert, Hall, and Olson 2001; Amundson 2005; Burian 2005; Müller 2005, 2006), Atkinson (1992) suggested that concepts such as *bauplan* (Riedl 1978), *canalization* (the reduced sensitivity of a phenotype to changes or perturbations in the underlying genetic and nongenetic factors that determine its expression: Wagner, Booth, and Bagheri-Chaichian 1997; Wilkins 1997; Gibson and Wagner 2000), and *developmental constraints* (see below) may serve in such a capacity.

Today we must add several more concepts to this list. One is *inherency*, the propensity of biological materials to assume preferred forms (Newman and Comper 1990; Newman and Müller 2006). Another is *evolvability*, the intrinsic potential of certain lineages to change during the course of evolution (Wagner and Altenberg 1996; Gerhart and Kirschner 1997; Conrad 1998; Kirschner and Gerhart 1998; Newman and Müller 2000, Schank and Wimsatt 2001; Hansen 2003; Earl and Deem 2004; Wagner and Laubichler 2004; Deacon 2006).[16] Still another crucial concept is *developmental modularity* (Wagner 1996; von Dassow et al. 2000; Gilbert and Bolker 2001; Stelling et al. 2001; Schlosser and Wagner 2002; Hall 2003; Burian 2005; Callebaut and Rasskin-Gutman 2005). Developmental systems are decomposable into components that operate largely according to their own, intrinsically determined principles. These components can be selected phenotypically. A developmental module consists of a set of genes, their products, their developmental interactions, and their interactions with epigenetic factors, including the resulting character complex and its functional effect. The genes affecting the modular character complex are typically characterized by a high degree of internal integration and a low degree of external connectivity—that is, pleiotropic relations are largely within-module. These modules, Brandon (1999, 177) claims, are *the* units of evolution by natural selection. Whether this verdict will bring the units and levels-of-selection debate to a stop remains to be seen. At any rate, as a principle connecting the genetic and epigenetic components of evolving developing repertoires, modularity has the potential to assume a central role in the evo-devo framework (von Dassow and Munro 1999).

Most important, from our perspective, we should add *evolutionary origination*, *innovation*, and *novelty*, a set of interrelated concepts collectively called the "innovation triad" (Müller and Newman 2005). In this domain, it is suggested, evo-devo could make its most pertinent contribution to evolutionary theory. We return in a later section to these concepts and the central role *homology* plays in them.

It should be noted that Wagner et al. (2000, 829) distinguish between the contributions evo-devo makes to the agenda of established research programs, such as the evolution of adaptations, and the genuine contributions of evo-devo, such as developmental constraints, evolutionary innovations, and the evolution of development in general.[17]

This much being said, it should be granted that evo-devo is still in search of a clear conceptual framework in which the aforementioned concepts, as well as related ones such as *robustness*—the capacity of organisms to compensate for perturbations (Fontana 2002; de Visser et al. 2003; Kitano 2004; A. Wagner 2005; Hammerstein et al. 2006)—can be integrated in a logically coherent and cautiously parsimonious way. A first necessary step in this direction may be to elucidate what specifically constitutes a *developmental mechanism* (von Dassow and Munro 1999; Salazar-Ciudad, Newman, and Solé 2001, Salazar-Ciudad, Jernvall, and Newman 2003). Increased clarity on what the proper questions and goals of evo-devo should be is another urgent desideratum, and may actually be easier to reach than one might expect (Wagner et al. 2000; Robert, Hall, and Olson 2001; Fontana 2002; Müller 2005).[18]

Evo-Devo Methods

Comparative developmental biology, experimental developmental biology (Müller 1991), and evolutionary developmental genetics are the methods most commonly relied on in evo-devo. These methods clearly are not unique to evo-devo (Müller 2006), and one should not assume that their combined application will automatically yield unproblematic novel results (see von Dassow and Munro 1999). When Haeckel proposed his views on recapitulation, the union of evolution and development he envisaged was primarily one of descriptive sciences. For Haeckel, as for Darwin before him, comparative embryology served as the source of knowledge in the construction of phylogenies (Richards 1992). Although the nexus between development and evolution may remain descriptive to a large extent (Nijhout, chapter 3, this volume), developmental biology (e.g., Gerhart and Kirschner 1997; Hall 1999; Gilbert 2003a; Carroll, Grenier, and Weatherbee 2004) has come a long way and today offers a wealth of causal-mechanistic theories and models—also, but not only, because it has become molecular (Robert, Hall, and Olson 2001; Müller 2005, 2006).

Biologists have long recognized that the evolvability of development is the key to morphological evolution. But molecular genetics is only now producing the substantial characterizations of developmental regulatory processes that make it possible to analyze them comparatively with regard to common molecular characteristics across a wide range of species (Wilkins 2001; Carroll, Grenier, and Weatherbee 2004).

In an attempt to zero in on the role of developmental mechanisms in evolutionary processes, and in particular evolutionary innovation, Wagner, Chiu, and Laubichler (2000) propose a checklist for establishing a causal link between molecular developmental evolution and phenotypic evolution. First, what is the developmental mechanism that accounts for the derived character (or character state)? More precisely, does the identified mechanism account specifically for the derived character state, or is it responsible for more fundamental and more ancient, plesiomorphic, character states within the lineage? Second, does the developmental mechanism for the derived character (or character state) map to the same node on the phylogeny as the derived character (or character state)? *Coincidence* is the first test of the causal efficacy of the mechanism in the evolutionary process. Only mechanisms that were active in the ancestor can account for the evolutionary origin of the character. The test for this may be difficult because of the recency bias built into cladistic character reconstruction methods: by the very nature of the cladistic method applied to recent species, the deepest node that in principle can be reconstructed is the most recent common ancestor of the extant clade (Wagner, Chiu, and Laubichler 2000, 824). Third, what are the developmental changes that occurred at the origin of the derived character (or character state)? As Wagner and colleagues see it, the study of the sequence evolution of developmental regulatory genes is a powerful tool for detecting candidate mutations that may be responsible for developmental and phenotypic transformations, yet this method has inherent limitations. Fourth, are the genetic differences sufficient to cause the derived character (or character state)? Advances in transgenic technology "make the thought of performing these experiments less crazy than [it was] a few years ago" (p. 828).

If the answers to all these questions support a hypothesis about the developmental mechanism for the origin of a novel phenotypic character, the conclusion is "all but unavoidable that this mechanism in fact was instrumental in causing the origin of the derived character (or character state) (Wagner, Chiu, and Laubichler 2000, 829)." What we call *evolutionary developmental epigenetics* is as relevant here as the more familiar evolutionary-developmental genetics (see below for examples).

The Explanatory Strategies of Evo-Devo

From an epistemological point of view, what current evo-devo seems to be largely lacking is a way to get from the knowledge of parts (entities and their properties) and what they do (process, activity)—knowledge that a field like developmental genetics increasingly provides—to a full-fledged, formal understanding of developmental phenomena. "Mechanism, per se, is an explanatory mode in which we describe what are the parts, how they behave intrinsically, and how those intrinsic behaviors of parts are coupled to each other to produce the behavior of the whole" (Von Dassow and Munro 1999, 309). This commonsense definition of mechanism implies an *inherently hierarchical decomposition* (Von Dassow and Munro 1999, 309; cf. Glennan 1996, 2002a, 2002b, on mechanism as interaction of parts; Machamer, Darden, and Craver 2000, 13–18, on mechanism as activity; Callebaut 2005a on the ubiquity of nearly decomposable systems in nature), which has historically been unattainable to developmental biologists because their knowledge of the parts was insufficient.[19] Instead, for explanation we still rely primarily on a sort of *local causation*: "The operational approach of developmental genetics, for instance, takes for granted the organism as a working whole, with no general assumptions about the nature of developmental mechanisms beyond the empirical fact that some fraction of mutations have discrete, visible effects. Developmental genetics begins with an induced anomaly (a mutation) and a (hopefully discrete) consequence, then proceeds to decipher a perturbation-to-consequence chain" (Von Dassow and Munro 1999, 309). This kind of account allows a description of what parts (genes and their products) do to each other, but it "doesn't articulate any sense of the mapping from genotype to phenotype, which is what we ultimately want" (p. 309; see also Hall 2003).[20] An alternative explanatory strategy focuses on the epigenetic factors that are causal in the evolution and organization of phenotypes. Here the generic physical properties of cells and tissue masses and their self-organizational properties are given priority over the molecules used in these processes (Müller and Newman 2003a, 2005, Newman 2005; Newman and Müller 2000, 2005).

Evo-Devo Packages

Our discussion up to now may have suggested that evo-devo is more monolithic than it actually is. In fact, various accounts of evo-devo exist, which in part reinforce one another, but in part also compete (box 2.1).

Gene selectionism (e.g., Cronin 1991; Williams 1966, 1992) provides the antiposition for all these views, as well as for a number of additional, complementary perspectives (box 2.2). It holds that the organism is but "the realization of a programme prescribed by its heredity" (Jacob 1970, 2) and that "the details of the embryonic developmental process, interesting as they may be, are irrelevant to evolutionary considerations" (Dawkins 1976, 62; but see Maynard Smith 1998). Recent work on niche construction (Laland, Odling-Smee, and Feldman 1999; Odling-Smee, Laland, and Feldman 2003), to the extent that it only pays lip service to development (Wimsatt and Griesemer, chapter 7, this volume), would also seem to be-

Box 2.1
Main evo-devo packages

> The main contributors to evo-devo combine theoretical, methodological, and epistemological perspectives with a genuine research *practice*. Following Gilbert 2003b, we view *core evo-devo* as encompassing both the program to explain evolution through changes in development and the program to reframe evolutionary biology more drastically along developmental lines (e.g., Raff 1996, 2000; Hall 1999, 2000; Gilbert 2003a; see figure 2.2). *Gene regulatory evolution* is a research program that arises from the methodical application of comparative molecular genetics to developmental biology. Its principal aim is the elucidation of the "developmental genetic machinery" (Arthur 2002) and the evolution of gene regulation. Presently it represents the most widespread and empirically productive approach to evo-devo (e.g., Akam 1995; Holland 2002; Carroll et al. 2004; Holland and Takahashi 2005; Davidson 2006). *Epigenetic evo-devo* is the main subject of the fourth section of this chapter. *Process structuralism* (e.g., Ho and Saunders 1979; Goodwin et al. 1983; Goodwin 1984, 1994; Webster and Goodwin 1996; cf. Depew and Weber 1995, Griffiths 1996, and Eble and Goodwin 2005) is a largely ahistorical approach that aims to explain form in terms of morphogenetic laws and generative principles. It regards organisms as fitting into timeless categories ("natural kinds"), similar to the periodic table of elements. Approaches to the *self-organization of biological complexity* abound and are multifarious, ranging from Heinz von Foerster's "order through noise," elaborated by Henri Atlan (e.g., Atlan and Koppel 1989; Fogelman Soulié 1991), to Stuart Kauffman's "adaptation on the edge of chaos" (e.g., Kauffman 1985, 1992, 1993, 1995) to the Prigogine school in far-from-equilibrium thermodynamics (e.g., Prigogine and Nicolis 1971; Prigogine and Goldbeter 1981; Nicolis 1995; cf. Camazine et al. 2001). Depew and Weber (1995) provide an excellent review (see also Callebaut 1998; Richardson 2001).

long to the gene selectionist paradigm, and so does the extended replicator perspective of Sterelny, Smith, and Dickison (1996).

A complication arises from the circumstance that not just competing theories—broad intellectual perspectives—are at stake. Process structuralism, self-organizational approaches to complexity, the dialectical account, or DSP—to name but a few examples—are typically presented as "packages" that not only include their own methodology and explanatory strategies, but are embedded in a broader "philosophy."

In this sense, evo-devo perspectives may be likened to (certain components of) paradigms, in which a characteristic set of metaphors play

Box 2.2
Additional evo-devo perspectives

> *Structural modeling (SM)* extends the explanation of phenotypic evolution from fitness considerations alone—the classical approach—to the topological structure of phenotype space as induced by the genotype-phenotype map (e.g., Fontana and Schuster 1998; Stadler and Stadler 2006). SM focuses on key aspects of evo-devo, but does not offer a representation of organismal development per se. The regulatory networks of gene expression and signal transduction that coordinate the spatiotemporal unfolding of complex molecules in organismal development have no concrete analogue in the RNA sequence-to-structure map. Yet the RNA folding map implements concepts like epistasis and phenotypic plasticity, thus "enabling the study of constraints to variation, canalization, modularity, phenotypic robustness and evolvability" (Fontana 2002, 1164). The *dialectical account* of biology urges thinking in terms of the organic construction of environments and the interpenetration between organisms and environments rather than adaptation (Levins 1968; Rose 1982a, 1982b, 1998; Rose et al. 1984; Levins and Lewontin 1985, 2006; Lewontin 1996, 2000a; Rose and Rose 2000; cf. Godfrey-Smith 2001b). *Systems biology* aims at explaining life mechanistically at the system (primarily cell) level. Whereas older work in this area, most notably biological cybernetics, had to concentrate on phenomenological analysis of physiological processes (cf. also Piaget, below), systems biology can take advantage of the rapid progress in molecular biology, furthered by technologies for making comprehensive measurements on DNA sequence, gene expression profiles, protein-protein interactions, and so on (e.g., Kitano 2001, 2002, 2004; Bruggeman et al. 2002; Hood and Galas 2003; Krohs and Callebaut 2007). Extant systems biology calls for a better awareness of the importance of organizing principles, in particular hierarchical organization (Mesarovic 1968; Newman 2003b; Mesarovic et al. 2004). Jean Piaget's attempt at a *cybernetic synthesis* (Cellérier et al. 1968; Piaget 1971, 1978, 1980), aptly summarized by Parker (2005) (see also Hendriks-Jansen 1996), is listed here as early attempt to articulate a phenotype-centered theory of biological and cognitive development and evolution—presumably flawed by Piaget's endorsement of Lamarckian inheritance (Deacon 2005; but see Hooker 1994). Griesemer's *reproducer perspective (RP)* is a process perspective that aims at understanding development, inheritance, and evolution by way of analyzing the processes that take place throughout an evolutionary hierarchy of levels of productive organization. The core idea of RP is *material overlap*: offspring form from organized parts of parents, such as cells, rather than by the copying of traits (Griesemer 2000a, 2000b, 2005, 2006b; Wimsatt and Griesemer, chapter 7, this volume). Reproducers can be important units of culture and cultural evolution, despite the

Box 2.2
(continued)

> common assumption that culture depends on transmission of units that are in some other way nonbiological. The *developmental systems perspective (DSP)* is characterized in notes 10 and 11, and its "engine," Wimsatt's *generative entrenchment*, in note 34. Caporael's (1997) *core configurations* model, based on human morphology and ecology in human evolutionary history, substitutes the concept of *repeated assembly* (of organisms, groups, habits, and so forth) for nature-nurture dualism. For Caporael, core configurations of face-to-face groups are the selective context for uniquely human mental systems (cf. also Wimsatt and Griesemer, chapter 7, this volume). A comparison of Caporael's and other recent calls for a "cognitive evo-devo" (e.g., Hernández Blasi and Bjorklund 2003) with OSA falls beyond the scope of this chapter.

exemplary heuristic and unificatory roles (see Jablonka 2000), and typically also a negative role—demarcation from other approaches.[21] In Lewontin's (1996, 1) words,

All sciences, but especially biology, have depended on dominant metaphors to inform their theoretical structures and to suggest directions in which the science can expand and connect with other domains of inquiry. Science cannot be conducted without metaphors. Yet, at the same time, these metaphors hold science in an eternal grip and prevent us from taking directions and solving problems that lie outside their scope.

(See also Nijhout 1990; Lewontin 2000a, chap. 1.) Metaphors also mediate between specific scientific cultures and culture at large (see, e.g., Schlanger 1971; Haraway 1976; Tibon-Cornillot 1992).[22]

Although one may be tempted to cluster the various approaches to evo-devo even further—thinking of, for instance, Goodwin's (1994) vindication of Kauffman's (1985, 1993) nonlinear models, or the tendency of advocates of DSP to portray the critical work of Lewontin and his associates as belonging to it (e.g., Oyama 2000, 8; Gray 2001, 200; Griffiths and Gray 2005, 418–419), or the suggestion that generative entrenchment (Wimsatt 2001) provides DSP with an "engine" (Robert 2004, 126)—the conceptual and methodological differences between these approaches remain considerable. We feel that in this incipient stage of the articulation of evo-devo it is wiser to cherish and cultivate conceptual and social variety than to apply Occam's razor.

Our distinction between the main evo-devo packages and additional perspectives on evo-devo is mostly based on considerations such as the fit between ideas professed and actual research practice, which is most obvious in the former case (with "practice" typically preceding "theory"). In contrast, the holistic incantations of the advocates of a dialectical biology, to take this example, often seem quite disconnected from their own research practice, which may be decidedly reductionistic. Or, in the case of DSP, most of the important research that its advocates invoke as endorsing their perspective (Griffiths and Gray 2005, 417–418), such as Gilbert Gottlieb's work in developmental psychobiology and Timothy Johnston's in learning theory, predates the articulation of their own perspective (see, e.g., Johnston 1982), which makes it difficult if not impossible to assess DSP's constructive (e.g., heuristic) potential at present (see Godfrey Smith 2001a). Other considerations relevant to our distinction are straightforwardly sociological: intellectual distance to core evo-devo (quantifiable by means of bibliometric methods), perspectives that are at present advocated by a single person or small group, or those that are currently taken by many to be obsolete (the Piagetian cybernetic synthesis). Evo-devo has incorporated many of the arguments of the critics of the synthetic theory but seems more successful at marrying its theory and practice than most of its contenders.

Griesemer (2000a, 2002) usefully distinguishes between theories, perspectives, and images. Following Levins (1968), he views *theories* as "collections of models and their robust consequences" that represent and interpret phenomena (Griesemer 2000a, 350; cf. Wimsatt 1997 and Giere 1999). Theoretical *perspectives* coordinate models and phenomena; such coordination is necessary because phenomena are complex, our scientific interests in them are heterogeneous, and the number of possible ways of representing them in models is too large. Adequate theorizing may require a variety of perspectives and models—a point worth keeping in mind in discussing what the "right" account of evo-devo is. Genetic determinism is an instance of a perspective. Theoretical perspectives are expressed in *images* that specify preferred lines of abstraction from phenomena of interest and prioritize principles in terms of which models may be constructed to represent phenomena. Weismannism has been a historically immensely powerful image of the genetic-determinist perspective, as was Watson's central-dogma diagram in the second half of the twentieth century. Many biologists (e.g., Hood and Galas 2003) today still think of the genetic-program image as heuristically fruitful.

What is an image, specifically? One function of Kuhnian exemplars is to make phenomena intelligible to students, researchers, and the public at large. (That the genetic-program image is catchy is plain enough.) Images that display a mechanism, especially if they are visual, are particulary apt to engender intelligibility, because they "show *how possibly, how plausibly*, or *how actually* things work" (Machamer, Darden, and Craver 2000, 21). The intelligibility of an image must not be confounded with the correctness of an explanation: images often will convince us even if, as (shorthand for) explanations, they are misleading; see de Regt's (2004) discussion of understanding. Intelligibility typically also evokes emotions, which in turn may act as positive or negative motivators.

Coming back to the general issue of evo-devo packages, we must ask to what extent the close links that are often suggested between specific theoretical notions and specific metalevel views are really robust or even inevitable or rather the result of the contingencies of one's scientific and philosophical education, social background, or standing. This question is important because, as, for example, Maienschein (2000) has shown for several of the preformation-versus-epigenesis debates in the eighteenth and nineteenth centuries, "epistemology (can) actually... drive the science" (p. 123). Smith (1992, 433) exemplifies a tendency to posit the existence of a neorationalist, "European view of metatheory," which is taken to be "more holistic and less empirical than its American counterpart" (p. 449). He associates this view with Riedl's and Wagner's systems-theoretical analyses (e.g., Riedl 1978; Wagner 1986) as well as with Goodwin's process structuralism (e.g., Goodwin 1984, 1994). Although there is some truth in Smith's view (Callebaut 1998), his picture is too undifferentiated. For instance, Riedl's evolutionary epistemology, which he viewed as an integral part of his theoretical biology (Riedl 1978; Wagner and Laubichler 2004), ultimately comes closer to Humean empiricism, with its emphasis on regularity rather than causation-as-production, than it resonates with the scientific-realist view that now dominates Anglo-American philosophy of biology (Callebaut 2005b). On the other hand, Hooker (1995, 227) and Mahner and Bunge (1997, 294–299), among others, have argued convincingly that there is no need to accept such a dichotomy as suggested by Smith. OSA assumes that one can resituate neo-Darwinian processes within a complex systems framework without going to either empiricist or rationalist extremes (see Depew and Weber 1995; Robert 2004).[23]

Against the background of evo-devo packages, the issue of the modesty or ambition of the various perspectives is not merely a tactical question, let alone a rhetorical flourish. At stake are the actors' deepest methodological, epistemological, and metaphysical convictions, their scientific ethos, and perhaps even their politics and ideology. (See, for instance, Ruse 1999, who discerns a steady increase in the influence of epistemic values and a concomitant diminution of wider cultural influences in the history of evolutionary theorizing since Erasmus Darwin.) Contrast, for instance, George Williams's or Edward Wilson's gene-selectionist austerity (Williams 1966, 1985, 1992; Wilson and Lumsden 1991) with Gould's pluralist playfulness.[24] Compare Maynard Smith's (1998, 45) "Reductionists to the Right, Holists to the Left," whose moral is that "it pays to be eclectic in our choice of theories," being "reductionist in one context and holist in another." As scientific realists, we are tempted to add that one should also realize that in the absence of any direct access to the Truth, it is ultimately on the success or failure of scientific careers as proxies for truth that bets are taken (see Hull in Callebaut 1993, 303; Hull 2001, 171, 182–183). No wonder that debates will at times be harsh! Difficulties in communication between the contending parties stem in part from the boundary work that accompanies scientific discipline building (Galison and Stump 1996; Harré 2005). An important task incumbent on philosophers and historians of biology will be to probe how deeply entrenched the metacommitments of the advocates of the various perspectives on evo-devo really are.

Reversing Foreground and Background: From Genetics to Epigenetics

In the introduction, Sansom and Brandon see the exclusion of development from the study of evolution for much of the twentieth century as the result of regarding evolution as a change in the genomes of a population over time.[25] On the selfish-gene view, "natural selection looked through the organism right to the genome," rendering the process of development epiphenomenal to the process of evolution. On this construal, the evolution of development of a full-fledged organism was but a Ruse (in both meanings of the word) of natural selection to increase the fitness of individual genes. This view has many flaws—the "beanbag" assumption, the assumption that genes are self-replicating, and so on—which have been sufficiently exposed that we can dispense with their discussion here (see, e.g., Lewontin 2000a, the papers in sections A and B of Singh et al. 2001, or Robert 2004). For our purposes in this chapter it will suffice to elaborate on the keywords *causation* (agency), *emergence*, and *pluralism* in the sense of a multilevel, hierarchical theory of natural selection, taking Gould's (2001) account as our point of departure. Together, we contend, these concepts force evo-devo to take epigenetic considerations—in the sense of "epigenesis"—as primordial for the organismic perspective, leaving it to others (there are plenty) to speak for the gene.

Causation

Replication must be displaced from the prominent role it plays in the gene-centric view of Williams, Dawkins, and Co. because it is not, and cannot be, the locus of agency. In Gould's (2001, 213) words, "Units of selection must be actors within the guts of the mechanism, not items in the calculus of results." Griesemer's reproducer perspective (see box 2.2) begins to pave the way for a full-fledged philosophical account of evolutionary-developmental processes as causal-mechanical processes, which can be based on genetic, epigenetic (figure 2.1), and environmental causation.

The philosophical view that best fits the causal-mechanistic account of explanation endorsed by OSA is a variety of philosophical naturalism,[26] *scientific realism*, which holds that (1) the world has a definite, mind-independent structure (*ontological* or *metaphysical realism*), (2) scientific theories are descriptions of their intended domain (whether observable or not) that are capable of being true or false (*semantic realism*), and (3) mature and predictively successful theories are well confirmed and approximately true of the world (*epistemic optimism*) (Psillos 2003, 60ff.; cf. Salmon 1984; Bhaskar 1989; Callebaut 1993; and Hooker 1995). Scientific realism combines the modest claim that there is a world that is independent from us,

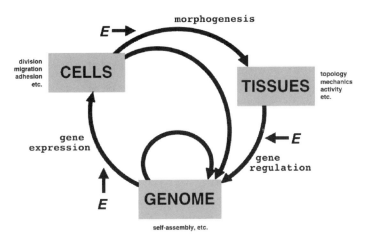

Figure 2.1
Epigenetic interactions between different levels of organization (with autonomous properties) and their environments (E) (after Müller 2001).

largely unobservable, which science attempts to map, with the more "presumptuous" claim that science can nonetheless succeed in arriving at a more or less faithful representation of the world, enabling us to know some (not necessarily "the") truth about it (Psillos 2003, 61)—a position that cannot be argued for from idealist, empiricist, and most constructivist premises (constructive realism being an exception; Giere 1988).

The causal-mechanistic account of explanation relinquishes the long-dominant *covering-law* account that viewed explanation as "subsumption" under one or more universal laws. (For example, that Newton's proverbial apple fell from the tree was taken to be explainable as an instantiation of his law of gravity in conjunction with appropriate initial and boundary conditions.) Instead, explanation is now accounted for in terms of mechanistic models (Salmon 1984, 1989, 1994, 1998; Bechtel and Richardson 1993; Glennan 1996, 2002a, 2002b; Machamer, Darden, and Craver 2000), capacities, powers, or propensities (Fisk 1973; Cartwright 1989; Ellis 1999), and, taking into account human cognitive limitations, "major factors" (Wimsatt 1974, 1976). This is not to say that functional/teleological explanations can be dispensed with at all levels, but the hope is that they can be rephrased so as to fit the general causal explanatory mold (see Sober 1993, 82–87, on naturalizing teleology and the distinction between ontogenetic and phylogenetic adaptation). Similarly, when, say, OSA interprets homologues as "attractors" of morphological design, this begs for an explanation, not of how the event to be explained was in fact produced, but of "how the event would have occurred regardless of which of a variety of causal scenarios actually transpired" (Sober 1983, 202ff.); also see our discussion of inherency below. Such "equilibrium explanations" (Sober) present disjunctions of possible causal scenarios, but the explanation does not specify which of them is the actual cause. Another complication with causal explanation, which we can only mention here without discussion, considers the desideratum that biological research should substitute for past causes the "traces"—state variables; see, for example, Padulo and Arbib 1974—left in the present by the operation of those causes (cf. Elster 1983, 33, and note 29).

Emergence

Gene-centrism counterfactually supposes that genes "build organisms" entirely in an additive fashion without nonlinear interactions among genes and their products, or for that matter, the physical processes that govern the material systems of which gene products are only one portion (Newman 2002). However, the fact of emergence is indisputable: "This aspect of the question is empirical but also entirely settled (and never really controversial): organisms are replete with emergent properties; our sense of organismic functionality and intentionality arises from these emergent features" (Gould 2001, 213; see Schlanger 1971; Pluhar 1978; Minati and Pessa 2002; Hooker 2004). Thus Dawkins's colorful metaphors of selfish genes and manipulated organisms "could not be more misleading, because he has reversed nature's causality: organisms are active units of selection; genes, although lending a helpful hand as architects, remain stuck within" (Gould 2001, 213–214). Hull's (1982, 1988) apt replacement of Dawkins's replicator-vehicle distinction by replicators and interactors, respectively, fine-tunes Lewontin's (1970) formalization of evolution as "heritable variation in fitness," but needs to be complemented by an account of what is materially transferred in evolution and development—Griesemer's "reproducer perspective," inspired by Maynard Smith's account of units of evolution in the 1980s, whose key principle is multiplication (see box 2.2). Individuals need not replicate themselves to be units of selection; it suffices that they contribute to the next generation by hereditary passage, and "magnify" their contributions relative to those of other individuals (Gould 2001; cf. Lewontin 1996 and Griesemer 2000a). Selection, then, occurs "when this magnification results *from the causal interaction of an evolutionary individual* (a unit of selection) *with the environment* in a matter that enhances the differential reproductive success of the individual" (Gould 2001, 216, emphasis in the original). Emergence, by contrast, can be seen as a by-product of dynamic systems (development) under evolutionary modification, regardless of whether selection or other factors initiate the change.

Pluralism

The modern synthesis was neither a consensus in favor of, nor a consensus against, a hierarchical concept of evolution. However, among the pioneers of the synthesis, Sewall Wright's thinking was paradigmatically hierarchical (Gayon 1989, 36). Wright's influence has clearly been instrumental in the comeback of group selection, which recently "has risen from the ashes to receive a vigorous rehearing" (Gould 2001, 216; see, e.g., Wade 1978; Wimsatt 1980; Sober and Wilson 1998; and Griesemer 2000c). This revival rests on two proposals that Gould and others believe can serve as centerpieces for a general theory of macroevolution: (1) the identification of evolutionary individuals as interactors, causal agents, and units of selection, and (2) the validation of a hierarchical theory of natural selection based on the recognition that evolutionary individuals exist at several levels of organization, including genes, cell lineages, organisms, demes, species, and clades (Gould 2001, 216). A multilevel view of evolution was also vindicated quite independently by the biologists and founders of general systems theory such as Ludwig von Bertalanffy (e.g., 1968) and Paul Weiss (e.g., 1970), who in turn influenced Riedl (1978), Wagner (1986), and OSA, and contributed to the current rise of systems biology (see box 2.2).

Reclaiming the phenotype, understood as a multilevel, hierarchical system, as the central concern of biology, we must qualify Smith and Sansom's (2001) contention that "evolution has resulted in organisms that develop." Relocating the causal nexus in epigenetic processes, we call for equal consideration of the proposal that *development has resulted in populations of organisms that evolve*. The next section will be an elaboration of this reversal of foreground and background. We propose to call the approach to evolution that seeks to incorporate the issues of causation, emergence, and pluralism with respect to levels into the formal framework of evolutionary theory the *organismic systems approach*.

The Organismic Systems Approach (OSA)

Evo-devo, eco-devo (Gilbert 2001), systems biology, and other concepts of organismic evolution contribute to OSA. At present the core conceptual framework is being provided by evo-devo. Contrary to unilinear, quasi-monocausal concepts of evolution such as genetic determinism, evo-devo as the empirical and theoretical analysis of the causal connection between embryological/developmental and evolutionary processes is immanently dialectical (figure 2.2). On the one hand, evo-devo is interested in the factors that together explain the origination of ontogenetic systems and the mechanisms that account for their subsequent modification. On the other hand, it investigates the ways in which the properties of ontogenetic processes influence the course of morphological evolution (see Raff 2000; Müller 2005).

How can OSA contribute to an extended evolutionary synthesis? A number of conceptual and methodological issues were mentioned above as part of the evo-devo packages. In this section we concentrate on epigenetic evo-devo concepts, leaving a treatment of and comparison with other work for another occasion. In our view, the most salient contributions of OSA are plausible theoretical and, where possible, also empirical accounts of: (1) the generation of primordial organismal form and structure (origination), (2) how novel structures arise in phenotypic evolution (innovation), (3)

The Organismic Systems Approach

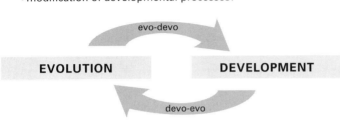

Figure 2.2
The dual structure of evo-devo. The two subagendas, evo-devo and devo-evo, address different kinds of questions (from Müller 2000b).

why not all the design options of a phenotypic space are realized (constraints), and (4) the causal processes that together make for the organization of the integrated phenotype (homology). In this chapter we can only discuss these issues in an abbreviated form; for a more complete survey, see Müller 2005.

Origination of Primordial Organ Forms and Body Plans

According to OSA, the correlation of an organism's form with its genotype, rather than being a defining condition of morphological evolution, is a highly derived property. This implies that other, nongenetic causal determinants of biological morphogenesis have been active over the course of evolution. Here it is necessary to distinguish between those mechanisms that are involved in the primary generation of first metazoan structures and body plans, origination (discussed in this section), the later modification of these arrangements by the processes of innovation (discussed in the following section), and variation, the mode of change dealt with most successfully by the neo-Darwinian paradigm.

Plasticity of form and developmental trajectory in modern-day organisms (West-Eberhard 2003) provides the starting point for this novel perspective on the origination of more ancient forms. Protists, fungi, plants, and animals such as arthropods and mollusks, may exhibit radically different forms in different environments or ecological settings. *Candida albicans*, for example, a fungal pathogen in humans, does not seem to have a "default morphology" (Magee 1997); rather, it is able to switch among forms ranging from single budding cells, to threadlike hyphae, to strings of yeastlike cells plus long separated filaments ("pseudohyphae"), depending on the local environment. In mice, the number of vertebrae can depend on the uterine environment: fertilized eggs of a strain with five lumbar vertebrae preferentially develop into embryos with six vertebrae when transferred into the uteri of a six-vertebra strain (McLaren and Michie 1958). Tadpoles of the frog *Rana temporaria* undergo significant changes in body and tail morphology within four days when subjected to changed predation environment (Van Buskirk 2002).

Neo-Darwinian interpretations of these phenotypic polymorphisms present them as specifically evolved adaptations and therefore sophisticated products of evolution. The different phenotypes of an organism with a given genotype are thus considered to be outcomes of subroutines of an overall genetic program that evolved as a result of distinct sets of selective pressures at different life-history stages. Alternatively, they are seen as manifestations of an evolutionary fine-tuned "reaction norm," or as products of evolution for evolvability. According to OSA, in contrast, much morphological plasticity is a reflection of the influence of external physicochemical parameters on any material system and is therefore primitive and inevitable rather than programmed (figure 2.3 and box 2.3).

Viscoelastic materials such as clay, rubber, lava, and jelly (soft matter: de Gennes 1992), are subject (by virtue of inherent physical properties) to being molded, formed, and deformed by the external physical environment. Most living tissues are soft matter, and all of them are also excitable media—materials that employ stored chemical or mechanical energy to respond in active and predictable ways to their physical environments (Mikhailov 1990). Much organismal plasticity results from such material properties.

Ancient organisms undoubtedly exhibited less genetic redundancy, metabolic integration, homeostasis, and developmental canalization than modern organisms and were thus more subject than the latter to external molding forces. Hence it is likely that in earlier multicellular forms morphological determination based on an interplay of intrinsic physical properties and external conditions was even more prevalent than it is today. In the scenario envisioned by OSA, then, morphological variation in response to the environment is a primitive, physically based property, carried over to a limited extent into modern organisms from the inherent plasticity and responsiveness to the external physical environment of the viscoelastic cell aggregates that constituted the first multicellular organisms.

The Organismic Systems Approach

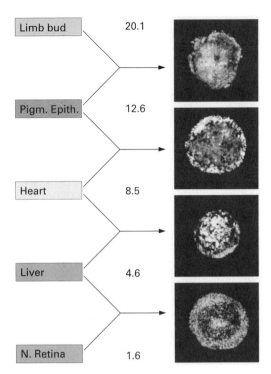

Figure 2.3
Spheres and tissue layering from in vitro combinations of cells from different vertebrate tissues. Cells sort out according to their surface tensions (dyne/cm), with the lower tension cells always surrounding the higher tension cells (after Foty et al. 1996).

Box 2.3
From the pre-Mendelian to the Mendelian world

Pre-Mendelian

Interchangable, generic forms resulting from multicellular aggregates, with no fixed, heritable relation between genotype and phenotype

Mendelian

Stabilized, heritable forms resulting from an increasingly fixed matching between genotype and phenotype established by natural selection

Source: Newman and Müller 2000.

The correspondence of a given genotype to one morphological phenotype, as typically seen in higher animals (but less so in multicellular protists, fungi, and plants), will be exceptional, then. Such close mapping can result from an evolutionary scenario in which the developmental mechanism by which a phenotype is generated changes from being sensitive to external conditions to being independent of such conditions. If modern organisms are "Mendelian," in the sense that genotype and phenotype are inherited in close correlation, and that morphological change is most typically dependent on genetic change, then the organismic systems hypothesis includes the postulate that there was a "pre-Mendelian world" of polymorphic organisms at the earliest stages of metazoan evolution whose genotypes and morphological phenotypes were connected in only a loose fashion (box 2.3; see also Newman 2005).

In this exploratory period of organismal evolution, the mapping of genotype to morphological phenotype would have been one-to-many, rather than one-to-one. This changed as the subsequent evolution of genetic redundancies (e.g., A. Wagner 1996) and other mechanisms supporting reliability of developmental outcome forged a closer linkage between genetic change and phenotypic change. In particular, natural selection that favored the maintenance of morphological phenotype in the face of environmental or metabolic variability, produced organisms characterized by a closer mapping of genotype to phenotype—that is, the familiar Mendelian world. But even as body plans and other major morphological features, such as the bauplan of the vertebrate limb (Newman and Müller 2005), became locked in by the accumulation of reinforcing genetic circuitry, fine-tuning of details of organismal, and particularly organ morphology continued (and continues) to occur through an interplay, albeit diminished in effect, of genetic and nongenetic factors.

Cell-cell adhesion is an evolutionary novelty that was the sine qua non of the earliest multicellular forms. Adhesivity is unlikely to have been stringently regulated with regard to expression levels on the occasion(s) when it first appeared. Subpopulations of cells in multicellular aggregates with adhesive differentials exceeding a threshold level will sort out, leading to multilayered structures (Steinberg and Takeichi 1994)—morphological novelties that would have been objects of natural selection. The stabilization of differential adhesion was thus probably among the earliest examples of biochemical differentiation to have become established in the metazoan world. Differential adhesion was coupled variously with cell polarity, inherent physical attributes and behaviors of cell aggregates (including diffusion and the formation of gradients), biochemical oscillation, and

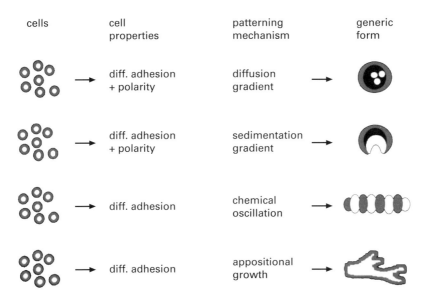

Figure 2.4
Repertoire of generic forms. Virtually all the features seen in early development of modern metazoan organisms can be attributed to the composition of one or more of the pattern-forming mechanisms with differential adhesion (Newman 1994). The combined effects of the various physical properties that were generic to the earliest multicellular aggregates considered as chemically excitable, viscoelastic soft matter, ensured the production of a profusion of forms.

standing wave formation. It would thus have generated a panoply of morphologically distinct forms—multilayered, hollow, segmented, separately and in combination (figure 2.4)—that is, novel organisms with the capacity to populate correspondingly new ecological niches (see Newman et al. 2006 for a more detailed account).

Evolutionary Innovation and Morphological Novelty

Once primordial structures and body plans had originated, phenotypic evolution proceeded through variation and innovation. OSA recognizes a distinction between these two processes. As a consequence, morphological novelties, such as feathers, eyes, or skeletal elements (box 2.4), represent a specific class of phenotypic change that differs from adaptations (Müller and Wagner 2003; Müller and Newman 2005; Wagner and Lynch 2005). It also holds that the processes underlying the generation of novelties are qualitatively different from the standard mechanisms of variation—that is,

Box 2.4
Examples of morphological novelties in the vertebrate skeleton

Panda's "thumb"	(panda bears)
Panda's "7th digit"	(panda bears)
snout bone	(boars)
calcar	(bats)
falciforme	(moles)
third forearm bone	(golden mole)
naviculare	(horses)
fibular crest	(theropod dinosaurs, birds)
preglossale	(passerine birds)
patella	(birds, mammals)
Source: After Müller and Wagner 1991.	

variation (plus selection) is considered to produce adaptation, whereas the outcome of innovation (plus selection) is novelty.

The treatment of the innovation and novelty problem depends to a great extent on how phenotypic novelties are defined (Müller 2002; Love 2003). Several definitions have been proposed,[27] but an operational definition favors the use of a morphological-character concept and of homology criteria: a morphological novelty is "a structure that is neither homologous to any structure in the ancestral species nor homonomous to any other structure of the same organism" (Müller and Wagner 1991, 243). This definition is narrow, excluding characters that deviate only quantitatively from the ancestral condition and that may often be regarded as novelties in a functional sense, but it permits identification of unambiguous cases of morphological novelty. Examples that satisfy the definition exist in all tissues and organ systems of plants and animals, such as, in the latter case, the skeletal system (box 2.4).

Focusing on novelties selected in accordance with this definition, it becomes possible for evo-devo to address empirical questions concerning the initiating conditions and the specific mechanistic processes responsible for the generation of a particular novelty. Here OSA does not give priority to the notion of genetic program that tacitly or overtly underlies the traditional account. Mutational change unavoidably accompanies any kind of phenotypic evolution, if only as a consolidator of change, but is not necessarily causal for the inception of new complex traits, except probably in rare cases of mutations with large and coordinated pleiotropic effects. OSA argues that the majority of novelties result from higher-level organizational properties, epigenetic interactions, and environmental influences that arise during the modification of established developmental systems. In this sense, novelties would be epigenetic by-products of the systems properties of development.

Among the initiating causes of innovation feature: natural selection, behavioral change (Mayr 1958, 1960), symbiotic combination (Margulis and Fester 1991), and environmental induction (Gilbert 2001; West-Eberhard 2003). OSA holds that, whatever the initiating conditions, the specificity of the resulting novelty is determined by the systems properties of development. In the case of natural selection, for instance, it is clear that selection cannot act on characters that are not yet in existence, and hence selection cannot directly cause novelty (see the subsection on "Selection and Emergence" below). Rather, as has been argued in several different contexts, novelties often arise as developmental side effects of selection, which acts on such parameters as body shape, size, or proportions, through the alteration of developmental processes—for example, the modification of cell behaviors or of developmental timing (Hanken 1985; Müller 1990; Müller and Wagner 1991; Newman and Müller 2000). In these and earlier views (Schmalhausen 1949) selection is regarded as a facilitating factor or as a general boundary condition rather than a direct cause of novelty.

The specific developmental mechanisms of novelty generation can also be multiple. Morphological novelties can be based on innovations at the molecular or cellular levels of development, including the establishment of new developmental interactions, or on the redeployment of existing developmental modules at a new location in the embryo. Other possibilities include the subdivision or combination of developmental modules, or the developmental individualization of serial structures such as teeth, vertebrae, or hair (Müller and Wagner 2003). Currently the bulk of empirical research is focused on the evolution of new gene regulatory interactions and the recruitment of genes and gene circuits into new developmental functions, as seen in a growing number of examples (Shapiro et al. 2004; Colosimo et al. 2005). Understanding the kinetics (Bolouri and Davidson 2003), dynamics (Cinquin and Demongeot 2005), and topological aspects (von Dassow et al. 2000; Salazar-Ciudad, Newman, and Solé 2001, Salazar-Ciudad, Solé, and Newman 2001) of developmental gene regulation and their correlation with morphogenetic events will be central in this endeavor. But as argued above, it will be equally necessary to elucidate the specific epigenetic conditions of novelty generation such as the role of biomechanics.

In summary, by considering the emergence and transformation of developmental mechanisms as an evolutionary problem in its own right, OSA arrives at the view that epigenetic mechanisms, rather than genetic changes, are the major sources of morphological novelty in evolution.

Constraints

Phenotypic evolution is limited and biased by developmental constraints that are contingent for all phylogenetic lineages (Maynard Smith et al. 1985). Constraints represent a central evo-devo principle (Streicher and Müller 1992; Amundson 1994; Eberhard 2001; Mayr 2001, 198–200; Wagner and Müller 2002), and hence are covered by OSA, but need not be discussed in detail here, because their role is widely accepted and now belongs to the standard picture (Schwenk and Wagner 2003). The relaxation or breaking up of constraints, which may be required in instances of innovation, is an important phenomenon that is readily accommodated by OSA. Possible mechanisms of constraint relaxation include mutational events or neutral phenotypic drift (Mayr 1960). But even in closely studied cases the interpretation of whether or not developmental constraints had to be overcome for the origination of an innovation can be difficult (cf. Eberhard 2001 versus Wagner and Müller 2002). Certainly constraints should not be understood as mere limitations to phenotypic variation, since they can also provide taxon-specific opportunities for novelties to arise. A novelty can result from a constrained developmental system, not because genetic or developmental variation is relaxed, but precisely because the system is unable to respond by variation and is forced to transcend a developmental threshold (Müller 1990; Streicher and Müller 1992). This can provide heightened potentialities for innovation in particular areas of phenotypic character space (Roth and Wake 1989; Arthur 2001).

Organization

The three features of OSA discussed above all pertain to the phenotype in evolution. Consequently, OSA takes the organization of the phenotype as its conceptual hallmark, in contrast to evolutionary theories that concentrate on gene frequencies in populations. The explanation of the genotype-phenotype relationship, not as an abstract statistical correspondence but in terms of mechanistic processes of structural organization, is the concern of OSA. In this it takes the evolution of characters and character states as realities and proposes that once generic morphological templates and novelties emerged through the interplay of epigenetic and genetic factors, they served as organizers in the further evolution of the phenotype and, at the same time, of their regulatory mechanisms in development.

This summary of OSA's position regarding the evolution of organization requires further specification. The kind of organization most relevant to this perspective is the generation, integration, and fixation of morphological building elements that result in evolutionarily stable body plans. These are the terms in which OSA represents the homology problem in evolutionary theory. It assumes that the constructional components have an organizing role in organismal evolution, a position that has been labeled the "organizational homology concept" (Müller 2003; Love and Raff 2005). The concept holds that, as more and more constructional detail is added, the already fixated and developmentally individualized parts (homologues) serve as accretion points for new components added in the evolving body plans. Furthermore, the homologues that initially arise from generic properties of cell masses, and later from conditional interactions between cells and tissues, provide "morphogenetic templates" for an increasing biochemical sophistication of cell and tissue interactions (Newman and Müller 2000). Although their co-option by the evolving genome results in the so-called genetic "programs" of development, the dynamics of these genetic control systems would involve stabilizing and fine-tuning the morphogenetic templates already present. Hence the sequence of gene expression changes associated with the generation of pattern and form during embryogenesis would be a consequence, not a cause, of phenotypic organization. At their final stage, the organizational homology concept asserts, homologues become autonomized—that is, the constructional building elements of the phenotypes become independent from their molecular and developmental underpinnings and come to act as attractors in body plan evolution (Striedter 1998).

The organizational homology concept briefly summarized here is a biological homology concept (Roth 1984; Wagner 1989), because it refers to the biological mechanisms that underlie the origination of homology, rather than to the genealogical and taxonomic aspects that are emphasized in historical homology concepts. Organizational homology recognizes the role of developmental constraints, but gives priority to the active contributions of organizing processes rather than to passive limitations.

The empirical approach to organization prompts the use of new computational tools that can accurately represent the relationship between gene activation, cell behavior, and morphogenesis. A host of such tools are being developed (e.g., Jernvall 2000; Streicher et al. 2000; Streicher and Müller 2001; Costa et al. 2005; Weninger et al. 2006). The data obtained by these and other techniques can be used to derive formal models of the principles relating genotype and phenotype, with the capacity to make testable predictions about the evolution of organization (Salazar-Ciudad, Newman, and Solé 2001; Salazar-Ciudad, Solé, and Newman 2001; Britten 2003; Bissell et al. 2003; Nijhout 2003; Rasskin-Gutman 2003).

Major Evolutionary Problems Addressed by OSA

OSA addresses a number of issues that are sidestepped by the received theory. OSA's explanatory focus is on phenotypic evolution. Homology, homoplasy, novelty, modularity—to name the principal ones—are phenotypic phenomena that are not specifically dealt with in the neo-Darwinian framework and by its population-genetic methods. In OSA these problems are regarded as central and are explained on the basis of mechanistic processes rather than statistical associations with abstract gene frequencies. Therefore, OSA-based models can be predictive not about which characters are going to be maintained in a population (the neo-Darwinian selection focus) but about what is likely to arise structurally. In addition, OSA addresses a number of long-standing issues in the neo-Darwinian discourse.

Tempo and Mode of Phenotypic Evolution The punctualism-gradualism debate receives renewed attention in OSA because development has nonlinear system properties that can result in punctuated events even when submitted to continuous change. Taking these properties into consideration, instances of rapid change of form are not only possible but are to be expected. The burst of morphological diversification recorded in the early Cambrian fossil record, for instance, rather a conundrum for any gradualistic theory, becomes much more easy to explain in a developmental framework that considers generic mechanisms of form generation. The framework of OSA suggests a pre-Mendelian phase of organismal evolution that conceivably laid the basis for this "Cambrian explosion" and the rapid elaboration (on the geological time scale) of the modern metazoan body plans that followed (Newman and Müller 2000; Newman 2005). The first morphologically complex multicellular organisms, represented by the Vendian fossil deposits, appear to have been flat, often-segmented, but apparently solid-bodied creatures. The concurrence of polarized cells with the multicellular state has the inevitable physical consequence of producing organisms with distinct interior cavities. Body cavities thus were a topological innovation permitting new geometric relationships among tissues and facilitating the origination of tissue multilayering, including triploblasty.

The mobilization of additional physical processes (differential adhesion, molecular diffusion, biochemical oscillation, Turing-type reaction-diffusion coupling) would have spurred the rapid elaboration of all the modern body plans within a space of perhaps 25 million years during the 100 million years following the advent of the Vendian forms (Newman 1994, 2005). That this could have occurred with little or no genetic evolution may also provide a resolution to the paradox (within the neo-Darwinian framework) of the conservation of the "developmental-genetic toolkit" (Carroll et al. 2004) across all animal phylogeny (Newman 2006).

Selection and Emergence Neo-Darwinism is about the change in organismal properties due to the dynamics of alleles in populations as determined by mutation, drift, and selection. But selection cannot set in until there are organismal properties to select. This is the problem of innovation, which was previously often referred to as "emergence." Although frequently mentioned as a property of complex systems such as development (e.g., Boogerd et al. 2005), emergence remained without representation in evolutionary theory. OSA's mechanistic concept of innovation fills this void by moving beyond the neo-Darwinian focus on variation-selection dynamics.

In this context, Fontana and Buss's (1994a, 1994b) modeling of an abstract chemistry implemented in a λ-calculus-based modeling platform is highly relevant. In their model world (see also box 2.2) the following features are generic, and hence would be expected to reappear if the "tape" of evolution were run twice: (1) hypercycles of self-reproducing objects arise; (2) if self-replication is inhibited, self-maintaining organizations arise; and (3) self-maintaining organizations, once established, can combine into higher-order self-maintaining organizations. Darwinian selection presupposes the existence of self-reproducing entities; but *self-maintaining organizations can arise in the absence of self-reproducing entities*. According to Fontana and Buss (1994a, 761),

> self-reproduction and self-maintenance are shared features of all extant organisms, barring viruses. It is not surprising, then, that there has been little attention paid to the generation of one feature independently of the other. While selection was surely ongoing when transitions in organizational grade occurred in the history of life [Griesemer 2000c], our model universe provides us the unique opportunity to ask whether selection played a necessary role.

Their findings indicate that it need not. They conclude that "separating the problem of the emergence of self-maintenance from the problem of self-reproduction leads to the realizazion that there exist routes to the generation of biological order other than that of natural selection" (cf. Goodwin 1984, 1994).

Integration In a theoretical framework that does not account for the generative and the emergent, there is no major requirement to deal with integration, and consequently such proposals were dismissed (e.g., Waddington's assimilation). In OSA innovation and novelty play a central role, and, consequently, integration is reemphasized. Here the problem is how novelties are accommodated into the preexisting, tightly coupled, constructional, developmental, and genetic systems of a taxon, to ensure their functionality and inheritance. Integration is likely to take place through different mechanisms, acting among and between different levels of organization—genetic, developmental, phenotypic, and functional. Waddington (1956, 1962) anticipated genetic integration, following from selection acting on the genetic variation that will arise with the spreading of a novel character. Today this corresponds in part to the notion of co-option during which orthologous and paralogous regulatory circuits acquire new developmental roles over the course of evolution (Wray 1999; Wray and Lowe 2000; Carroll et al. 2004). The consequential genetic integration will increasingly stabilize and overdetermine the generative processes, resulting in an ever-closer mapping between genotype and phenotype. Such transitions can be interpreted as a change from emergent to hierarchical gene networks (Salazar-Ciudad, Newman, and Solé 2001; Salazar-Ciudad, Solé, and Newman 2001).

OSA provides for the possibility that not all integration necessarily rests on early genetic fixation. Many phenotypic characters can, on the one hand, be experimentally suppressed by changing the epigenetic conditions of their formation and, on the other hand, experimentally induced modifications in one subsystem of development can be accommodated by other subsystems even when the initial change is dramatic. On the basis of this evidence it is argued that epigenetic integration will usually come first, and may suffice to maintain a novelty for long evolutionary periods, without any need for genetic "hardwiring" of the new interactions. But the epigenetic patterns of integration may also provide the templates for subsequent genetic integration (Newman and Müller 2000; Müller 2003b). Epigenetic integration "locks in" the novel characters that arose as a consequence of the mechanisms discussed earlier and thus will generate stability (and even heritability) of new building units in organismal body plans. Similar concepts emphasizing epigenetic integration are generative entrenchment (Wimsatt 1986, 2001) and epigenetic traps (Wagner 1989).

Inherency If the morphological organization of organisms is (in an evolutionary sense) predictable to a certain degree from the material properties and generative rules of their constituent tissues (compare Vermeij 2006 on the predictability of evolutionary innovations), then an additional principle is added to the external selectionism paradigm of neo-Darwinism. This principle we propose to call inherency (Newman and Müller 2006). "Something is inherent either if it will always happen (e.g., entropy) or if the potentiality for it always exists and actuality can only be obstructed" (Eckstein 1980, 138–139).

In the evolutionary context inherency means that the morphological motifs of modern-day organisms have their origins in the generic forms assumed by cell masses interacting with one another and their microenvironments, and were only later integrated into developmental repertoires by stabilizing and canalizing genetic evolution.[28] Therefore, in OSA, the causal basis of phenotypic evolution is not reduced to gene regulatory evolution and population genetic events, but includes the formative factors inherent in the evolving organisms themselves, such as their physical material properties, their self-organizing capacities, and their reactive potential to external influence. Whereas neo-Darwinism from Monod (1971) to Williams (1985, 1992) and Gould (2002) emphasized the contingency of evolution[29] (as does DSP with respect to development: Oyama, Griffiths, and Gray 2001[30]), the watchword of OSA becomes inherency.

Regarding individual development, inherency defies the "blueprint" or "program" notions that abound in present accounts. Cell collectives and tissue masses take on form not because they are instructed to do so but because of the inherent physical and self-organizational properties of interacting cells. This does not mean that genes have no role or that gene regulation is unimportant in development. Of course cell properties and cell interaction of all organisms, ancient and modern, depend on the molecules that genes specify, but the resulting biological forms and specific cell arrangements are not encoded in any deterministic fashion in the genome (Neumann-Held and Rehmann-Sutter 2006). Inherency locates the causal basis of morphogenesis in the dynamics of interaction between genes, cells, and tissues—each endowed with their own "autonomous" physical and functional properties (figure 2.1).

Epistemological and Pragmatic Aspects of Evo-Devo

Will developmental biology only (re)supplement evolutionary biology, or could evo-devo eventually prompt a fundamental theoretical rethinking of evolution itself (see Depew and Weber 1995)? The honest but easy historian's answer is, of course, that it is too early to tell (Hull 2006), but as participating scientists and philosophers we have a different ax to grind. Scenarios in terms of "modesty" versus "ambition" have been suggested. A caveat seems in order here. The task the proponents of an extended, "evo-devo synthesis" face may be so formidable that even achieving more modest aims may turn out herculean. A historical perspective is useful here. Amundson (1994, 576) reminds us that "evolutionary biology was built on a huge black box—Darwin could never have written the *Origin of Species* if he had not wisely bracketed the mechanism of inheritance." One of the great advantages of Mendelian genetics in the early decades of the twentieth century, it is often thought, was that it "bypassed the uncharted swamp of development" (Hull 1998, 89). But today the situation looks radically different:

Pre-Synthesis Darwinians at least realized the need for a theory of inheritance, although they doubted that Mendelism was that theory. Most post-Synthesis neo-Darwinians do not require developmental biological contributions to evolution theory [e.g., Wallace 1986]. Developmentalists may or may not be able to demonstrate that a knowledge of the processes of ontogenetic development is essential for the explanation of evolutionary phenomena. (Amundson 1994, 576)

Provided developmentalists can demonstrate this, and offer well-founded developmental evolutionary explanations, the result will be "a dramatic synthesis of divergent explanatory and theoretical traditions," Amundson concluded. We feel the time to attempt such a synthesis has come, although its contours are most likely going to differ greatly from the received view of explanatory theory reduction that centered on "imperialism and competition for primacy and fundamentality" (Griesemer 2006).

What kind of explanatory unification[31] can evo-devo in general and OSA in particular reasonably be expected to contribute to?[32] Arthur (1987, 182), pondering whether biologists can realistically aspire to a "biological Grand Unified Theory" as a long-term aim, suggested that "biology can transcend its biospheric origins and develop principles that will apply to life wherever we encounter it." He offered principles of self-organization and homeostatic regulation as good candidates. In a similar vein, Kauffman (1993, 22–23) envisages a synthesis of evolutionary biology and developmental biology that would be based on the identity of generative mechanisms (see Goodwin 1984; Eble and Goodwin 2005):

The task of enlarging evolutionary theory would be far more complete...if we could show that fundamental aspects of evolution and ontogeny had origins in some mea-

The Organismic Systems Approach

Box 2.5
Possible interrelations between natural selection (NS) and self-organization (SO)

> - NS, not SO, drives evolution.
> - SO constrains NS.
> - SO is the null hypothesis against which evolutionary change is to be measured.
> - SO is an auxiliary to NS in causing evolutionary change.
> - SO drives evolution, but is captured and fixated by NS.
> - NS is itself a form of SO.
> - NS and So are two aspects of a single evolutionary process.
>
> *Source*: Adapted from Depew and Weber 1995; cf. Kauffman 1993, 1995; Newman and Müller 2000; and Richardson 2001.

sure reflecting self-organizing properties of the underlying systems. The present paradigm is correct in its emphasis on the richness of historical accident, the fact of drift, the many roles of selection, and the uses of design principles in attempts to characterize the possible goals of selection.... the task must be to include self-organizing properties in a broadened framework, asking what the effects of selection and drift will be when operating on systems which have their own rich and robust self-ordered properties.

Kauffman's vision provides one attractive interpretation of what the long-term unifying function of evo-devo might be. But at present, the actual and conceivable interrelations between the dominant selectionist paradigm and the paradigm of self-organization are still quite unclear (box 2.5).

It is fair to say that the contribution to a unified theoretical biology most workers in the field of evo-devo currently envision is more modest. Rather than some grand unification scheme, what they have in mind can perhaps be rendered best in terms of Darden and Maull's (1977) concept of an "interfield." In their analysis, a scientific field is an area of science consisting of a central *problem*, a *domain* consisting of items taken to be *facts* related to that problem, *general explanatory factors and goals* providing expectations as to how the problem is to be solved, *techniques and methods*, and often (but not necessarily) also *concepts, laws, and theories* that are related to the problem and attempt to realize the explanatory goals. If developmental biology and evolutionary biology—or more correctly, a number of their subdisciplines that are relevant to evo-devo[33]—are regarded as fields, evo-devo might be regarded as an "interfield" that borrows resources from one field (say, a developmental analysis of homology) to solve a problem in another field (in our example, the evolution of homology). This may look modest enough, yet, as some of the historical case studies of Darden and her group document (e.g., Darden 1991; Darden and Craver 2002; cf. also Darden 1986), such humble beginnings can engender quite revolutionary breakthroughs. (On a much grander scale, one could think here of the transition of biochemistry to molecular biology in the 1930s, and the migration of physicists and chemists into biology as stimulated by the Rockefeller Foundation; e.g., Abir-Am 1982.) Evo-devo, as we have presented it here, will require nothing less than a rethinking of the fundamentals of developmental as well as evolutionary biology—not in any philosophically "foundational" sense, but in the much more down-to-earth terms of Wimsatt's "generative entrenchment" of ideas.[34]

Most historians and at least some philosophers of biology hold that evolutionary biology is not a unified science, and that the modern synthesis was not so much a conceptual/theoretical as an institutional synthesis (e.g., Smocovitis 1996; Burian 2005). Moreover, as already indicated, the ideal of "the" unity of science as envisaged by the logical empiricists and epitomized in Schaffner's (1993) account of reduction/replacement is no longer taken for granted, to put it mildly (Dupré 1983, 1993; Rosenberg 1994; Galison and Stump 1996). If Cartwright (1999, 1) is right, "The laws that describe this world are a patchwork, not a pyramid." In biology as elsewhere, reduction does not work in practice (Hull 1973; cf. Longino 2000); physicalist reduction, namely, the reduction of all the theories of complex objects to physics, is humanly impossible due to our bounded rationality.[35] Further obstacles to unification are related to the data-driven nature of much contemporary research. "Since World War II," Patrick Suppes (1979, 17) wrote almost thirty years ago, "the engines of empiricism have vastly outrun the horse-drawn carriages of theory." See, for example, Henry (2003) on the tensions between data- and hypothesis-driven approaches to systems biology. The current "reign of theory pluralism" (Suppes 1978) is in part a consequence of this "Baconian" bias toward inductive knowledge acquisition.

If the postmodernist fashion to celebrate the end of the unity (singular) of science (singular) was all there was to it, the rhetoric of evo-devo enthusiasts calling for a developmental synthesis would be shallow indeed. But we take it that the one-sided ideology of unification dear to the positivists has only been replaced by an equally misplaced, exclusive focus on the factors in scientific practice that tend to diversity and separate scientific discourses (Radnitzky 1987, 1988; Reisch 2005). A more balanced view should also take into account the unifying tendencies that are a consequence of regimes and practices of experimentation and instrumentation. These, according to Hacking (1996, 69), "have been more powerful as a source of unity among diverse sciences than have grand unified theories." Instruments (Clarke and Fujimura 1992) are speedily transformed from one discipline to another, not according to theoretical principles but in order to interface with phenomena. As an example, Hacking mentions the tunneling electron microscope: at first thought suitable only for metallurgy, it has expanded into cell biology "in ways not all that well thought-out, and sometimes by accident" (p. 69). We propose that techniques like rheological analysis of tissues and "model tissues" that address the physical aspects (Forgacs and Newman 2005), time-resolved quantitative PCR that addresses the temporal gene expression aspects (Zhang et al. 2006), and "deconstructive" comparative knockout analysis in in vivo and in vitro systems (Lall and Patel 2001; Liu and Kaufman 2005) could play similar unifying roles with respect to the "archeology" of evo-devo genetics—that is, the record of the primitive genetic toolkit.[36]

Another topic for philosophical investigation that could deepen and enrich evo-devo's self-understanding is the vexing role of laws in biology and, intimately related to that, the function of natural kinds in biological modeling and theorizing (Griffiths 1996; Wagner 1996; Laporte 2004); see also the recent debate on the "myth of essentialism" in the history and philosophy of biology (Amundson 2005, 2006; Hull 2006). According to Cartwright (1999, 1), the laws of nature "do not take after the simple, elegant and abstract structure of a system of axioms and theorems." One extreme position wants to dispense with biological laws altogether on the ground that even in the practice of physics they play negligible roles only (see Giere 1999). The other extreme position, represented by certain process structuralists, is that developmental biology will deliver the goods. Thus Eble and Goodwin (2005) write: "The successes of evolutionary developmental biology suggest that the comparison of historical and developmental kinds in developmental morphospace may soon generate more immediate insight into the nature of evolution than the comparison of historical and adaptive kinds in fitness landscapes." OSA's view of inherency clearly situates it more toward the latter end of the spectrum, but exploring these fascinating issues transcends the bounds of this chapter.

Here again, as for most of the themes we have explored, a pluralism of approaches seems desirable. The historian of biology Evelyn Fox Keller (2000a) has plausibly characterized the previous century as "The Century of the Gene." We are currently witnessing a growing awareness among biologists that some of the foundational features of the reigning evolutionary paradigm, in particular its genetic determinism and adaptationism, require substantial revision and need to be complemented by other concepts and theories. We hope to have convinced readers that evo-devo, and more specifically our organismic systems approach to evolution and development, including its naturalistic philosophical outlook, have the potential to contribute to a truer picture of life on this earth.

Notes

1. It is worth noting that when Treviranus published his *Biologie, oder Philosophie der lebenden Natur für Naturforscher und Ärzte* in 1802, the possibility of integration he envisaged was based on the eighteenth-century change in generation theory from preformationism to epigeneticism. "Biology and epigenesis were born together and grew up together" (Depew 2005).

2. "A biological explanation should invoke no factors other than the laws of physics, natural selection, and the contingencies of history. The idea that an organism has a complex history through which natural selection has been in constant operation imposes a special constraint on evolutionary theorizing" (Williams 1985, 1–2). See note 29.

3. "The functional biologist deals with all aspects of the decoding of the programmed information contained in the DNA of the fertilized zygote. The evolutionary biologist, on the other hand, is interested in the history of these programs of information and in the laws that control the changes of these programs from generation to generation. In other words, he is interested in the causes of these changes" (Mayr [1961] 1988, 26). According to Weismann's doctrine, all causality other than that due to environments, which is ignored at the level of the individual cell or organism, can be traced to germ or genes; "the body or phenotype is a causal dead end" (Griesemer 2002, 98).

4. In principle, that is; in practice, adaptationists in behavior and cognition studies tend to shun proximate questions (Callebaut 2003).

5. In the 1950s, when the phrase "synthetic theory" was not as common as it would become later, quite a few authors used to refer to the "selection theory of evolution," "neo-Darwinism," or sometimes "neo-Mendelism" instead. According to Gayon (1989, 4), "Such a terminological hesitation clearly indicates that there is indeed a major theoretical commitment in the synthetic theory: It is fundamentally a general consensus on a genetic theory of natural selection." The issue of adaptationism cannot be pursued in this chapter; see, for example, Antonovics 1987; Amundson 1994; Rose and Lauder 1996; Ahouse 1998; Orzack and Sober 2001; Andrews, Gangestad,

and Matthews 2002. We have found useful Sober's (1998, 72) characterization of adaptationism as the claim that "natural selection has been the only important cause of the phenotypic traits found in most species."

6. Thus, on Lewontin's (2000b, 193–194) view, "It is *not* within the problematic of population geneticists to discover the basic biological phenomena that govern evolutionary change.... The basic phenomena are already provided... by biological discoveries in classical and molecular genetics, cell biology, developmental biology, and ecology. Nor is it within the problematic of *observational* population genetics to discover the ways in which the operation of these causal phenomena can interact to produce effects. The elucidation of the structure of the network of causal pathways, and of the relation between the magnitudes of these elementary forces and their effects on evolution, is an entirely analytic problem." Because we lack the necessary observational power (and, in practice, will always lack it), it also "cannot be the task of population genetics to fill in the particular quantitative values in the basic structure that will provide a correct and testable detailed explanation in any arbitrary case.... Rather, the task of population genetics is to make existential claims about outcomes of evolutionary processes and about significant forces that contribute to these outcomes." Lewontin's pessimistic conclusion is that "there is an inverse relation between the degree of specificity of these existential claims and the size of the domain to which they apply."

7. Gayon (1989, 4–5) noted that there are basically two ways of criticizing the synthetic theory as a whole: (1) to attack natural selection as the ultimate factor controlling all evolutionary processes; (2) to contest the tacit subordination between the diverse fields of research. With a few exceptions such as OSA, which points to the problems of origination of traits and qualitative phenotypic change as the major blind spots of the synthetic theory (Müller and Newman 2003a), biologists engaged in the evo-devo enterprise adopt the second strategy.

8. If not vitalism: As Varela (1979, 5) once pointed out, pushing all the properties pertaining to a coherent, cooperative whole such as the functioning cell into the DNA, which now contains some abstract description of the "teleogenic project" of the cell, brings one close to the vitalists, who insisted on a simular reduction of the characteristics of life to some component other than the cooperative relations of the cellular unity.

9. In developmental biology this notion has gained seeming confirmation in findings that successive developmental steps are typically triggered by episodes of new gene expression and that experimental alteration of gene expression frequently leads to changed developmental outcome. In evolutionary biology the tenet that genes determine form is essentially equivalent to the reigning neo-Darwinian paradigm: (1) evolution is the hereditary transmission of phenotypic change; (2) genes are the medium of heredity; (3) the sum of selected genes therefore specifies and determines the phenotypic differences between organisms (Newman and Müller 2005).

10. We prefer this label (following Griesemer 2000a) to the more common "developmental systems theory" (DST) because, as Griffiths and Gray (2005, 417) acknowledge, what we are talking about here is really "a general theoretical perspective on development, heredity and evolution" intended to facilitate the study of the many factors that influence development without reviving "dichotomous" debates over nature or nurture, and so on. "While theories yield models for explaining, theoretical perspectives yield guidelines for theorizing and for modeling" (García Deister 2005, 28). We will elaborate on the notion of "perspectives" later.

11. Griffiths and Gray (2005, 420–424) argue that the criticism of the parity thesis coming from evo-devo is based on a misunderstanding (see also Gray 2001). But their own whiggish rendering of the history of evo-devo ("EDB [evolutionary developmental biology], whose growth as a discipline has been closely tied to discoveries in developmental genetics, has embraced a conception of the developmental system as an emergent feature of the genome" (p. 421); cf. Arthur 2002) contributes little to solving this and other disagreements between DSP and evo-devo.

12. Another influential interpretation of causation, the *manipulationist* or *interventionist* account, views causal and explanatory relationships as potentially exploitable for purposes of manipulation and control (Woodward 2003; cf. Hacking 1983; Pearl 2000). Exploring these and other, often competing views of the nature of causation for a better understanding of causal processes in evo-devo remains a task for the future.

13. Goodwin (1984) accused the synthetic theory of having neglected the important lessons of developmental mechanics for evolution. Neo-Darwinism treated embryogenesis as an aspect of inheritance, considering that a certain category of occasional causes (genes) ultimately determine form. Thus no room was left for the general principle of biological organization. On Goodwin's alternative view, the process of natural selection acting on genes is subordinated to an "intrinsic dynamics of the living state" (see Gayon 1989, 24).

14. For example, Dawkins's (1976) account of evolution in terms of his replicator-vehicle distinction "does not provide resources to identify empirically the physical avatars of his functional entities. We know that the informational genes are tied to matter and structure, but if evolutionary theory—to be generally enough to cover cultural and conceptual change—must be devoid of *all* reference to concrete mechanisms, it cannot follow from the theory, for example, that genes are inside organisms or are even parts of organisms, as Dawkins's language suggests. Strictly, only the *correlations* between replicator and vehicle due to causal connections of a completely unspecified sort can be implied by such a theory. Striving to get matter and specific structure out of the theory in order to make it apply to immaterial realms may thus leave it bankrupt as an account of causal connection for the material, biological cases" (Griesemer 2005, 79).

15. One standard way to characterize how sciences differ from each other is to specify the modes of explanation they require. Physics is "the standard instance and

model of a science using causal explanation" (Elster 1983, 18), but the realm of causal explanation, which invokes antecedent causes, extends to the life sciences, psychology, social science, and even the humanities. Whereas physics is not normally viewed as having room for explanation in terms of (actual) future consequences of a phenomenon, functional explanation is the hallmark of evolutionary biology and plays a more limited role in other biological disciplines, including biochemistry (Rosenberg 1985), as well as in the social sciences. (Notice that "functional biology" as referred to at the beginning of this section uses causal, not functional explanation.) Intentional explanation is crucial in the social sciences and humanities only; its role in cognitive ethology is controversial (see, e.g., Sterelny 2000).

16. For Gerhart and Kirschner (1997), evolvability is the capacity of a process to generate nonlethal functional variation on which selection can act. In our own epigenetic framework, evolvability is interpreted in an entirely different fashion. It represents the *continued efficacy of epigenetic processes in a lineage*—some of them quite ancient, others of more recent origin—and as such is tied to primitive morphogenetic plasticity. Genetic evolution will tend to suppress such evolvability and buffer the development of form (Newman and Müller 2000, 306).

17. Although one can say with Wagner et al. (2000, 821) that development "can be seen as another set of biological characters that evolve," our emphasis on the primacy of epigenetic mechanisms suggests that the evolution of development cannot be adequately captured within the reigning, gene-centric neo-Darwinian paradigm and should therefore be at the heart of the evo-devo enterprise.

18. As Robert, Hall, and Olson (2001, 957) note, "The aims of evo-devo are not carved in stone," which is exactly what one would expect of an evolutionary account of science à la Hull (1982, 1988). Yet the lists of aims of evo-devo of Hall (2000) and Wagner et al. (2000) map almost one-to-one onto one another.

19. But see Brandon (1990, 185; 1996, 192–202), who disagrees that mechanism implies an inherently hierarchical decomposition. Although we are largely sympathetic with Brandon's (1996, 179ff.) attempt to construe a mechanistic philosophy that would allow biologists to avoid the "false choice" between reductionism and holism, we would prefer not to equate mechanism with "causal pattern" (p. 194), but rather leave some flesh on the bones such as a specification of the material(s) out of which a mechanism is constructed (cf. Griesemer 2005, 62, n. 1). See also Kauffman (1971) and Wimsatt (1974, 1980) on "articulation-of-parts" explanation. Tabery (2004) suggests that the aforementioned notions of mechanism as interaction and as activity may complement each other.

20. See Britten's (2003) "(only) details determine" thesis, according to which development is entirely determined by self-assembling individual genes and molecules ("details") that interact as a result of a long process of natural selection.

21. Given the inflationary use of the term *paradigm* that followed the popularization of Thomas Kuhn's views, many historians and philosophers of science have become wary of the term, but biologists (e.g., Wilkins 1996; Strohman 1997) are reempowering it. The feature we have in mind was captured best in the so-called disciplinary matrix (Kuhn 1970). As a kind of scientific Weltanschauung, disciplinary matrices are acquired largely implicitly during scientific education, in which the learning of "exemplars"—archetypal applications of symbolic generalizations or theories to phenomena—is of paramount importance. Commonness of vocabulary and "symbolic generalizations" notwithstanding, scientific communities with different exemplars will hold different theories, and hence will "view the world differently." Still, according to Kuhn, depending on one's exemplars one will also tend to ask different questions and hold different values. (For a critical assessment, see Suppe 1977, 138–151.) At the level of Weltanschauungen, biological discourse is heavily imbued with metaphors.

22. "The reason why competition, selfish genes, struggle, adaptation, climbing peaks in fitness landscapes, doing better and making progress, are so important as metaphors in neo-Darwinism is because they make sense of evolution in terms that are familiar to us from our social experience in this culture" (Goodwin 1994, 168).

23. Hooker (1994) has argued for this middle-of-the-road position in painstaking detail in the case of Piaget's biology and genetic epistemology. In a similar vein, Amundson (1993) has shown for the debate between behaviorists and cognitivists in psychology that paradigmatic packages are not necessarily as monolithic as they are often presented as being. More generally, Bhaskar (e.g., 1978) has made a case for a critical realism as "an appropriate metaphilosophy for an adequate account of science" on which philosophy will "play two essential roles—namely as a Lockean underlabourer and occasional midwife, and as a Leibnizian conceptual analyst and potential critic" (Bhaskar 1989, 82).

24. "My memory of Steve [Gould] is indelibly tied to the celebration of diversity—diversity of approaches, of explanations, of organisms, and of people" (Wake 2002, 2346). However, see Newman 2003a for a discussion of Gould's tendency to overprivilege genetic causation at the ontogenetic level and nongenetic causation at the behavioral level. Longino (2000, 282) points out the political desirability of epistemological pluralism as "a philosophical epistemology that recognizes the local character of prescriptive epistemologies associated with particular approaches." Such a sensibility for local epistemic cultures is lacking in E. O. Wilson's (1998) call for the unification of all human knowledge by means of *consilience*, the (inductive) proof that everything in our world is organized in terms of a small number of fundamental natural laws—his deep concern for the conservation of biodiversity notwithstanding.

25. Epigenetics is here meant as a collective term for all nonprogrammed factors of development (Newman and Müller 2000). Note that there are fundamental differences in usage between the "epigenesis" of developmental biology and the "epige-

netics" of gene regulation (Griesemer 2002; Müller and Olsson 2003; Jablonka and Lamb 2005).

26. Naturalism has many roots that reach back to Aristotle and include the philosophical traditions of empiricism and materialism, Spinoza, and the American pragmatists. Despite the multiplicity and variety of its formulations, a concise and consistent characterization of ("streamlined") naturalism seems possible today. Naturalism regards philosophy as continuous with science (Callebaut 1993), requires philosophical assertions to be testable (Hull 2001, part III), shuns a priori and transcendental arguments (Callebaut 2003, 2005b), and considers scientific explanations as paradigmatic naturalistic explanations (Giere 1988, 2006; Hull 1988, 2001). A consistent naturalism is a methodological rather than an ontological monism and can best be understood in terms of methodological maxims rather than metaphysical doctrines, which makes it difficult to define naturalism positively (Giere 2006). The open-ended character of naturalism dovetails nicely with mechanism, which on a rather common understanding is also agnostic with respect to ontological commitments (Brandon 1996, 192ff.; cf. Bechtel and Richardson 1993 or Callebaut 1993).

27. For a more detailed discussion see Müller and Newman 2005. The uses of homology and the organizational homology concept are discussed in Müller 2003.

28. See Conway Morris 2003, 505, which refers to the "still unresolved" problem of inherency "whereby much of the potentiality of structures central to evolutionary advancement, e.g. mesoderm, neural crest, are already 'embedded' in more primitive organisms."

29. This is not to be confounded with contingency as understood in logic or, for that matter, sociology (e.g., Luhmann 1984), where something is considered contingent if it is neither necessary nor impossible—which is not identical to "possible," for what is necessary must also be possible. As regards historical contingency, we should be aware that we are entering a conceptual minefield. Historical contingency has been associated with chance (Monod 1971), irreversibility (Georgescu-Roegen 1971) or "lock-in" (W. B. Arthur 1989), and nonrepeatability (Elster 1976). It has been equated with randomness and stochasticity (Chaisson 2001), and has even been conflated with the epistemological notion of unpredictability (Gould 1989, 2002; but see Vermeij 2006). Gould described historical explanations as taking "the form of narrative: E, the phenomenon to be explained, arose because D came before, preceded by C, B, and A. If any of these stages had not occurred, or had transpired in a different way, then E would not exist (or would be present in a substantially altered form, E', requiring a different explanation" (Gould 1989, 283; all further quotes in this note are from this page). But, given A–D, E *"had to arise"* (italics ours), and is in this sense nonrandom. Yet "no law of nature enjoined E"; any variant E' arising from altered antecedents would have been equally explainable, "though massively different in form and effect." For Gould, then, contingency means that the final result is "dependent... upon everything that came before—the unerasable and determining

[sic] signature of history." Superficially, one could think that Gould combined an epistemological notion of contingency, unpredictability, with an ontological one, which Oyama (2000, 116) dubs "causal dependency." But to us, "unerasable and determining signature of history" suggests that the past leaves its traces in the present, which takes us back to epistemology or at least makes for a more complicated picture ("*How much* of the past do we have to know to understand the present?"). It seems to us that interpreting contingency in terms of stochasticity (Chaisson) is equally problematic—viewed from Gould's perspective, that is—because one would now have to rule out stochastic processes that are memoryless (the Markov property: the present state of the system predicts future states as well as the whole history of past and present states). We are assuming here that models and theories that exhibit time lags have a more prominent role to play in biology than in the physical sciences, because the structural knowledge that would enable us to disregard the more ancient causes (A–C in Gould's schema) has not been attained (yet?).

30. Oyama (2000, 116) argues for "a notion of development in which contingency is central and constitutive, not merely secondary alteration of more fundamental, 'preprogrammed' forms."

31. If, as scientific realists, we consider the covering-law account of explanation flawed because mere subsumption is not explanation yet, we must also point out that not all the nonreductionistic accounts of unification that are currently in vogue are necessarily explanatory (see Halonen and Hintikka 1999).

32. As R. L. Carroll (2000) argues, an expanded evolutionary synthesis needs to integrate new concepts and information not only from developmental biology, but from systematics, geology, and the fossil record as well.

33. We agree with Gilbert (2003b, 348) that "not all parts of developmental biology and not all parts of evolutionary biology are involved in these new unions," although we do not share his emphasis on developmental genetics and population genetics. The situation is actually more complicated since "paleontology, morphometrics, QTL mapping, ecology, life history strategy research, and functional morphology" (and, we would add, biological physics) are also involved. If evo-devo is taken to include ecological considerations (Gilbert 2001; Gilbert and Bolker 2003; Hall et al. 2004), this list must be expanded even further.

34. Von Baer's "laws of development" may be summarized under the general principle "Differentiation proceeds from the general to the particular." Wimsatt's "developmental lock" model of *generative entrenchment* (GE) formalizes this principle (Wimsatt 1986). The model sustains, among other tenets, that features expressed earlier in development (1) have a higher probability of being required for features that will appear later; (2) will, on average, have a larger number of "downstream" features dependent on them; (3) are phylogenetically older; (4) are more likely to be widely distributed taxonomically than features expressed later in development. Features

that are deeply generatively entrenched, if "mutated," are likely to cause major developmental abnormalities. GE, which allows for environmental information to be generatively entrenched, has been applied to the innate-acquired distinction (Wimsatt 1986, 1999), evo-devo (Schank and Wimsatt 2001; Wimsatt 2001), conceptual evolution (Griesemer and Wimsatt 1989), and the "evo-devo of culture" (Wimsatt and Griesemer, chapter 7, this volume), among other domains; see also table 2.2.

35. The physical correlate of a higher-level property or kind is typically massively disjunctive, making it vanishingly unlikely that the properties or kinds of higher levels will enter into the kinds of universal laws characteristic of physics or chemistry.

36. For Hacking, mathematical tools have played a similar role as pragmatic unifiers since the day of Descartes, Newton, and Leibniz. A nice contemporary example is the steady expansion of game theory since its inception in the 1940s, from a theory of human decision making to a tool for modeling evo-devo (Ross 2006) and the evolution of biochemical systems (Pfeiffer and Schuster 2005).

References

Abir-Am, P. G. 1982. The discourse of physical power and biological knowledge in the 1930s: A reappraisal of the Rockefeller Foundation's "policy" in molecular biology. *Social Studies of Science* 12: 341–382.

Ahouse, J. C. 1998. The tragedy of a priori selectionism: Dennett and Gould on adaptationism. *Biology and Philosophy* 13: 359–391.

Akam, M. 1995. Hox genes and the evolution of diverse body plans. *Philosophical Transactions of the Royal Society London* B349: 313–319.

Amundson, R. 1993. On the plurality of psychological naturalisms: A response to Ringen. *New Ideas in Psychology* 11: 193–204.

Amundson, R. 1994. Two concepts of constraint: Adaptationism and the challenge from developmental biology. *Philosophy of Science* 61: 556–578.

Amundson, R. 2005. *The Changing Role of the Embryo in Evolutionary Thought*. Cambridge: Cambridge University Press.

Amundson, R. 2006. Evo Devo as cognitive psychology. *Biological Theory* 1: 10–11.

Andrews, P. W., S. W. Gangestad, and D. Matthews. 2002. How to carry out an exaptationist program. *Behavioral and Brain Sciences* 25: 489–553.

Antonovics, J. 1987. The evolutionary dys-synthesis: What bottles for which wine? *American Naturalist* 129: 321–331.

Arthur, W. B. 1989. Competing technologies, increasing returns, and lock-in by historical events. *Economic Journal* 99: 116–131.

Arthur, W. 1987. *Theories of Life: Darwin, Mendel and Beyond.* Harmondsworth: Penguin Books.

Arthur, W. 2001. Developmental drive: An important determinant of the direction of phenotypic evolution. *Evolution and Development* 3: 271–278.

Arthur, W. 2002. The emergent conceptual framework of evolutionary developmental biology. *Nature* 415: 757–764.

Atkinson, J. W. 1992. Conceptual issues in the reunion of development and evolution. *Synthese* 91: 93–110.

Atlan, H., and M. Koppel. 1989. The cellular computer DNA: Program or data? *Bulletin of Mathematical Biology* 52: 335–348.

Autumn, K., M. J. Ryan, and D. B. Wake. 2002. Integrating historical and mechanistic biology enhances the study of adaptation. *Quarterly Review of Biology* 77: 383–408.

Bhaskar, R. 1978. *A Realist Theory of Science.* 2nd ed. Atlantic Highlands, NJ: Humanities Press.

Bhaskar, R. 1989. *Reclaiming Reality: A Critical Introduction to Contemporary Philosophy.* New York: Verso.

Bechtel, W., and R. C. Richardson. 1993. *Discovering Complexity: Decomposition and Localization as Strategies in Scientific Research.* Princeton, NJ: Princeton University Press.

Bissell, M. J., I. S. Mian, D. Radisky, and E. Turley. 2003. Tissue specificity: Structural cues allow diverse phenotypes from a constant genotype. In G. B. Müller and S. A. Newman, eds., *Origination of Organismal Form,* 103–117. Cambridge, MA: MIT Press.

Bolouri, H., and E. H. Davidson. 2003. Transcriptional regulatory cascades in development: Initial rates, not steady state, determine network kinetics. *Proceedings of the National Academy of Sciences USA* 100: 9371–9376.

Boogerd, F. C., F. J. Bruggeman, R. C. Richardson, A. Stephan, and H. V. Westerhoff. 2005. Emergence and its place in nature: A case study of biochemical networks. *Synthese* 145: 131–164.

Brandon, R. N. 1990. *Adaptation and Environment.* Princeton, NJ: Princeton University Press.

Brandon, R. N. 1996. *Concepts and Methods in Evolutionary Biology.* Cambridge: Cambridge University Press.

Brandon, R. N. 1999. The units of selection revisited: The modules of selection. *Biology and Philosophy* 14: 167–180.

Britten, R. J. 2003. Only details determine. In G. B. Müller and S. A. Newman, eds., *Origination of Organismal Form*, 75–86. Cambridge, MA: MIT Press.

Bruggeman, F. J., H. V. Westerhoff, and F. C. Boogerd. 2002. BioComplexity: A pluralist research strategy is necessary for a mechanistic explanation of the "live" state. *Philosophical Psychology* 15: 411–440.

Burian, R. M. 1988. Challenges to the Evolutionary Synthesis. In M. K. Hecht and B. Wallace, eds., *Evolutionary Biology*, 247–269. New York: Plenum Press.

Burian, R. M. 2005. *The Epistemology of Development, Evolution, and Genetics: Selected Essays*. Cambridge: Cambridge University Press.

Callebaut, W. 1993. *Taking the Naturalistic Turn, or How Real Philosophy of Science Is Done*. Chicago: University of Chicago Press.

Callebaut, W. 1998. Self-organization and optimization: Conflicting or complementary approaches? In G. Van de Vijver, S. N. Salthe, and M. Delpos, eds., *Evolutionary Systems*, 79–100. Dordrecht: Kluwer.

Callebaut, W. 2003. Lorenz's philosophical naturalism in the mirror of contemporary science studies. *Ludus Vitalis* 11: 27–55.

Callebaut, W. 2005a. The ubiquity of modularity. In W. Callebaut and D. Rasskin-Gutman, eds., *Modularity*, 3–28. Cambridge, MA: MIT Press.

Callebaut, W. 2005b. Again, what the philosophy of biology is not. *Acta Biotheoretica* 53: 93–122.

Callebaut, W., and D. Rasskin-Gutman, eds. 2005. *Modularity: Understanding the Development and Evolution of Natural Complex Systems*. Cambridge, MA: MIT Press.

Camazine, S., J.-L. Deneubourg, N. R. Franks, J. Sneyd, G. Theraulaz, and E. Bonabeau. 2001. *Self-Organization in Biological Systems*. Princeton, NJ: Princeton University Press.

Caporael, L. R. 1997. The evolution of truly social cognition: The core configurations model. *Personality and Social Psychology Review* 1: 276–298.

Carroll, R. L. 2000. Towards a new evolutionary synthesis. *Trends in Ecology and Evolution* 15: 27–32.

Carroll, S., J. Grenier, and S. Weatherbee. 2004. *From DNA to Diversity: Molecular Genetics and the Evolution of Animal Design*. 2nd ed. Malden, MA: Blackwell.

Cartwright, N. 1989. *Nature's Capacities and Their Measurement*. Oxford: Clarendon Press.

Cartwright, N. 1999. *The Dappled World: A Study of the Boundaries of Science*. Cambridge: Cambridge University Press.

Cellérier, G., S. Papert, and G. Voyat. 1968. *Cybernétique et épistémologie*. Paris: Presses Universitaires de France.

Chaisson, E. J. 2001. *Cosmic Evolution: The Rise of Complexity in Nature*. Cambridge, MA: Harvard University Press.

Cinquin, O., and J. Demongeot. 2005. High-dimensional switches and the modeling of cellular differentiation. *Journal of Theoretical Biology* 233: 391–411.

Clarke, A., and J. Fujimura, eds. 1992. *The Right Tools for the Job: At Work in Twentieth Century Life Sciences*. Princeton, NJ: Princeton University Press.

Collier, J. D. 1999. Causation is the transfer of information. In H. Sankey, ed., *Causation and Laws of Nature*, 215–245. Dordrecht: Kluwer.

Colosimo, P. F., K. E. Hosemann, S. Balabhadra, G. Villarreal Jr., M. Dickson, J. Grimwood, J. Schmutz, R. M. Myers, D. Schluter, and D. M. Kingsley. 2005. Widespread parallel evolution in sticklebacks by repeated fixation of ectodysplasin alleles. *Science* 307: 1928–1933.

Conrad, M. 1998. Towards high evolvability dynamics. In G. Van de Vijver, S. N. Salthe, and M. Delpos, eds., *Evolutionary Systems*, 33–43. Dordrecht: Kluwer.

Conway Morris, S. 2003. The Cambrian "explosion" of metazoans and molecular biology: Would Darwin be satisfied? *International Journal of Developmental Biology* 47: 505–515.

Costa, L. da F., B. A. N. Travençolo, A. Azeredo, M. E. Beletti, G. B. Müller, D. Rasskin-Gutman, G. Sternik, M. Ibañes, and J. C. Izpisúa Belmonte. 2005. Field approach to three-dimensional gene expression pattern characterization. *Applied Physics Letters* 86: 143901–143903.

Cronin, H. 1991. *The Ant and the Peacock: Altruism and Sexual Selection from Darwin to Today*. Cambridge: Cambridge University Press.

Darden, L. 1986. Relations among fields in the Evolutionary Synthesis. In W. Bechtel, ed., *Integrating Scientific Disciplines*, 113–123. Dordrecht: Nijhoff.

Darden, L. 1991. *Theory Change in Science: Strategies from Mendelian Genetics*. New York: Oxford University Press.

Darden, L., and C. Craver. 2002. Strategies in the interfield discovery of the mechanisms of protein synthesis. *Studies in History and Philosophy of Biology and Biomedical Sciences* 33C: 1–28.

Darden, L., and N. Maull. 1977. Interfield theories. *Synthese* 44: 43–64.

Davidson, E. H. 2006. *The Regulatory Genome: Gene Regulatory Networks in Development and Evolution*. New York: Academic Press.

Dawkins, R. 1976. *The Selfish Gene*. Oxford: Oxford University Press.

Deacon, T. W. 2005. Beyond Piaget's phenocopy: The baby in the Lamarckian bath. In S. T. Parker, J. Langer, and C. Milbrath, eds., *Biology and Knowledge Revisited: From Neurogenesis to Psychogenesis*, 87–122. Mahwah, NJ: Erlbaum.

Deacon, T. W. 2006. Reciprocal linkage between self-organizing processes is sufficient for self-reproduction and evolvability. *Biological Theory* 1: 136–149.

de Gennes, P. G. 1992. Soft matter. *Science* 256: 495–497.

Depew, D. J. 2005. Darwin's evolution and "Baldwin effect." International Colloquium, Epigenetic Developments: In Search of a New Model for Complexity in Sciences and Philosophy, Rome, September 13–14.

Depew, D. J., and B. H. Weber, eds. 1985. *Evolution at a Crossroads: The New Biology and the New Philosophy of Science*. Cambridge, MA: MIT Press.

Depew, D. J., and B. H. Weber. 1995. *Darwinism Evolving: System Dynamics and the Genealogy of Natural Selection*. Cambridge, MA: MIT Press.

de Regt, H. 2004. Making sense of understanding. *Philosophy of Science* 71: 98–109.

de Visser, J. A. G. M., J. Hermisson, G. P. Wagner, L. A. Meyers, H. Bagheri-Chaichian, J. L. Blanchard, L. Chao, J. M. Cheverud, S. F. Elena, W. Fontana, G. Gibson, T. F. Hansen, D. Krakauer, R. C. Lewontin, C. Ofria, S. H. Rice, G. von Dassow, A. Wagner, and C. Whitlock. 2003. Evolution and detection of genetic robustness. *Evolution* 57: 1959–1972.

Dowe, P. 1992. Wesley Salmon's process theory of causality and the conserved quantity theory. *Philosophy of Science* 59: 195–216.

Dowe, P. 2000. *Physical Causation*. Cambridge: Cambridge University Press.

Dupré, J. 1983. The disunity of science. *Mind* 92: 321–346.

Dupré, J. 1993. *The Disorder of Things*. Cambridge, MA: Harvard University Press.

Earl, D. J., and M. W. Deem. 2004. Evolvability is a selectable trait. *Proceedings of the National Academy of Sciences USA* 101: 11531–11536.

Eberhard, W. G. 2001. Multiple origins of a major novelty: Moveable abdominal lobes in male sepsid flies (Diptera: Sepsidae) and the question of developmental constraints. *Evolution and Development* 3: 206–222.

Eble, G. J., and B. C. Goodwin. 2005. Natural kinds and the dialectic between evolution and development. Unpublished manuscript.

Eckstein, H. 1980. Theoretical approaches to explaining collective violence. In T. R. Gurr, ed., *Handbook of Political Conflict*, 135–167. New York: Free Press.

Ehrlich, P. R. 2000. *Human Natures: Genes, Cultures, and the Human Prospect*. Washington, DC: Island Press.

Eldredge, N. 1985. *Unfinished Synthesis: Biological Hierarchies and Modern Evolutionary Thought*. New York: Oxford University Press.

Ellis, B. 1999. Causal powers and laws of nature. In H. Sankey, ed., *Causation and Laws of Nature*, 19–34. Dordrecht: Kluwer.

Elster, J. 1976. A note on hysteresis in the social sciences. *Synthese* 33: 371–391.

Elster, J. 1983. *Explaining Technical Change: A Case Study in the Philosophy of Science*. Cambridge: Cambridge University Press.

Fisk, M. 1973. *Nature and Necessity: An Essay in Physical Ontology*. Bloomington: Indiana University Press.

Fogelman Soulié, F., ed. 1991. *Les théories de la complexité: Autour de l'oeuvre d'Henri Atlan*. Paris: Seuil.

Fontana, W. 2002. Modeling "evo-devo" with RNA. *BioEssays* 24: 1164–1177.

Fontana, W., and L. Buss. 1994a. "The arrival of the fittest": Toward a theory of biological organization. *Bulletin of Mathematical Biology* 56: 1–64.

Fontana, W., and L. Buss. 1994b. What would be conserved if "the tape were played twice"? *Proceedings of the National Academy of Sciences USA* 91: 757–761.

Fontana, W., and P. Schuster. 1998. Shaping space: The possible and the attainable in RNA genotype-phenotype mapping. *Journal of Theoretical Biology* 194: 491–515.

Forgacs, G., and S. A. Newman. 2005. *Biological Physics of the Developing Embryo*. Cambridge: Cambridge University Press.

Foty, R. A., C. M. Pfleger, G. Forgacs, and M. S. Steinberg. 1996. Surface tensions of embryonic tissues predict their mutual envelopment behavior. *Development* 122: 1611–1620.

Fry, L. 2000. *The Emergence of Life on Earth: A Historical and Scientific Overview*. New Brunswick, NJ: Rutgers University Press.

Futuyma, D. J. 1988. Sturm und Drang and the Evolutionary Synthesis. *Evolution* 42: 217–226.

Galison, P., and D. J. Stump, eds. 1996. *The Disunity of Science: Boundaries, Contexts, and Power*. Stanford, CA: Stanford University Press.

García Deister, V. 2005. *Resisting Dichotomies: Causal Images and Causal Processes in Development*. Posgrado en Filosofie de la Ciencia, Facultad de Filosofía y Letras, Instituto de Investigaciones Filosóficas, Universidad Nacional Autónoma de México. www.stv.umb.edu/n05garcia.pdf.

Gayon, J. 1989. Critics and criticisms of the Modern Synthesis: The viewpoint of a philosopher. *Evolutionary Biology* 24: 1–49.

Georgescu-Roegen, N. 1971. *The Entropy Law and the Economic Process*. Cambridge, MA: Harvard University Press.

Gerhart, J., and M. Kirschner. 1997. *Cells, Embryos, and Evolution: Toward a Cellular and Developmental Understanding of Phenotypic Variation and Evolutionary Adaptability*. Oxford: Blackwell Science.

Gibson, G., and G. P. Wagner. 2000. Canalization in evolutionary genetics: A stabilizing theory? *BioEssays* 22: 372–380.

Giere, R. N. 1988. *Explaining Science: A Cognitive Approach*. Chicago: University of Chicago Press.

Giere, R. N. 1999. *Science Without Laws*. Chicago: University of Chicago Press.

Giere, R. N. 2006. Modest evolutionary naturalism. *Biological Theory* 1: 50–58.

Gilbert, S. F. 2001. Ecological developmental biology: Developmental biology meets the real world. *Developmental Biology* 233: 1–12.

Gilbert, S. F. 2003a. *Developmental Biology*. 7th ed. Sunderland, MA: Sinauer Associates.

Gilbert, S. F. 2003b. Evo-Devo, Devo-Evo, and DevGen-PopGen. *Biology and Philosophy* 18: 347–362.

Gilbert, S. F., and J. A. Bolker. 2001. Homologies of process and modular elements of embryonic construction. *Journal of Experimental Zoology (Molecular and Developmental Evolution)* B291: 1–12.

Gilbert, S. F., and J. A. Bolker. 2003. Ecological developmental biology: Preface to the symposium. *Evolution and Development* 5: 3–8.

Gilbert, S. F., J. M. Opitz, and R. A. Raff. 1996. Resynthesizing evolutionary and developmental biology. *Developmental Biology* 173: 357–372.

Glennan, S. S. 1996. Mechanisms and the nature of causation. *Erkenntnis* 44: 49–71.

Glennan, S. S. 2002a. Contextual unanimity and the units of selection problem. *Philosophy of Science* 69: 118–137.

Glennan, S. S. 2002b. Rethinking mechanistic explanation. *Philosophy of Science* 69: S342–S353.

Godfrey-Smith, P. 2001a. On the status and explanatory structure of Developmental Systems Theory. In S. Oyama et al., eds., *Cycles of Contingency*, 283–297. Cambridge, MA: MIT Press.

Godfrey-Smith, P. 2001b. Organism, environment, and dialectics. In R. S. Singh et al., eds., *Thinking about Evolution*, vol. 2, 253–266. Cambridge: Cambridge University Press.

Goodwin, B. C. 1984. Changing from an evolutionary to a generative paradigm in biology. In J. W. Pollard, ed., *Evolutionary Theory: Paths into the Future*, 99–120. Chichester: Wiley.

Goodwin, B. C. 1994. *How the Leopard Changed Its Spots: The Evolution of Complexity*. London: Weidenfeld and Nicholson.

Goodwin, B. C., N. Holder, and C. C. Wylie, eds. 1983. *Development and Evolution*. Cambridge: Cambridge University Press.

Gould, S. J. 1989. *Wonderful Life: The Burgess Shale and the Nature of History*. New York: Norton.

Gould, S. J. 2001. The evolutionary definition of selective agency, validation of the theory of hierarchical selection, and fallacy of the selfish gene. In R. S. Singh et al., eds., *Thinking about Evolution*, vol. 2, 208–234. Cambridge: Cambridge University Press.

Gould, S. J. 2002. *The Structure of Evolutionary Theory*. Cambridge, MA: Belknap Press of Harvard University Press.

Gould, S. J., and R. C. Lewontin. 1979. The spandrels of San Marco and the Panglossian paradigm: A critique of the adaptationist programme. *Proceedings of the Royal Society of London* B: 581–598.

Gray, R. D. 1992. Death of the gene: Developmental Systems strike back. In P. Griffiths, ed., *Trees of Life: Essays in the Philosophy of Biology*, 165–209. Dordrecht: Kluwer.

Gray, R. D. 2001. Selfish genes or developmental systems? In R. S. Singh et al., eds., *Thinking about Evolution*, vol. 2, 184–207. Cambridge: Cambridge University Press.

Griesemer, J. R. 2000a. Development, culture, and the units of inheritance. *Philosophy of Science* 67: 348–368.

Griesemer, J. R. 2000b. Reproduction and the reduction of genetics. In P. Beurton, R. Falk, and H.-J. Rheinberger, eds., *The Concept of the Gene in Development and Evolution: Historical and Epistemological Perspectives*, 240–285. Cambridge: Cambridge University Press.

Griesemer, J. 2000c. The units of evolutionary transition. *Selection* 1: 67–80.

Griesemer, J. 2002. What is "epi" about epigenetics? *Annals of the New York Academy of Sciences* 981: 97–110.

Griesemer, J. R. 2005. The informational gene and the substantial body: On the generalization of evolutionary theory by abstraction. In M. R. Jones and N. Cartwright, eds., *Idealizations XII: Correcting the Model: Idealization and Abstraction in the Sciences*, 59–115. Amsterdam: Rodopi.

Griesemer, J. R. 2006a. Theoretical integration, cooperation, and theories as tracking devices. *Biological Theory* 1: 4–7.

Griesemer, J. R. 2006b. Genetics from an evolutionary process perspective. In E. M. Neumann-Held and C. Rehman-Sutter, eds., *Genes in Development*, 199–237. Durham, NC: Duke University Press.

Griesemer, J. R., and W. C. Wimsatt. 1989. Picturing Weismannism: A case study of conceptual evolution. In M. Ruse, ed., *What the Philosophy of Biology Is*, 75–137. Dordrecht: Kluwer.

Griffiths, P. E. 1996. Darwinism, process structuralism, and natural kinds. *Philosophy of Science* 63: 51–59.

Griffiths, P. E., and R. Gray. 1994. Developmental systems and evolutionary explanation. *Journal of Philosophy* 91: 277–304.

Griffiths, P. E., and R. Gray. 2005. Discussion: Three ways to misunderstand developmental systems theory. *Biology and Philosophy* 20: 417–425.

Hacking, I. 1983. *Representing and Intervening: Introductory Topics in the Philosophy of Natural Science*. Cambridge: Cambridge University Press.

Hacking, I. 1996. The disunities of the sciences. In P. Galison and D. J. Stump, eds., *The Disunity of Science*, 37–74. Stanford, CA: Stanford University Press.

Hahlweg, K., and C. A. Hooker. 1989. Evolutionary epistemology and philosophy of science. In K. Hahlweg and C. A. Hooker, eds., *Issues in Evolutionary Epistemology*, 21–150. Albany: State University of New York Press.

Hall, B. K. 1999. *Evolutionary Developmental Biology*. 2nd ed. Dordrecht: Kluwer.

Hall, B. K. 2000. Evo-devo or devo-evo—Does it matter? *Evolution and Development* 2: 177–178.

Hall, B. K. 2003. Unlocking the black box between genotype and phenotype: Cell condensations as morphogenetic (modular) units. *Biology and Philosophy* 18: 219–247.

Hall, B. K., R. D. Pearson, and G. B. Müller, eds. 2004. *Environment, Development, and Evolution*. Cambridge, MA: MIT Press.

Halonen, I., and J. Hintikka. 1999. Unification—It's magnificent but is it explanation? *Synthese* 120: 27–47.

Hamburger, V. 1980. Embryology and the Modern Synthesis in evolutionary theory. In E. Mayr and W. Provine, eds., *The Evolutionary Synthesis*, 97–112. Cambridge: Cambridge University Press.

Hammerstein, P., E. H. Hagen, A. V. M. Herz, and H. Herzel. 2006. Robustness. *Biological Theory* 1: 88–91.

Hanken, J. 1985. Morphological novelty in the limb skeleton accompanies miniaturization in Salamanders. *Science* 229: 871–874.

Hansen, T. F. 2003. Is modularity necessary for evolvability? Remarks on the relationship between pleiotropy and evolvability. *BioSystems* 69: 1–12.

Haraway, D. 1976. *Crystals, Fabrics, and Fields: Metaphors of Organicism in Twentieth-Century Developmental Biology*. New Haven, CT: Yale University Press.

Harré, R. 2005. Transcending the emergence/reduction distinction: The case of biology. In A. O'Hear, ed., *Philosophy, Biology and Life*, 1–20. Cambridge: Cambridge University Press.

Hazen, R. M. 2005. *Genesis: The Scientific Quest for Life's Origins*. Washington, DC: Joseph Henry Press.

Hendriks-Jansen, H. 1996. *Catching Ourselves in the Act: Situated Activity, Interactive Emergence, and Human Thought*. Cambridge, MA: MIT Press.

Henry, C. M. 2003. Systems biology. *Chemical and Engineering News* 81(29): 45–55.

Hernández Blasi, C., and D. F. Bjorklund. 2003. Evolutionary developmental psychology: A new tool for better understanding human ontogeny. *Human Development* 46: 259–281.

Ho, M.-W., and P. T. Saunders. 1979. Beyond neo-Darwinism: An epigenetic approach to evolution. *Journal of Theoretical Biology* 78: 573–591.

Holland, P. W. H. 2002. The fall and rise of evolutionary developmental biology. In D. M. Williams and P. L. Forey, eds., *Milestones in Systematics*, 261–275. Boca Raton: CRC Press.

Holland, P. W. H., and T. Takahashi. 2005. The evolution of homeobox genes: Implications for the study of brain development. *Brain Research Bulletin* 66: 484–490.

Hood, L., and D. Galas. 2003. The digital code of DNA. *Nature* 421: 444–448.

Hooker, C. A. 1994. Regulatory constructivism: On the relation between evolutionary epistemology and Piaget's genetic epistemology. *Biology and Philosophy* 9: 197–244.

Hooker, C. A. 1995. *Reason, Regulation, and Realism: Toward a Regulatory Systems Theory of Reason and Evolutionary Epistemology*. Albany: State University of New York Press.

Hooker, C. A. 2004. Asymptotics, reduction and emergence. *British Journal for the Philosophy of Science* 55: 435–479.

Hull, D. L. 1973. Reduction in genetics—Doing the impossible? In P. Suppes et al., eds., *Logic, Methodology and Philosophy of Science IV*, 619–635. Amsterdam: Elsevier.

Hull, D. L. 1982. The naked meme. In H. C. Plotkin, ed., *Learning, Development, and Culture: Essays in Evolutionary Epistemology*, 273–327. Chichester: Wiley.

Hull, D. L. 1988. *Science as a Process: An Evolutionary Account of the Social and Conceptual Development of Science*. Chicago: University of Chicago Press.

Hull, D. L. 1998. Introduction to Part II. In D. L. Hull and M. Ruse, eds., *The Philosophy of Biology*, 89–92. Oxford: Oxford University Press.

Hull, D. L. 2001. *Science and Selection: Essays on Biological Evolution and the Philosophy of Science*. Cambridge: Cambridge University Press.

Hull, D. L. 2006. The essence of scientific theories. *Biological Theory* 1: 16–18.

Jablonka, E. 2000. Lamarckian inheritance systems in biology: A source of metaphors and models in technological evolution. In J. Ziman, ed., *Technological Innovation as an Evolutionary Process*, 27–40. Cambridge: Cambridge University Press.

Jablonka, E., and M. J. Lamb. 2005. *Evolution in Four Dimensions: Genetic, Epigenetic, Behavioral, and Symbolic Variation in the History of Life*. Cambridge, MA: MIT Press.

Jacob, F. 1970. *The Logic of Life: A History of Heredity*. Princeton, NJ: Princeton University Press.

Jernvall, J. 2000. Linking development with generation of novelty in mammalian teeth. *Proceedings of the National Academy of Sciences USA* 97: 2641–2645.

Johnston, T. D. 1982. Learning and the evolution of developmental systems. In H. C. Plotkin, ed., *Learning, Development, and Culture*, 411–442.

Kauffman, S. 1971. Articulation of parts explanation in biology and the rational search for them. In R. Buck and R. S. Cohen, eds., *PSA 1970*, 257–272. Dordrecht: Reidel.

Kauffman, S. 1985. Self-organization, selective adaptation, and its limits: A new pattern of inference in evolution and development. In D. J. Depew and B. H. Weber, eds., *Evolution at a Crossroads*, 169–207. Cambridge, MA: MIT Press.

Kauffman, S. 1992. The sciences of complexity and "origins of order." In J. Mitten-thal and A. Baskin, eds., *Principles of Organization in Organisms*, 309–391. Reading, MA: Addison-Wesley.

Kauffman, S. 1993. *The Origins of Order: Self-Organization and Selection in Evolution*. Oxford: Oxford University Press.

Kauffman, S. 1995. *At Home in the Universe: The Search for Laws of Complexity*. Oxford: Oxford University Press.

Keller, E. F. 2000a. *The Century of the Gene*. Cambridge, MA: Harvard University Press.

Keller, E. F. 2000b. Making sense of life: Explanation in developmental biology. In R. Creath and J. Maienschein, eds., *Biology and Epistemology*, 244–260. Cambridge: Cambridge University Press.

Kirschner, M., and J. Gerhart. 1998. Evolvability. *Proceedings of the National Academy of Sciences USA* 95: 8420–8427.

Kitano, H. 2001. *Foundations of Systems Biology*. Cambridge, MA: MIT Press.

Kitano, H., ed. 2002. Systems biology: A brief overview. *Science* 295: 1662–1668.

Kitano, H. 2004. Biological robustness. *Nature Genetics* 5: 826–837.

Krohs, U., and W. Callebaut. 2007. Data without models merging with models without data. In F. C. Boogerd, F. J. Bruggeman, J.-H. S. Hofmeyer, and H. V. Westerhoff, eds., *Systems Biology: Philosophical Foundations*, 181–213. Amsterdam: Reed-Elsevier.

Kuhn, T. S. 1970. *The Structure of Scientific Revolutions*. 2nd ed. Chicago: University of Chicago Press.

Lall, S., and N. H. Patel. 2001. Conservation and divergence in molecular mechanisms of axis formation. *Annual Review of Genetics* 35: 407–437.

Laland, K. N., F. J. Odling-Smee, and M. W. Feldman. 1999. Niche construction, biological evolution and cultural change. *Behavioral and Brain Sciences* 23: 131–175.

Laporte, J. 2004. *Natural Kinds and Conceptual Change*. Cambridge: Cambridge University Press.

Laubichler, M. D. 2000. Homology in development and the development of the homology concept. *American Zoologist* 40: 777–788.

Laubichler, M. D., and J. Maienschein, eds. 2007a. *From Embryology to Evo-Devo: A History of Embryology in the 20th Century*. Cambridge, MA: MIT Press.

Laubichler, M. D., and J. Maienschein. 2007b. Embryos, cells, genes, and organisms: A few reflections on the history of evolutionary developmental biology. In R. Sansom and R. N. Brandon, eds., *Integrating Evolution and Development*, 1–92. Cambridge, MA: MIT Press.

Levins, R. 1968. *Evolution in Changing Environments*. Princeton, NJ: Princeton University Press.

Levins, R., and R. C. Lewontin. 1985. *The Dialectical Biologist*. Cambridge, MA: Harvard University Press.

Levins, R., and R. C. Lewontin. 2006. A program for biology. *Biological Theory* 1: 333–335.

Lewontin, R. C. 1970. The units of evolution. *Annual Review of Ecology* 1: 1–23.

Lewontin, R. C. 1996. Evolution as engineering. In J. Collado-Vides, B. Magasanik, and T. F. Smith, eds., *Integrative Approaches to Molecular Biology*, 1–10. Cambridge, MA: MIT Press.

Lewontin, R. C. 2000a. *The Triple Helix: Gene, Organism, and Environment*. Cambridge, MA: Harvard University Press.

Lewontin, R. C. 2000b. What do population geneticists know and how do they know it? In R. Creath and J. Maienschein, eds., *Biology and Epistemology*, 191–215. Cambridge: Cambridge University Press.

Liu, P. Z., and T. C. Kaufman. 2005. *even-skipped* is not a pair-rule gene but has segmental and gap-like functions in *Oncopeltus fasciatus*, an intermediate germband insect. *Development* 132: 2081–2092.

Lolle, S. J., J. L. Victor, J. M. Young, and R. E. Pruitt. 2005. Genome-wide non-mendelian inheritance of extra-genomic information in *Arabidopsis*. *Nature* 434: 505–509.

Longino, H. E. 2000. Toward an epistemology for biological pluralism. In R. Creath and J. Maienschein, eds., *Biology and Epistemology*, 261–286. Cambridge: Cambridge University Press.

Love, A. 2003. Evolutionary morphology, innovation, and the synthesis of evolutionary and developmental biology. *Biology and Philosophy* 18: 309–345.

Love, A., and R. A. Raff. 2005. Larval ectoderm, organizational homology, and the origins of evolutionary novelty. *Journal of Experimental Zoology (Molecular and Developmental Evolution)* B306: 18–34.

Løvtrup, S. 1984. Ontogeny and phylogeny from an epigenetic point of view. *Human Development* 27: 249–261.

Luhmann, N. 1984. *Soziale Systeme: Grundriss einer allgemeinen Theorie*. Frankfurt am Main: Suhrkamp.

Machamer, P., L. Darden, and C. F. Craver. 2000. Thinking about mechanisms. *Philosophy of Science* 67: 1–25.

Magee, P. T. 1997. Which came first, the hypha or the yeast? *Science* 277: 52–53.

Mahner, M., and M. Bunge. 1997. *Foundations of Biophilosophy*. Berlin: Springer.

Maienschein, J. 2000. Competing epistemologies and developmental biology. In R. Creath and J. Maienschein, eds., *Biology and Epistemology*, 122–137. Cambridge: Cambridge University Press.

Margulis, L., and R. Fester, eds. 1991. *Symbiosis as a Source of Evolutionary Innovation.* Cambridge, MA: MIT Press.

Maynard Smith, J. 1998. *Shaping Life: Genes, Embryos, and Evolution.* New Haven, CT: Yale University Press.

Maynard Smith, J., and E. Szathmáry. 1995. *The Major Transitions in Evolution.* Oxford: Oxford University Press.

Maynard Smith, J., and E. Szathmáry. 1998. *The Origins of Life: From the Birth of Life to the Origins of Language.* Oxford: Oxford University Press.

Maynard Smith, J., R. M. Burian, S. A. Kauffman, P. A. Alberch, J. Campbell, B. C. Goodwin, L. Lande, D. Raup, and L. Wolpert. 1985. Developmental constraints and evolution: A perspective from the Mountain Lake conference on development and evolution. *Quarterly Review of Biology* 60: 265–287.

Mayr, E. 1958. Behavior and systematics. In A. Roe and G. G. Simpson, eds., *Behavior and Evolution*, 341–362. New Haven, CT: Yale University Press.

Mayr, E. 1960. The emergence of evolutionary novelties. In S. Tax, eds., *Evolution After Darwin: The University of Chicago Centennial, Vol. 1: The Evolution of Life: Its Origin, History and Future*, 349–380. Chicago: University of Chicago Press.

Mayr, E. 1961. Cause and effect in biology. *Science* 134: 1501–1506. Reprinted in E. Mayr, *Toward a New Philosophy of Biology*, 24–37. Cambridge, MA: Belknap Press of Harvard University Press.

Mayr, E. 1982. *The Growth of Biological Thought.* Cambridge, MA: Belknap Press of Harvard University Press.

Mayr, E. 2001. *What Evolution Is.* New York: Basic Books.

Mayr, E., and W. B. Provine, eds. 1980. *The Evolutionary Synthesis: Perspectives on the Unification of Biology.* Cambridge, MA: Harvard University Press.

McLaren, A., and D. Michie. 1958. An effect of the uterine environment upon skeletal morphology in the mouse. *Nature* 181: 1147–1148.

Mesarovic, M. D. 1968. Systems theory and biology: View of a theoretician. In M. D. Mesarovic, ed., *Systems Theory and Biology*, 59–87. Berlin: Springer.

Mesarovic, M. D., S. N. Sreenath, and J. D. Keene. 2004. Search for organising principles: Understanding in systems biology. *Systems Biology* 1: 19–27.

Mikhailov, A. S. 1990. *Foundations of Synergetics I.* Berlin: Springer.

Minati, G., and E. Pessa, eds. 2002. *Emergence in Complex, Cognitive, Social, and Biological Systems.* Dordrecht: Kluwer.

Monod, J. 1971. *Chance and Necessity.* New York: Knopf. (French original 1970.)

Moss, L. 2002. *What Genes Can't Do.* Cambridge, MA: MIT Press.

Müller, G. B. 1990. Developmental mechanisms at the origin of morphological novelty: A side-effect hypothesis. In M. H. Nitecki, ed., *Evolutionary Innovations*, 99–130. Chicago: University of Chicago Press.

Müller, G. B. 1991. Experimental strategies in evolutionary embryology. *American Zoologist* 31: 605–615.

Müller, G. B. 2002. Novelty and key innovation. In M. Pagel, ed., *Encylopedia of Evolution*, 827–830. New York: Oxford University Press.

Müller, G. B. 2003. Homology: The evolution of morphological organization. In G. B. Müller and S. A. Newman, eds., *Origination of Organismal Form*, 51–69. Cambridge, MA, MIT Press.

Müller, G. B. 2005. Evolutionary developmental biology. In F. M. Wuketits and F. J. Ayala, eds., *Handbook of Evolution*, vol. 2, 87–115. San Diego: Wiley.

Müller, G. B. 2007. Six memos for Evo-devo. In M. D. Laubichler and J. Maienschein, eds., *From Embryology to Evo-Devo: A History of Embryology in the 20th Century*, 499–524. Cambridge, MA: MIT Press.

Müller, G. B., and S. A. Newman. 1999. Generation, integration, autonomy: Three steps in the evolution of homology. In G. Cardew and G. R. Bock, eds., *Homology*, 65–79. Chichester: Wiley.

Müller, G. B., and S. A. Newman, eds. 2003a. *Origination of Organismal Form: Beyond the Gene in Development and Evolution.* Cambridge, MA: MIT Press.

Müller, G. B., and S. A. Newman. 2003b. Origination of organismal form: The forgotten cause in evolutionary theory. In G. B. Müller and S. A. Newman, eds., *Origination of Organismal Form*, 3–10. Cambridge, MA, MIT Press.

Müller, G. B., and S. A. Newman. 2005. The innovation triad: An Evo-Devo agenda. *Journal of Experimental Zoology (Molecular and Developmental Evolution)* 304: 487–503.

Müller, G. B., and L. Olsson. 2003. Epigenesis and epigenetics. In B. K. Hall and W. M. Olson, eds., *Keywords and Concepts in Evolutionary Developmental Biology*, 114–123. Cambridge, MA: Harvard University Press.

Müller, G. B., and G. P. Wagner. 1991. Novelty in evolution: Restructuring the concept. *Annual Review of Ecology and Systematics* 22: 229–256.

Müller, G. B., and G. P. Wagner. 2003. Innovation. In B. K. Hall and W. M. Olson, eds., *Keywords and Concepts in Evolutionary Developmental Biology*, 218–227. Cambridge, MA: Harvard University Press.

Neumann-Held, E. M., and C. Rehmann-Sutter, eds. 2006. *Genes in Development: Rereading the Molecular Paradigm.* Durham, NC: Duke University Press.

Newman, S. A. 1994. Generic physical mechanisms of tissue morphogenesis: A common basis for development and evolution. *Journal of Evolutionary Biology* 7: 467–488.

Newman, S. A. 2002. Developmental mechanisms: Putting genes in their place. *Journal of Biosciences* 27: 97–104.

Newman, S. A. 2003a. Nature, progress and Stephen Jay Gould's biopolitics. *Rethinking Marxism* 15: 479–496.

Newman, S. A. 2003b. The fall and rise of systems biology: Recovering from a half-century gene hinge. *GeneWatch* (July/August) 8–12.

Newman, S. A. 2005. The pre-Mendelian, pre-Darwinian world: Shifting relations between genetic and epigenetic mechanisms in early molecular evolution. *Journal of Biosciences* 30: 75–85.

Newman, S. A. 2006. The developmental-genetic toolkit and the molecular homology-analogy paradox. *Biological Theory* 1: 12–15.

Newman, S. A., and W. D. Comper. 1990. "Generic" physical mechanisms of morphogenesis and pattern formation. *Development* 110: 1–18.

Newman, S. A., G. Forgacs, and G. B. Müller. 2006. Before programs: The physical origination of multicellular forms. *International Journal of Developmental Biology* 50: 289–299.

Newman, S. A., and G. B. Müller. 2000. Epigenetic mechanisms of character origination. *Journal of Experimental Zoology (Molecular and Developmental Evolution)* B288: 304–317. Reprinted in G. P. Wagner and M. D. Laubichler, eds., *The Character Concept in Evolutionary Biology*, 559–579. San Diego: Academic Press, 2001.

Newman, S. A., and G. B. Müller. 2005. Origination and innovation in the vertebrate limb skeleton: An epigenetic perspective. *Journal of Experimental Zoology (Molecular and Developmental Evolution)* B304: 593–609.

Newman, S. A., and G. B. Müller. 2006. Genes and form: Inherency in the evolution of developmental mechanisms. In E. M. Neumann-Held and C. Rehmann-Sutter, eds., *Genes in Development*, 38–73. Durham, NC: Duke University Press.

Nicolis, G. 1995. *Introduction to Nonlinear Science*. Cambridge: Cambridge University Press.

Nijhout, H. F. 1990. Metaphors and the roles of genes in development. *BioEssays* 23: 441–446.

Nijhout, H. F. 2003. Gradients, diffusion, and genes in pattern formation. In G. B. Müller and S. A. Newman, eds., *Origination of Organismal Form*, 165–181. Cambridge, MA: MIT Press.

Nijhout, H. F. 2007. Complex traits: Genetics and evolution. In R. Sansom and R. N. Brandon, eds., *Integrating Evolution and Development*, 93–112. Cambridge, MA: MIT Press.

Odling-Smee, J., K. N. Laland, and M. W. Feldman. 2003. *Niche Construction: The Neglected Process in Evolution*. Princeton, NJ: Princeton University Press.

Orzack, S., and E. Sober. 2001. *Adaptationism and Optimality*. Cambridge: Cambridge University Press.

Oyama, S. 2000. *Evolution's Eye: A Systems View of the Biology-Culture Divide*. Durham, NC: Duke University Press.

Oyama, S., P. E. Griffiths, and R. D. Gray. 2001. *Cycles of Contingency: Developmental Systems and Evolution*. Cambridge, MA: MIT Press.

Padulo, L., and M. Arbib. 1974. *System Theory*. Philadelphia: Saunders.

Parker, S. T. 2005. Piaget's phenocopy model revisited: A brief history of ideas about the origins of adaptive genetic variations. In S. T. Parker, J. Langer, and C. Milbrath, eds., *Biology and Knowledge Revisited: From Neurogenesis to Psychogenesis*, 33–86. Mahwah, NJ: Erlbaum.

Pearl, J. 2000. *Causality: Models, Reasoning, and Inference*. Cambridge: Cambridge University Press.

Pfeiffer, T., and S. Schuster. 2005. Game-theoretical approaches to studying the evolution of biochemical systems. *Trends in Biochemical Sciences* 30: 20–25.

Piaget, J. 1971. *Biology and Knowledge: An Essay on the Relations Between Organic Regulations and Cognitive Processes*. Chicago: University of Chicago Press.

Piaget, J. 1978. *Behavior and Evolution*. New York: Pantheon Books.

Piaget, J. 1980. *Adaptation and Intelligence, Organic Selection and Phenocopy*. Chicago: University of Chicago Press.

Pluhar, E. 1978. Emergence and reduction. *Studies in the History and Philosophy of Science* 9: 279–289.

Prigogine, I., and A. Goldbeter. 1981. Nonequilibrium self-organization in biochemical systems. In J. Poortmans, ed., *Biochemistry of Exercise*, vol. 3, 3–12. Baltimore, MD: University Park Press.

Prigogine, I., and G. Nicolis. 1971. Biological order, structure and instabilities. *Quarterly Review of Biophysics* 4: 107–148.

Psillos, S. 2003. The present state of the scientific realism debate. In P. Clark and K. Hawley, eds., *Philosophy of Science Today*, 59–82. Oxford: Clarendon Press.

Radnitzky, G., ed. 1987–1988. *Centripetal Forces in the Sciences*. vol. 1 (1987); vol. 2 (1988). New York: Paragon House.

Raff, R. A. 1996. *The Shape of Life: Genes, Development, and the Evolution of Animal Form*. Chicago: University of Chicago Press.

Raff, R. A. 2000. Evo-devo: The evolution of a new discipline. *Nature Reviews Genetics* 1: 74–79.

Rasskin-Gutman, D. 2003. Boundary constraints for the emergence of form. In G. B. Müller and S. A. Newman, eds., *Origination of Organismal Form*, 305–322. Cambridge, MA: MIT Press.

Richards, R. J. 1992. *The Meaning of Evolution: The Morphological Construction and Ideological Reconstruction of Darwin's Theory*. Chicago: University of Chicago Press.

Richards, R. J. 2002. *The Romantic Conception of Life: Poetry, Philosophy, and Organism in the Age of Goethe*. Chicago: University of Chicago Press.

Richardson, R. C. 2001. Complexity, self-organization and selection. *Biology and Philosophy* 16: 655–683.

Reid, R. 1985. *Evolutionary Theory: The Unfinished Synthesis*. London: Croon Helm.

Reisch, G. A. 2005. *How the Cold War Transformed Philosophy of Science*. Cambridge: Cambridge University Press.

Riedl, R. 1978. *Order in Living Organisms: A Systems Analysis of Evolution*. New York: Wiley. (German original 1975.)

Robert, J. S. 2004. *Embryology, Epigenesis, and Evolution: Taking Development Seriously*. Cambridge: Cambridge University Press.

Robert, J. S., B. K. Hall, and W. M. Olson. 2001. Bridging the gap between developmental systems theory and evolutionary developmental biology. *BioEssays* 23: 954–962.

Rose, H., and S. Rose, eds. 2000. *Alas, Poor Darwin: Arguments against Evolutionary Psychology*. London: Jonathan Cape.

Rose, M. R., and G. V. Lauder, eds. 1996. *Adaptation*. San Diego: Academic Press.

Rose, S., ed. 1982a. *Against Biological Determinism*. London: Allison and Busby.

Rose, S., ed. 1982b. *Towards a Liberatory Biology*. London: Allison and Busby.

Rose, S. 1998. *Lifelines: Biology beyond Determinism*. Oxford: Oxford University Press.

Rose, S., L. J. Kamin, and R. C. Lewontin. 1984. *Not in Our Genes: Biology, Ideology and Human Nature*. Harmondsworth: Penguin Books.

Rosen, R. 1991. *Life Itself: A Comprehensive Inquiry into the Nature, Origin, and Fabrication of Life*. New York: Columbia University Press.

Rosenberg, A. 1985. *The Structure of Biological Science*. Cambridge: Cambridge University Press.

Rosenberg, A. 1994. *Instrumental Biology, or the Disunity of Science*. Chicago: University of Chicago Press.

Ross, D. 2006. Game theory in studies of evolution and Development: Prospects for deeper use. *Biological Theory* 1: 29–30.

Roth, G., and D. B. Wake. 1989. Conservatism and innovation in the evolution of feeding in vertebrates. In D. B. Wake and G. Roth, eds., *Complex Organismal Functions: Integration and Evolution in Vertebrates*, 7–21. Chichester: Wiley.

Roth, V. L. 1984. On homology. *Biological Journal of the Linnaean Society* 22: 13–29.

Ruse, M. 1999. *Mystery of Mysteries: Is Evolution a Social Construction?* Cambridge, MA: Harvard University Press.

Salazar-Ciudad, I., J. Jernvall, and S. A. Newman. 2003. Mechanisms of pattern formation in development and evolution. *Development* 130: 2027–2037.

Salazar-Ciudad, I., S. A. Newman, and R. Solé. 2001. Phenotypic and dynamical transitions in model genetic networks. I. Emergence of patterns and genotype-phenotype relationships. *Evolution and Development* 3: 84–94.

Salazar-Ciudad, I., R. Solé, and S. A. Newman. 2001. Phenotypic and dynamical transitions in model genetic networks. II. Application to the evolution of segmentation mechanisms. *Evolution and Development* 3: 95–103.

Salmon, W. C. 1984. *Scientific Explanation and the Causal Structure of the World*. Princeton, NJ: Princeton University Press.

Salmon, W. C. 1989. *Four Decades of Scientific Explanation*. Minneapolis: University of Minnesota Press.

Salmon, W. C. 1994. Causality without counterfactuals. *Philosophy of Science* 61: 297–312.

Salmon, W. C. 1998. *Causality and Explanation*. New York: Oxford University Press.

Sarkar, S. 2000. Information in genetics and developmental biology: Comments on Maynard Smith. *Philosophy of Science* 67: 208–213.

Schaffner, K. F. 1993. Theory structure, reduction, and disciplinary integration in biology. *Biology and Philosophy* 8: 319–347.

Schaffner, K. F. 1998. Genes, behavior and developmental emergentism: One process, indivisible? *Philosophy of Science* 65: 209–252.

Schank, J. C., and W. C. Wimsatt. 2001. Evolvability: Adaptation and modularity. In R. S. Singh et al., eds., *Thinking about Evolution*, vol. 2, 322–335. Cambridge: Cambridge University Press.

Schlanger, J. E. 1971. *Les métaphores de l'organisme*. Paris: J. Vrin.

Schlosser, G., and G. P. Wagner, eds. 2002. *Modularity in Development and Evolution*. Chicago: University of Chicago Press.

Schmalhausen, I. I. 1949. *Factors of Evolution*. Chicago: University of Chicago Press.

Schwenk, K., and G. P. Wagner. 2003. Constraint. In B. K. Hall and W. M. Olson, eds., *Keywords and Concepts in Evolutionary Developmental Biology*, 52–61. Cambridge, MA: Harvard University Press.

Shapiro, M. D., M. E. Marks, C. L. Peichel, B. K. Blackman, K. S. Nereng, B. Jonsson, D. Schluter, and D. M. Kingsley. 2004. Genetic and developmental basis of evolutionary pelvic reduction in threespine sticklebacks. *Nature* 428: 717–723.

Singh, R. S., C. B. Krimbas, D. B. Paul, and J. Beatty, eds. 2001. *Thinking about Evolution, Vol. 2: Historical, Philosophical, and Political Perspectives*. Cambridge: Cambridge University Press.

Smith, K. C. 1992. Neo-rationalism versus neo-Darwinism: Integrating development and evolution. *Biology and Philosophy* 7: 431–451.

Smith, K. C., and R. Sansom. 2001. Introductory statement for the ISHPSSB 1999 Evo-Devo sessions. http://www.ishpssb.org/oldmeetings/2001/program.htm#evo-devo1.

Smocovitis, B. V. 1996. *Unifying Biology: The Evolutionary Synthesis and Evolutionary Biology*. Princeton, NJ: Princeton University Press.

Sober, E. 1983. Equilibrium explanation. *Philosophical Studies* 43: 201–210.

Sober, E. 1993. *Philosophy of Biology*. Oxford: Oxford University Press.

Sober, E. 1998. Six sayings about adaptationism. In D. L. Hull and M. Ruse, eds., *The Philosophy of Biology*, 72–86. Oxford: Oxford University Press.

Sober, E., and D. S. Wilson. 1998. *Unto Others: The Evolution and Psychology of Unselfish Behavior*. Cambridge, MA: Harvard University Press.

Stadler, P. F., and B. M. R. Stadler. 2006. Genotype-phenotype maps. *Biological Theory* 1: 268–279.

Steinberg, M. S., and M. Takeichi. 1994. Experimental specification of cell sorting, tissue spreading, and specific spatial patterning by quantitative differences in cadherin expression. *Proceedings of the National Academy of Sciences USA* 91: 206–209.

Stelling, J., A. Kremling, M. Ginkel, K. Bettenbrock, and E. D. Gilles. 2001. Towards a virtual biological laboratory. In H. Kitano, ed., *Foundations of Systems Biology*, 189–208. Cambridge, MA: MIT Press.

Sterelny, K. 2000. Primate worlds. In C. Heyes and L. Huber, eds., *The Evolution of Cognition*, 143–162. Cambridge, MA: MIT Press.

Sterelny, K., K. C. Smith, and M. Dickison. 1996. The extended replicator. *Biology and Philosophy* 11: 377–403.

Streicher, J., M. A. Donat, B. Strauss, R. Sporle, K. Schughart, and G. B. Müller. 2000. Computer-based three-dimensional visualization of developmental gene expression. *Nature Genetics* 25: 147–152.

Streicher, J., and G. B. Müller. 1992. Natural and experimental reduction of the avian fibula: Developmental thresholds and evolutionary constraint. *Journal of Morphology* 214: 269–285.

Streicher, J., and G. B. Müller. 2001. 3D modelling of gene expression patterns. *Trends in Biotechnology* 19(4): 145–148.

Striedter, G. F. 1998. Stepping into the same river twice: Homologues as recurrent attractors in epigenetic landscapes. *Brain, Behavior, and Evolution* 52: 218–231.

Strohman, R. C. 1997. The coming Kuhnian revolution in biology. Nature Biotechnology 15: 194–200.

Suppe, F. 1977. Introduction. In F. Suppe, ed., *The Structure of Scientific Theories*, 2nd ed., 3–232. Urbana: University of Illinois Press.

Suppes, P. 1978. The plurality of science. In P. D. Asquith and I. Hacking, eds., *PSA 1978*, vol. 2, 3–16. East Lansing, MI: Philosophy of Science Association.

Suppes, P. 1979. The role of formal methods in the philosophy of science. In P. D. Asquith and H. E. Kyburg Jr., eds., *Current Research in Philosophy of Science*, 16–27. East Lansing, MI: Philosophy of Science Association.

Tabery, J. G. 2004. Synthesizing activities and interactions in the concept of a mechanism. *Philosophy of Science* 71: 1–15.

Tibon-Cornillot, M. 1992. *Les corps transfigurés: Mécanisation du vivant et imaginaire de la biologie*. Paris: Seuil.

Tinbergen, N. 1963. On aims and methods of Ethology. *Zeitschrift für Tierpsychologie* 20: 410–433.

Van Buskirk, J. 2002. Phenotypic lability and the evolution of predator-induced plasticity in tadpoles. *Evolution* 56: 361–370.

Van Speybroeck, L. 2002. From epigenesis to epigenetics: The case of C. H. Waddington. *Annals of the New York Academy of Sciences* 981: 61–81.

Varela, F. J. 1979. *Principles of Biological Autonomy*. New York: North-Holland.

Vermeij, G. J. 2006. Historical contingency and the purported uniqueness of evolutionary innovations. *Proceedings of the National Academy of Sciences USA* 103: 1804–1809.

von Bertalanffy, L. 1968. *General System Theory: Foundations, Development, Applications*. Harmondsworth: Penguin Books.

Von Dassow, G., and E. Munro. 1999. Modularity in animal development and evolution: Elements of a conceptual framework for Evo-devo. *Journal of Experimental Zoology (Molecular and Developmental Evolution)* B285: 307–325.

von Dassow, G., E. Meir, E. M. Munro, G. M. Odell. 2000. The segment polarity network is a robust developmental module. *Nature* 406: 188–192.

Waddington, C. H. 1956. *Principles of Embryology*. London: Allen and Unwin.

Waddington, C. H. 1962. *New Patterns in Genetics and Development*. New York: Columbia University Press.

Wade, M. J. 1978. A critical review of the models of group selection. *Quarterly Review of Biology* 53: 101–114.

Wagner, A. 1996. Genetic redundancy and its evolution in networks of transcriptional regulators. *Biological Cybernetics* 74: 559–569.

Wagner, A. 2005. *Robustness and Evolvability in Living Systems*. Princeton, NJ: Princeton University Press.

Wagner, G. P. 1986. The systems approach: An interface between development and population genetic aspects of evolution. In D. M. Raup and D. Jablonski, eds., *Patterns and Processes in the History of Life*, 149–166. Berlin: Springer.

Wagner, G. P. 1989. The biological homology concept. *Annual Review of Ecology and Systematics* 20: 51–69.

Wagner, G. P. 1996. Homologues, natural kinds, and the evolution of modularity. *American Zoologist* 36: 36–43.

Wagner, G. P., and L. Altenberg. 1996. Complex adaptations and the evolution of evolvability. *Evolution* 50: 967–976.

Wagner, G. P., G. Booth, and H. Bagheri-Chaichian. 1997. A population genetic theory of canalization. *Evolution* 51: 329–347.

Wagner, G. P., C.-C. Chiu, and M. D. Laublichler. 2000. Developmental evolution as a mechanistic science: The inference from developmental mechanisms to evolutionary processes. *American Zoologist* 40: 819–831.

Wagner, G. P., and M. D. Laubichler. 2004. Rupert Riedl and the re-synthesis of evolutionary and developmental biology: Body plans and evolvability. *Journal of Experimental Zoology (Molecular and Developmental Evolution)* B302: 92–102.

Wagner, G. P., and V. J. Lynch. 2005. Molecular evolution of evolutionary novelties: The vagina and uterus of placental mammals. *Journal of Experimental Zoology (Molecular and Developmental Evolution)* 304B: 580–592.

Wagner, G. P., and G. B. Müller. 2002. Evolutionary innovations overcome ancestral constraints: A re-examination of character evolution in male sepsid flies (Diptera: Sepsidae). *Evolution and Development* 4: 1–6.

Wake, D. B. 2002. On the scientific legacy of Stephen Jay Gould. *Evolution* 56: 2346.

Wallace, B. 1986. Can embryologists contribute to an understanding of evolutionary mechanisms? In W. Bechtel, ed., *Integrating Scientific Disciplines*, 149–163. Dordrecht: Nijhoff.

Weber, M. 1998. *Die Architektur der Synthese: Entstehung und Philosophie der modernen Evolutionstheorie.* Berlin: Walter de Gruyter.

Webster, G., and B. Goodwin. 1996. *Form and Transformation.* Cambridge: Cambridge University Press.

Weiss, P. A. 1970. Life, order, and understanding: A theme in three variations. *[University of Texas] Graduate Journal* 8 (Suppl.): 1–157.

Weninger, J. W., S. H. Geyer, T. J. Mohun, D. Rasskin-Gutman, T. Matsui, I. Ribeiro, L. da F. Costa, J. C. Izpisúa Belmonte, and G. B. Müller. 2006. High-resolution episcopic microscopy: A rapid technique for detailed 3D analysis of gene activity in the context of tissue architecture and morphology. *Anatomy and Embryology* 211: 213–221.

West-Eberhard, M. J. 2003. *Developmental Plasticity and Evolution.* Oxford: Oxford University Press.

Wilkins, A. S. 1996. Are there Kuhnian revolutions in biology? *BioEssays* 18: 695–696.

Wilkins, A. S. 1997. Canalization: A molecular genetic perspective. *BioEssays* 19: 257–262.

Wilkins, A. S. 2001. *The Evolution of Developmental Pathways.* Sunderland, MA: Sinauer Associates.

Williams, G. C. 1966. *Adaptation and Natural Selection: A Critique of Some Current Evolutionary Thought.* Princeton, NJ: Princeton University Press.

Williams, G. C. 1985. A defense of reductionism in evolutionary biology. *Oxford Surveys in Evolutionary Biology* 2: 1–21.

Williams, G. C. 1992. *Natural Selection: Domains, Levels and Challenges.* Oxford: Oxford University Press.

Wilson, E. O. 1998. *Consilience: The Unity of Knowledge.* London: Little, Brown and Company.

Wilson, E. O., and C. J. Lumsden. 1991. Holism and reductionism in sociobiology: Lessons from the ants and human culture. *Biology and Philosophy* 6: 401–411.

Wimsatt, C. 1974. Reductive explanation: A functional account. In *PSA 1974*, 671–710. Dordrecht: Reidel.

Wimsatt, W. C. 1976. Reductionism, levels of organization and the mind-body problem. In G. G. Globus, G. Maxwell, and I. Savodnik, eds., *Consciousness and the Brain: A Scientific and Philosophical Inquiry*, 199–267. New York: Plenum Press.

Wimsatt, W. C. 1980. Reductionistic research strategies and their biases in the units of selection controversy. In T. Nickles, ed., *Scientific Discovery, Vol. 2: Case Studies*, 213–259. Dordrecht: Reidel.

Wimsatt, W. C. 1986. Developmental constraints, generative entrenchment, and the innate-acquired distinction. In W. Bechtel, ed., *Integrating Scientific Disciplines*, 185–208. Dordrecht: Martinus Nijhoff.

Wimsatt, W. C. 1997. False models as means to truer theories. In M. H. Nitecki and A. Hoffman, eds., *Neutral Models in Biology*, 23–55. New York: Oxford University Press.

Wimsatt, W. C. 1999. Generativity, entrenchment, evolution, and innateness: Philosophy, evolutionary biology, and conceptual foundations of science. In V. G. Hardcastle, ed., *Where Biology Meets Psychology*, 139–179. Cambridge, MA: MIT Press.

Wimsatt, W. C. 2001. Generative entrenchment and the developmental systems approach to evolutionary processes. In S. Oyama et al., eds., *Cycles of Contingency*, 219–237. Cambridge, MA: MIT Press.

Wimsatt, W. C., and J. R. Griesemer. 2006. Re-producing entrenchments to scaffold culture: How to re-develop cultural evolution. In R. Sansom and R. N. Brandon, eds., *Integrating Evolution and Development*, 227–324. Cambridge, MA: MIT Press.

Woodward, J. 2003. *Making Things Happen: A Theory of Causal Explanation*. Oxford: Oxford University Press.

Wray, G. A. 1999. Evolutionary dissociations between homologous genes and homologous structures. *Novartis Foundation Symposium* 222: 189–203.

Wray, G. A., and C. J. Lowe. 2000. Developmental regulatory genes and echinoderm evolution. *Systematic Biology* 49: 151–174.

Zhang, T. J., B. G. Hoffman, T. Ruiz de Algara, and C. D. Helgason. 2005. SAGE reveals expression of Wnt signalling pathway members during mouse prostate development. *Gene Expression Patterns* 6: 310–324.

3 Complex Traits: Genetics, Development, and Evolution

H. Frederik Nijhout

The relationship between genotype and phenotype has long been of interest to geneticists, developmental biologists, evolutionary biologists, and epidemiologists. Yet in spite of this, our understanding of how genes affect complex traits is still fairly primitive. This is, in part, because the relationship between genotypes and phenotypes is ambiguous: different genotypes can produce the same phenotype (for instance if alleles are recessive), and one genotype can produce different phenotypes (for instance in social insects where morphologically distinct castes such as workers, soldiers, and queens can be genetically identical). In addition, phenotypes are remarkably well buffered against genetic and environmental variation, and this robustness tends to mask the effects of genes on traits. It is therefore difficult to detect the effects of single genes in a naturally varying complex trait. Much of our understanding of how genes affect complex traits comes from interpretations of the effects of artificial constructs of engineered genes that greatly over- or underexpress a given gene, or by studying the effects of rare alleles with large phenotypic effect in an artificially constant genetic background. Although this approach is sensible from an experimental point of view, it has given us a highly biased view of how genetic variation is related phenotypic variation. This bias, that specific properties of phenotypic traits can be attributed to specific genes, is expressed most clearly in the continued search for the role of single genes as the causes of complex diseases.

The relationship between genotype and phenotype is usually thought of as some kind of "mapping function" in which the phenotype is a mathematical function of the genes. Until recently, studies on the genetics and evolution of complex traits have used polynomial functions or simple mapping functions in which genes act additively or multiplicatively in determining the value of the phenotype. Although such simplified mapping functions have provided many important theoretical insights into evolutionary genetics, they do not describe the rather complex and nonlinear mechanisms by which genes interact, as revealed by developmental genetics. Because nonlinearities produce context-dependent relationships between dependent and independent variables, it is likely that simple linear models cannot capture important aspects of the relationship between genes and complex traits.

To understand the relationship between genotype and phenotype we need a good notion of what genes do, and we have actually known this for a long time. Genes code for proteins, and it is the properties of proteins and the way they interact with their surroundings that must somehow explain what we perceive to be the properties of genes. Proteins do not function in isolation, but always occur in a cellular or organismal context—embedded in membranes, attached to other proteins, or compartmentalized within organelles—and this context is critical for their normal function. The map of genotype to phenotype should thus somehow take account of what proteins do, and how they interact to produce things that we recognize as the phenotype. In the sections that follow I will outline one possible way of constructing a realistic and hopefully accurate genotype-to-phenotype mapping, and discuss a novel insight into the relationship between development and evolution produced by this approach.

Mathematical Models

One way to examine whether we understand a mechanism is to attempt a mathematical description. A quantitative description forces us to be explicit about the assumptions we make, and can therefore reveal some of the gaps in our understanding. I note at the outset a distinction between two kinds of mathematical models. The first is a description of a hypothetical mechanism designed to discover the necessary and sufficient conditions for a particular phenomenon. The second is simply a quantitative description of the processes that are known to operate. The first of these has a relatively poor history of success in biology, in no small measure due to the fact that biological systems do not tend to work by the simplest mechanism. The second approach is essentially descriptive, and thus relies on having a lot of information about the system we wish to model. It is this second approach I will explore here, by examining how we can achieve a quantitative description of what genes (actually proteins) do.

Networks of transcriptional regulation that control various aspects of early embryonic development are now well enough understood that several research groups have developed mathematical models of them (Novak and Tyson 1995; Reinitz and Sharp 1995; Bodnar 1997; Sharp and Reinitz 1998;

Von Dassow et al. 2000). These models are quantitative descriptions of the interactions among genes and gene products. Although these models have been used primarily to investigate the robustness of genetic networks, they can predict the phenotypic effects of mutations and are therefore useful for studying the relationships between genetic variation and phenotypic variation.

A mathematical model of a transcriptional regulation network is, in effect, a quantitative description of the actual mechanism that underlies the relationship between genotype and phenotype. In this case the genotype is represented by the values of the kinetic constants that describe the rates of gene expression and the phenotype is the final pattern of gene expression. This can be the pattern of gene expression within a cell, and if gene products can move from cell to cell, it can be a spatial pattern of gene expression.

Functionally different alleles for a gene will code for proteins that differ in the value of one or more kinetic constants. This can probably be best understood by reference to enzyme kinetics. In the Michaelis-Menten equation (rate = $V_{max}[S]/(K_m + [S])$), the rate of an enzymatic reaction is a function of V_{max}, K_m, and the concentration of substrate S. The V_{max} and K_m can be thought of as the genetic properties of the enzyme. The V_{max} is a function of the amount of enzyme present and the rate of release of the reaction product, and is thus determined in part by the rate of transcription and translation of the gene for the enzyme. The K_m is a measure of the catalytic efficiency of an enzyme and is a function of the structure of the enzyme. The K_m can thus be altered by mutations that change the amino acid sequence of the enzyme. Genetic variation that alters the degree of expression of a gene is therefore expressed as variation in the V_{max} and genetic variation that alters the structure of the catalytic portion of the enzyme is expressed as variation in the K_m and/or the V_{max}. Environmental factors such as pH can affect the shape of the enzyme and can therefore alter the V_{max} and K_m. The genetic parameters are thus defined only under specific environmental conditions, which illustrates that the relationship between genotype and phenotype is also a function of the environment. The concentration of the substrate can be thought of as an environmental parameter if it is supplied from outside the system. In many cases, however, the concentration of the substrate is determined in large part by the activities of other enzymes in the system and it is therefore a property of the system as a whole.

The equations that are used to describe the regulation of gene expression are variants of the Michaelis-Menten equation and produce a hyperbolic or sigmoid relationship between the strength of the regulatory input and the degree of activation of a gene. Unfortunately the values of the kinetic constants for developmental genes are not well known, and much effort has gone into devising methods for deducing them from time series of gene expression patterns and from the responses of expression patterns to experimental manipulations (Reinitz and Sharp 1995; Sharp and Reinitz 1998). Yet, even in the absence of precise parameter values we can learn a great deal about the relationship between genotypes and phenotypes in realistic genetic networks. The important point is that to understand the roles genes play in the makeup of phenotypes, it is necessary to understand the roles of gene products in the ontogeny and function of a trait, and to be able to represent those roles by mathematical equations. With an accurate mathematical model of a mechanism by which the phenotype arises, it is possible to examine the phenotypic effects of variation in each of the components of that mechanism.

Visualization

The effect of variation in the activity of one gene, or gene product, on the phenotype can be visualized by means of a graph that represents all possible values of gene activity as the independent variable, and the value of the phenotype as the dependent variable (figure 3.1). On such a graph, individual genotypes are points along the x-axis. The slope of the graph at such a point is a measure of the sensitivity of the phenotype to genetic variation around that genotype. If the slope is shallow or zero, then genetic variation (mutations) will have little or no effect on the phenotype, the mutation would be considered neutral, and the phenotype would be considered robust to variation at that gene. If the slope is steep, the same amount of genetic variation will have a large effect on the phenotype and one would draw different conclusions about neutrality and robustness. When the relationship between genotype and phenotype is nonlinear, the effect of a mutation will depend on the exact value of the genotype (figure 3.1).

Of course, an underlying assumption of such a univariate representation is that all other factors that affect the phenotype are held constant, which in real systems is not a reasonable assumption. Among the most useful things that a mathematical model of the phenotype allows us to do is to study the effects of simultaneous variation in several genes. The graph that describes the effects of variation in two genes is a two-dimensional surface (e.g., figure 3.2). A graph of the simultaneous variation in n genes would be an n-dimensional hypersurface with n orthogonal axes (one for each of the

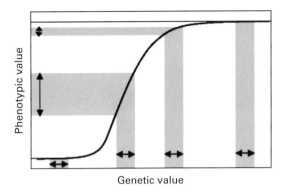

Figure 3.1
When the relationship between genotype and phenotype is nonlinear, a given amount of genetic variation can have a large effect, a small effect, or no effect at all on the phenotype, depending entirely on where the genotypic value lies. Genotypes on the right and left of this diagram would be considered robust because genetic variation (indicated by the width of the gray bands) in those regions would not alter the phenotype. The same amount of genetic variation in the middle region of the diagram will be associated with more or less large phenotypic effects.

independent variables) plus an orthogonal axis for the phenotype (equivalent to the y-axis in figure 3.1). Although such a surface can be easily computed, it cannot be represented on paper. Hence we will restrict our examples here to bivariate cases, with the understanding that real systems are likely to be multivariate.

Context Dependency

We take as our example a mitogen-activated protein kinase (MAPK) signaling cascade, which typically consists of a chain of three kinases. Suppose there is genetic variation in the phosphorylation-rate constants of the first and second kinases in the cascade. The phenotype is the activity of the last kinase (ERK in vertebrates), and a graph of this phenotype with respect to the two genetic variables is shown in figure 3.2. The multidimensional graph of phenotype as a function of genotype is called a *phenotypic surface*. Individuals can be thought of as points on this surface. The effect of mutations that alter the activity of the first two kinases will cause an individual to move to a different position on the surface.

In spite of its simplicity, this is a surprisingly rich system that illustrates most of the important concepts needed to understand the genetic

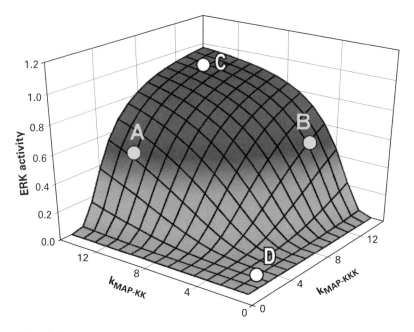

Figure 3.2
Phenotypic surface of a MAPK cascade. Here the phenotype is ERK activity and is graphed as a function of the phosphorylation rate constants of the first (MAP-KKK) and second (MAP-KK) kinases in the cascade. These rate constants are genetic properties of the kinases, and a broad range of values for different alleles are represented on the x- and y-axes. In this example, two different genotypes, with different rate constants for the two kinases, produce two identical phenotypes, A and B. The sensitivity of these two phenotypes to mutation is proportional to the local slopes of the surface at locations A and B. Thus, a small amount of genetic variation in the MAP-KK gene would have little effect on the phenotype of A, but a similar amount of variation would have a large phenotypic effect in B. The opposite is true for small to moderate genetic variation in the gene for MAP-KKK. Phenotypes on the relatively horizontal regions of the surface (C, D) would be perceived to be robust to genetic variation in the two genes because such variation would have little or no effect on the phenotype. All molecular and developmental mechanisms that produce robustness must have phenotypic surfaces with horizontal regions.

and evolutionary properties of a realistic genotype-to-phenotype mapping for a complex trait. What follows will be inferences and conclusions that can be drawn from any system that has a nonlinear phenotypic surface. This includes (but is not in the least restricted to) most biochemical pathways, genetic regulatory networks, signaling pathways, physiological processes, and morphogenetic mechanisms. In what follows I will use biochemical systems as examples, but the arguments will apply generally to any developmental, physiological, or morphological system.

The local slopes of the surface are a measure of the sensitivity to genetic variation at that location. For instance, the phenotype of an individual at location A would be sensitive to a modest amount of quantitative variation in the gene for the MAP-KKK but not in the gene for the MAP-KK, and such an individual would therefore experience MAP-KKK as the major gene for the trait, and MAP-KK as a gene of minor effect. In classical evolutionary genetics such a gene with minor effect might be called a modifier gene, and in quantitative genetics it would be detected as a quantitative trait locus (QTL).

The critical thing to recognize is that exactly the opposite would be true for an individual at location B (that is, such an individual would experience MAP-KKK as the major gene for the trait, and MAP-KK as a gene of minor effect), even though A and B possess identical genetic mechanisms and have identical phenotypes! This illustrates that the quantitative effect of a gene on a trait is not a property of the gene itself, but depends strongly on the system in which it is embedded, which we normally call the genetic background. In this case, the gene for MAP-KKK is the genetic background for MAP-KK, and vice versa.

The slopes of a phenotypic surface are therefore related in some way to the covariance (or the correlation, which is a scaled covariance) between genetic and phenotypic variation. To obtain the covariance for a particular population one would need to know not only the slope but also the amount of variation within a population along each axis. For a high correlation between genetic and phenotypic variation one would need both a high slope and a large genetic variance. The covariance of any gene with the trait, irrespective of the dimensionality and complexity of the underlying mechanism, and in the absence of linkage disequilibrium, is given by

$$Cov(g_1, p) = \frac{\beta_1 \, Var(g_1) \, Var(p)}{\sum_{i=1}^{n} \beta_i^2 \, Var(g_i)}$$

where g_1 is the gene of interest, p is the phenotype, β is the regression of genetic values on phenotypic values along the genetic axis of gene i (approximately the local slope of the phenotypic surface), and the summation is over all n genes whose variation affects the trait. For a bivariate case, like that illustrated in figure 3.2, in which all other factors are held constant, $n = 2$. This equation can also be expressed as a correlation, in which case it becomes

$$r_{(g_1, p)} = \sqrt{\frac{\beta_1^2 \, Var(g_1)}{\sum_{i=1}^{n} \beta_i^2 \, Var(g_i)}}$$

So the correlation between a given gene and a trait at any one location on a phenotypic surface is proportional to the relative slopes at that location in the direction of each of the axes and relative genetic variances in those directions (Nijhout 2002). All other things being equal, a gene with little or no allelic variation will be less correlated with a trait than a gene with a lot of allelic variation. But if the slope of the surface along the axis of variation of the first gene is very steep and the slope along the axis of variation of the second gene is small or zero, then the first gene may be more highly correlated with the trait, in spite of its small variance.

One interesting property of correlations is that if each of the genes varies independently from the others (i.e., if there is no linkage disequilibrium), then the coefficients of determination (the squared correlation coefficients) must add up to one ($\Sigma r^2 = 1$). This summation property implies that if the correlations between each of the genes and trait are not equal, then only one or a few genes can be highly correlated with the trait (another way of putting this is that they cannot all be highly correlated with the trait yet be uncorrelated with each other). If the relationship between genes and traits were linear, and all genes contributed equally to the trait, then the correlation of each gene with the trait would be simply inversely proportional to the number of variable genes. If there are many genes, then each will be weakly correlated with the trait. If not all genes contribute equally to variation of a trait, then only one or a few genes will be highly correlated with the trait and the rest will be very weakly correlated. When the relationship between genotype and phenotype is nonlinear, then the gene that is most highly correlated with the trait will vary with genetic background (for the reasons illustrated in figure 3.2), and therefore in different populations different genes will be most highly correlated with the trait, even though the mechanism through which they affect the trait is exactly the same.

Evolution

It is often stated that evolution comes about through changes in the developmental "program" of an organism, so that the evolution of form is effectively due to the evolution of the underlying developmental mechanism. With a phenotypic surface we can actually visualize two quite different modes of evolution. One of these involves a change in mechanism, the other actually does not, and both are equally effective at altering the phenotype.

Evolution can occur by moving from one location of a phenotypic surface to another. This mode would not require any change in the developmental mechanism, and can be accomplished entirely by mutations that alter the activity of gene products and selection for an optimal phenotype. Selection without mutation will also move a population around on the surface by changing allele frequencies.

Alternatively, mutations can change the underlying developmental mechanism, for instance by adding, removing, or retargeting regulatory interactions. This could be done by mutations in the regulatory region of a gene, so that it acquires or loses a binding site for a transcriptional regulator. Or it could happen by mutations in the coding regions of genes for transcriptional regulators, so that they no longer recognize the same DNA-binding sites in the genes they control. Such a mutation would change the shape of the phenotypic surface (recalling that the surface is a graphic representation of the interactions) and thus would produce new patterns of association between many genes and the phenotype.

Both of these modes would be considered microevolutionary in that they occur within populations, and do not necessarily generate new taxonomic entities. The first one is what population geneticists usually have in mind when they talk about microevolution. This mode could therefore be called Fisherian microevolution because it happens by changes in gene frequencies. By contrast, developmental biologists typically think of the second as the mechanism of developmental-evolutionary change, which we therefore will call Waddingtonian microevolution because it happens by genetic changes in the developmental mechanism. Both Fisherian and Waddingtonian modes can be gradualistic or saltatory, depending entirely on whether mutation introduces alleles with slight or with highly deviant physiological effects. In Fisherian evolution a highly deviant allele would cause individuals to jump to a distant location on a phenotypic surface, whereas in Waddingtonian evolution a highly deviant allele would cause a significant distortion of the phenotypic surface that will carry an individual to a new phenotypic value (e.g., figure 3.8 and associated discussion).

Natural selection acts on the phenotype, but heritable evolutionary change is due to changes in the genotype. Because phenotypic surfaces provide the links between genotype and phenotype it is possible to use these surfaces to visualize and predict the genetic response to selection on the phenotype. Under directional selection, a population will tend to move in the direction of the steepest local slope of the phenotypic surface (Rice 1998), assuming that the necessary genetic variation exists. The trajectory of a population on a phenotypic surface can thus be predicted from the gradient of the surface. A mathematical visualization program, such as MatLab (The MathWorks, Inc.) can be used to easily calculate the gradient and depict it as a vector field where the arrows show the direction of the gradient and their length is proportional to the slope (right-hand panel in figure 3.3).

Selection usually favors some optimal intermediate phenotype. In a phenotypic surface, an intermediate phenotype is a contour (an isophenotype curve), and any phenotypic surface can be converted into a fitness landscape by taking the absolute slope on either side of the optimal fitness contour (figure 3.4). The vector field on this new surface again predicts the direction in which the genotype will change from any given starting point. Interestingly, in a bivariate case, the optimal phenotype is a long ridge, so there is no single optimal genotype. Many genotypes can produce the same phenotypic value, so where a population ends up depends very much on where it started from. This may help explain why different populations with similar phenotypes do not necessarily have the same genetic makeup. Rice (1998) has shown that when a population arrives on a ridge it can evolve by moving along the ridge to a point where the slopes orthogonal to the ridge are smallest. This corresponds to a point where the phenotype is most strongly canalized (i.e., where the fitness surface is the flattest and thus where genetic variation has the smallest effect on the phenotype).

So when two genes are variable, the optimal phenotype is a one-dimensional ridge or line. Similarly, if three genes in a system are variable, then the optimum phenotype will be represented by a two-dimensional surface in 3-space. With many variable genes the optimal phenotype becomes a hypervolume with a very large number of allelic combinations that can give the identical optimal phenotype.

How Nonlinear Are Phenotypic Surfaces?

This question actually has two answers. One relates to the surface as a whole, the other relates to the surface as it is seen by a realistic population. The overall surface is likely to be highly nonlinear. The reason for this can be best understood if we focus on the shapes of one-dimensional sections through the surface parallel to the variable axes. Such a section illustrates

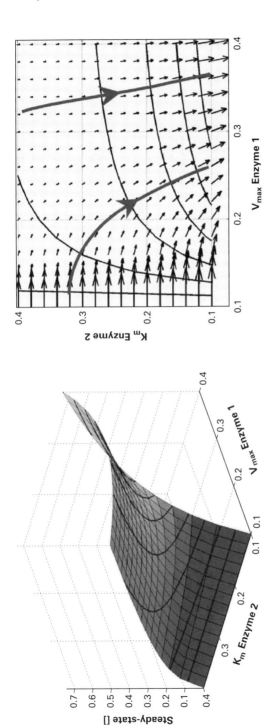

Figure 3.3

Phenotypic surface for a chain of enzymatic reactions in which two of the enzymes are polymorphic: there is allelic variation for the V_{max} of one enzyme and for the K_m of another. The phenotype is the steady-state concentration of the final product of the chain, as explained in the text. The slopes of the phenotypic surface predict the direction of evolution if selection is directional (e.g., for ever-larger phenotypic values), assuming there is no limitation on the number and diversity of alleles for the two parameters. The slopes can be calculated and are depicted as a vector field in the right panel, which also shows the predicted evolutionary trajectories of two populations starting from different initial conditions.

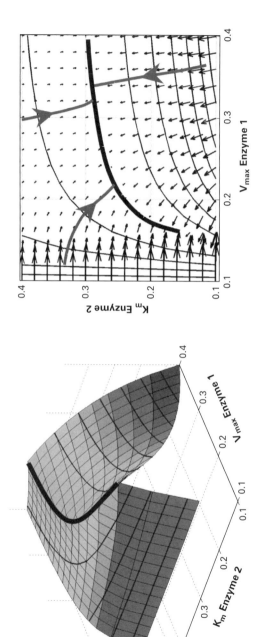

Figure 3.4

When an intermediate phenotype has maximal fitness, the highest-fitness phenotypes will be found along a contour of the surface, and evolution will be toward this optimal phenotype. The fitness landscape can be derived from the phenotypic surface by reflection through the maximal-fitness plane, so that the maximal-fitness contour now becomes a ridge of maximal fitness (illustrated here for the surface shown in figure 3.3). The vector field in the right panel indicates the direction and steepness of the slopes, and the predicted evolutionary trajectories from three different starting points are shown.

the effect that variation in one factor will have, when all other factors are kept constant at some value. There are limited possibilities for these shapes. In a few cases such a section will be an ever-increasing (or decreasing) function—for instance, if the axis represents the amount of gene product, such as an enzyme, then the rate of the reaction catalyzed would go up as the enzyme concentration goes up. Theoretically this could go on to infinity, but in practice there will be a limitation of substrate and a limitation on the amount of enzyme that can exist in a cell, so in reality the curve will level off. The range of shapes that sections through a phenotypic surface can have are illustrated in figure 3.5. Overall, these shapes are highly nonlinear, but they each have linear regions as well as nonlinear regions.

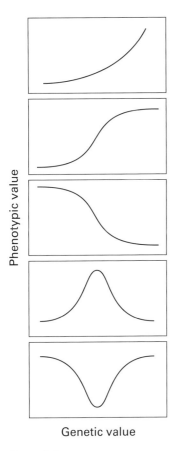

Figure 3.5
Diversity of shapes of one-dimensional sections through a phenotypic surface.

How much of such a surface a population actually experiences depends entirely on the range of allelic variation. If the range of allelic variation is small then the population will occupy only a small region of the surface, and for that population the surface will appear to be linear or flat. Thus, if the genetic variation in a population is small relative to the curvature of the surface, the surface will appear flat, or nearly so, but if the range of genetic variation is large relative to the curvature of the surface this will not be the case.

There is reason to believe that natural selection will act to keep the range of genetic variation small. The reason for this can be illustrated by reference to a two-dimensional fitness landscape (figure 3.6). Allelic values of individuals at the extremes of a distribution will combine to produce offspring that fall well away from the optimal phenotype and will therefore be selected against. Thus random breeding among members of a population will tend

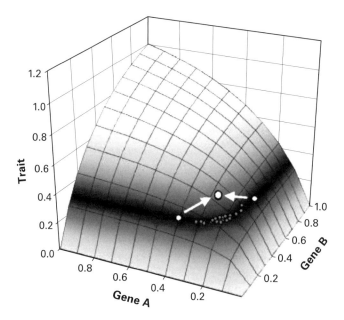

Figure 3.6
Nonlinear phenotypic landscape. The optimal fitness contour is indicated by darker shading. A population of individuals is indicated by small spots. If outliers in the genotypic distribution (large spots) interbreed (arrows), they would produce a phenotype that lies away from the optimal contour. The alleles contributing to such phenotypes would be eliminated. This process restricts the diversity of alleles in the population.

to disfavor the offspring of individuals that differ too much in the values of the alleles they carry. In the long term this will restrict the range of allelic diversity that can be supported in a population. The result is that a population will occupy only a relatively small region of a phenotypic surface, and for such a population the surface will be effectively flat.

It is likely that, by chance alone, different, noninterbreeding, populations of a species will end up in different regions of a phenotypic landscape. Each population will have limited genetic variation and thus perceive its landscape as being practically flat, so all its statistical genetic properties that depend on linearity will be perfectly consistent and unaffected by the overall nonlinearity of the surface. If the populations occur on the same fitness surface then their phenotypes will be the same, even though the ranges of their genotypes would be nonoverlapping and could, in fact, be extremely different. If two such populations should interbreed, they would produce offspring with highly nonadaptive phenotypes, because of admixture of alleles with very different values. Such populations would be effectively reproductively isolated.

Mutation and selection would move such populations along their fitness ridge, and they may end up in the same region, and thus become genetically compatible. This scenario resembles that of classical evolution on a fitness landscape, in that a population may, of course, become "stuck" at a local optimum if insufficient genetic variation is available. Continued genetic isolation of such populations could lead to the accumulation of different mutations in each and this would increase their incompatibility, eventually leading to speciation.

Constraints on Evolution

It is generally assumed that mutation can continually supply alleles of the right values to take the phenotype wherever selection and the phenotypic surface lead it. Given enough time, and a large enough population, this may well be the case. In the short term, and in realistically small populations, there is likely to be a limit to genetic diversity. The exact nature of phenotypically relevant genetic variation has not yet been determined for any natural population, but it seem reasonable to assume that the number and diversity of alleles are limited.

A phenotype cannot evolve in just any direction. The shape of the phenotypic surface, and the genetic diversity present in a population, each pose distinct constraints on the possible modes of evolution. Again, this is best explained visually. Figure 3.7 illustrates a population on a hypothetical

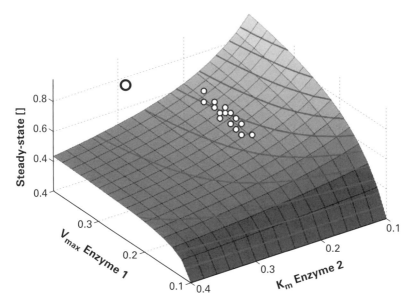

Figure 3.7
An illustration of genetic and developmental constraints on evolution. The phenotypic surface is the same as that shown in figure 3.3. The dots represent the distribution of genotypes and phenotypes in a population. Because the genetic variation is limited, the population mean can only evolve within the region represented by the dots. This is a genetic constraint on evolution. But even if there is a much broader range of genetic diversity, the only phenotypes that can be produced are those that are "on" the surface; the phenotype indicated by the large dot cannot be obtained with the particular genotype indicated. This is a developmental constraint on evolution. Changes elsewhere in the enzymatic pathway, however, could alter the surface so that it "rises" to the level of the desired phenotype without allelic changes in the two genes depicted here.

surface. First, a population can only evolve in the direction in which there is genetic variation, and only a small portion of the phenotypic surface may actually be available to a given population. This is a *genetic constraint* on evolution. Second, individuals can only be "on" the surface, and mutation and selection can move phenotypes around on the surface, but a phenotype that is not on the surface cannot be obtained. This is what is meant by a *developmental constraint* on evolution.

Only mutations that change the underlying mechanism, and therefore change the shape of the surface, have the potential of generating new phenotypic values for a given array of genotypes. For instance, figure 3.8 illus-

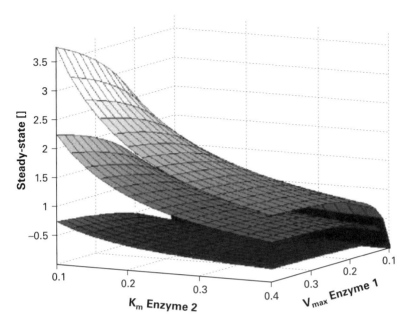

Figure 3.8
Three phenotypic surfaces that differ in the activity of a third enzyme. Variation in the third factor (enzyme) alters both the shape and the position of the phenotypic surface with respect to the first two factors. A phenotype (steady-state concentration of product in this case) that could not be obtained with a particular combination of values of enzymes 1 and 2 can, however, be obtained if a third enzyme is added to the system.

trates the effect of variation in three genes, two that vary continuously and the third represented by three different values. Mutations in the third gene have the effect of altering the level and the shape of the surface, and thus cause a particular combination of alleles in the other two genes to become associated with a new phenotypic value.

This illustrates context dependency once more, in that the relationship between the phenotype and any set of allelic values of the first two genes depends on the value of the third gene. It is a bit more difficult to depict a phenotype surface when all three genes have continuous variation. For three genes the phenotypic surface is a volume and the values of the phenotypes can be indicated by numbers (i.e., a 3-D matrix of numbers). The gradient in such a three-dimensional space can be visualized as a 3-D vector field that predicts the relative rates at which each determinant will change under directional selection (again, assuming an infinite-allele model).

The Nature of the Variable Factors

Until now we have assumed that the variables depicted on the axes of a phenotypic surface represent genes, or rather, the activities of gene products. There are, however, many determinants of the phenotype that are not genetic, plus in most systems we do not have sufficient information to reduce all heritable determinants to the gene level. Indeed, the relationship between genotype and phenotype can be extremely sensitive to environmental variables, as manifest by the ubiquity of reaction norms (Schmalhausen 1949; Schlichting and Pigliucci 1998; West-Eberhard 2003). Environmental variables like nutrients, temperature, and photoperiod are typical modulators of phenotypic plasticity and thus have important regulatory roles in development. These environmental variables can alter growth rates, can change patterns of hormone secretion and gene expression, and can thus have profound effects on the developing phenotype (temperature-dependent sex determination is a nice example).

The effect of environmental variables can also be captured in a mathematical model by describing the way they alter gene expression and protein activity. For instance, the effect of temperature on the rate of reaction can be explicitly introduced into a kinetic equation, and temperature then becomes an independent (ortogonal) axis of variation in the system that contains this equation and describes the phenotype. The effect of variation in other environmental variables can be graphed as additional independent (orthogonal) axes of variation of a phenotypic surface. In this way we can visualize the interaction of all potential environmental and genetic variation as a single multidimensional phenotypic surface.

Such a phenotypic surface in effect puts genetic and environmental variation on the same footing, and leads to a more critical understanding of how genes and environment interact to produce the phenotype. A phenotypic surface with multiple orthogonal genetic and environmental axes can be thought of as a multidimensional reaction norm. Thus, if a mutation takes an individual to a suboptimal phenotype, it should be possible by examination of the phenotypic surface (or by examination of the equations of which the surface is a description) to determine whether a change in one or more of the environmental variables could bring the phenotype back to an optimal value.

References

Bodnar, J. W. 1997. Programming the *Drosophila* embryo. *Journal of Theoretical Biology* 188: 391–445.

Gilchrist, M. A., and H. F. Nijhout. 2001. Non-linear developmental processes as sources of dominance. *Genetics* 159: 423–432.

Kacser, H., and J. A. Burns. 1981. The molecular basis of dominance. *Genetics* 97: 639–666.

Nijhout, H. F. 2002. The nature of robustness in development. *BioEssays* 24: 553–563.

Nijhout, H. F., A. M. Berg, and W. T. Gibson. 2003. A mechanistic study of evolvability using the mitogen activated protein kinase cascade. *Evolution & Development* 5: 281–294.

Nijhout, H. F., and S. M. Paulsen. 1997. Developmental models and polygenic characters. *American Naturalist* 149: 394–405.

Novak, B., and J. J. Tyson. 1995. Quantitative analysis of a molecular model of mitotic control in fission yeast. *Journal of Theoretical Biology* 173: 283–305.

Reinitz, J., and D. H. Sharp. 1995. Mechanisms of *eve*-stripe formation. *Mechanisms of Development* 49: 133–158.

Rice, S. H. 1998. The evolution of canalization and the breaking of Von Baer's laws: Modeling the evolution of development with epistasis. *Evolution* 52: 647–656.

Schlichting, C. D., and M. Pigliucci. 1998. *Phenotypic Evolution: A Reaction Norm Perspective*. Sunderland, MA: Sinauer.

Schmalhausen, I. I. 1949. *Factors of Evolution*. Philadelphia: Blakiston.

Sharp, D. H., and J. Reinitz. 1998. Prediction of mutant expression patterns using gene circuits. *BioSystems* 47: 79–90.

Von Dassow, G., E. Meir, E. Munro, and G. M. Odell. 2000. The segment polarity network is a robust developmental module. *Nature* 406: 188–192.

Wagner, G. P., G. Booth, and H. Bagheri-Chaichian. 1997. A population genetic theory of canalization. *Evolution* 51: 329–347.

West-Eberhard, M. J. 2003. *Developmental Plasticity and Evolution*. Oxford: Oxford University Press.

4 Functional and Developmental Constraints on Life-Cycle Evolution: An Attempt on the Architecture of Constraints

Gerhard Schlosser

Why are there no elephants with wings? Did the right kind of mutation just never happen? Or are wings just not among the kinds of structures that can be built in elephant embryos, not even in all possible kinds of mutants? Or could they maybe be built, but would just not serve elephants very well? These kinds of questions are at stake in an ongoing controversy about the role of constraints in evolution (see, e.g., Gould and Lewontin 1979; Dullemeijer 1980; Alberch 1980, 1982, 1989; Wake 1982, 1991; Cheverud 1984; Maynard Smith et al. 1985; Wake and Larson 1987; Gould 1989a; Antonovics and van Tienderen 1991; McKitrick 1993; Amundson 1994; Losos and Miles 1994; Schwenk 1994, 2001; Resnik 1996; Hall 1996; Raff 1996; Arthur 1997; Hodin 2000; Strathmann 2000; Wagner and Schwenk 2000; Sterelny 2000). The minimal consensus in this debate (e.g., Maynard Smith et al. 1985) seems to acknowledge the following. First, natural selection will occur in large populations when some trait actually exists in several variants, these variants differ in fitness and the variants and their fitness values are heritable. Second, the occurrence of variants is due to random mutation, but is nonetheless biased or constrained in certain ways due to the peculiar development of the organisms under consideration.

Regarding the relative importance of both factors in determining the course of evolution, however, evolutionary biologists appear to be split into two camps. These are often portrayed (e.g., Gould and Lewontin 1979; Wake and Larson 1987; Alberch 1989; Wake and Roth 1989) as an externalist/functionalist/adaptationist/selectionist camp on the one side, and an internalist/structuralist camp on the other side. Externalists/functionalists emphasize the importance of selection: evolutionary change is guided by functional demands of the environment, leading to adaptation to external environmental factors; the evolving organism is relatively free to vary and to respond in multiple ways to different environmental challenges. Internalists/structuralists on the other hand emphasize the importance of internal constraints: evolutionary change is channeled more by the internal generative or structural properties of the organism than by adaptation to external environmental factors; the scope of selection is limited, because it can only choose among the small number of possible variants that are permitted by developmental constraints.

For decades there has been little progress in resolving this debate. Our understanding of development has just been too incomplete to provide much insight into constraints on variation. However, the situation has begun to change with the recent breakthrough in developmental biology (see, e.g., Raff 1996; Gerhart and Kirschner 1997; Gilbert 2006). This developmental revolution has sparked hopes that neo-Darwinian evolutionary theory may finally be supplemented with a theory that allows us to predict constraints on variation, thus leading to a truly synthetic developmental evolutionary theory (Wagner and Altenberg 1996; Gilbert et al. 1996; Raff 1996; Gerhart and Kirschner 1997; Brandon 1999; Schlosser 2002).

However, despite some stunning successes of the burgeoning field of evolutionary developmental biology (e.g., new insights into homologies based on gene expression data), the implications of this research program for evolutionary biology have not yet been worked out in much detail (Wilkins 1998; Hall 2000). In particular, little attention has been paid to variability and intrapopulational variation of developmental processes (reviewed in Stern 2000; Purugganan 2000). To what extent can different developmental processes vary independently of each other and how may such variants spread in populations? Such questions, which lie at the heart of the debate on constraints, are currently difficult to answer. Partly this is certainly due to empirical deficits: few data about the variability and intrapopulational variation of developmental processes are available, because they are difficult and cumbersome to collect. But part of the difficulty is due to two conceptual deficits that are evident in the misleadingly dichotomous portrait of the two camps sketched above. These conceptual issues are the topic of this chapter. They need to be resolved, in order to know what to look for in empirical studies of constraints.

First, constraints are often construed as being somehow opposed to selection in their focus on internal versus external factors or their emphasis on structural versus functional explanation. For some authors, constraints are virtually identical with developmental constraints—that is, biased variation due to structural or generative internal properties of a system regardless of selection for external functional demands (e.g., Gould and Lewontin 1979; Alberch 1980, 1982, 1989; Gould 1989a; Schwenk 1994; Webster and Goodwin 1996). Other authors admit the existence of functional constraints, but retain the close association of the notion of constraints with

Functional and Developmental Constraints on Life-Cycle Evolution 115

internal determinants of evolutionary change (involved in internal selection), while external determinants of evolutionary change are always considered components of (external) selection (e.g., Whyte 1965; Dullemeijer 1980; Wake 1982; Wagner and Schwenk 2000; Schwenk 2001). In the first part of this chapter (sections 1 and 2), I wish to suggest an alternative perspective that dissolves this apparent dichotomy between constraints and selection (see also Arthur 1997). Rather, constraints, which arise from the necessity to maintain a stable/functional organization after variation, may be developmental or functional, internal or external (section 1 and 2), and simply describe the boundary conditions on a process whose dynamics (under certain conditions) are described by selection (section 2).

Second, too much attention has been devoted to the question of the relative contribution of constraints versus selection in modulating the direction of evolutionary transformations, neglecting the much more interesting question concerning the architecture of constraints (but see Cheverud 1984, 1996; Wagner 1995; Wagner and Altenberg 1996; Wagner and Schwenk 2000; Wagner and Laubichler 2000; Wagner and Mezey 2000; Hansen and Wagner 2001; Stadler et al. 2001; Wagner and Stadler 2003; Schlosser 2002). This will be the topic of the second part of this chapter (sections 3–7). I suggest that constraints on evolutionary transformations of a given system should to a certain degree reflect the pattern of generative and functional couplings among its essential components, here termed constituents (section 3). As a consequence, evolutionary transformations of a particular constituent of a given system will usually not be constrained equally strongly by every other constituent. Rather there may be units comprising a limited number of constituents that act as *units of evolution* and reciprocally constrain evolutionary changes predominantly among themselves (section 4). Such units of evolution, however, do not necessarily act as units of selection in particular selection processes (section 5). Units of evolution can be considered modules of evolutionary transformations. These may under certain conditions coincide with modular units of development or behavior during a life cycle (section 6). Therefore, experimental studies of development and physiology can be used to frame hypotheses about units of evolution, which can then be tested in a comparative framework (section 7).

4.1 What Are Constraints?

Constraints are invoked to explain deviations from the expected. More specifically, constraints are conditions that prevent from happening what would otherwise have been possible or that make more improbable what would otherwise have been more probable. Two points are important here. First, this characterization presupposes a process-oriented view of constraint. When structures are said to be constrained, this has to be read as an elliptic formulation of the claim that certain processes in which these structures participate are constrained. Second, the recognition of constraints requires that we have an idea of the expected, a theory of the possible.

Constraints Are Restrictions within the Realm of the Physically Possible
At a first glance, it may be tempting to identify the possible very broadly with the conceivable. We certainly can imagine all kind of strange worlds, with different laws, where more kinds of things are possible than in our universe (Alexander 1985). For example, we may imagine worlds in which the evolution of elephant-sized mice were physically possible. In contrast, elephant-sized mice cannot evolve in our world, because when any object of a given shape is sized up in our world, volume and weight grow more rapidly than surface area. Hence an elephant-sized mouse would not be able to live, because for instance its bones would break and its oxygen supply would be insufficient. Relative to all imaginable worlds, therefore, our world allows only for a limited set of processes and in that sense may be termed "constrained." Such "constraints" are sometimes called "universal constraints," because they apply to all processes in our world (Alexander 1985; Maynard Smith et al. 1985; Gould 1989a; Thomas and Reif 1993; Hall 1996).

I doubt, however, that such a wide and arbitrary concept of constraint will lead to very useful insights (see also Schwenk 1994). Instead, I propose that we should confine the possible to the physically possible (i.e., physically possible in our universe). What is physically possible is determined by a set of universal and immutable laws of transformation ("laws of nature") and a set of initial and boundary conditions permitted by our theories of the universe. Constraints, then, refer to conditions that prohibit the realization of certain states or events, even though they are physically possible.

From this perspective, it is dangerous to infer the presence or absence of constraints from "gaps in morphospace," where the latter is defined by all conceivable combinations of some parameters (e.g., Raup 1966; Alberch 1980, 1989; Thomas and Reif 1993; Rasskin-Gutman 2001). On the one hand, real gaps may not be noticeable as gaps in a theoretical morphospace, because the latter considers too few relevant parameters; on the other hand, some gaps in a theoretical morphospace may be due to physical impossibilities rather than being due to constraints.

Generative Constraints and Stability Constraints

The notion of constraint can be further specified, by noting that the laws of transformation that enter into the definition of the physically possible also impose temporal order on it: from any possible state only a subset of all the other possible states can be reached in a specified number of steps or period of time. Given certain immutable laws of transformation, this reachable subset—let us call it the causally reachable subset—is in principle defined by (1) a set of conditions specifying the initial state of a system or process, and (2) a set of conditions specifying possible perturbations (e.g., by specifying permitted kinds, intensities, and frequencies of external influences on the constituents and their interactions). Given these sets of conditions, many physically possible states cannot be reached, or cannot be reached in a specified period of time. Therefore, these conditions impose constraints on the transformations of the system in question. Such constraints, imposed by the organization of any system on the states it can generate after perturbation, will be called *generative constraints*.

However, we almost never talk about constraints when systems and processes are only subject to generative constraints. While volcanic eruptions can only have certain outcomes, we usually do not invoke constraints to explain this. Generative constraints are only invoked in cases where an additional class of constraints comes into play, namely, constraints that I will refer to as *stability constraints*. These constraints apply to transformations of the organization of a particular stable or persistent system or process relative to a set of possible perturbations. Only a subset of transformations of any state of such a system that can be generated (i.e., that are permitted by generative constraints) will be nondisruptive (nonlethal in the case of organisms)—that is, will again result in a stable state of a system with the same type of organization.[1] Hence, stability constraints are constraints on the nondisruptive transformations of a given stable system; they simply reflect the fact that only some of the generatively possible transformations of a system preserve its integrity.

For instance, steering a car, stepping on the gas, applying the brakes are transformations of the state of a car that will not destroy the car, when kept within certain limits. These transformations are permitted by generative as well as by stability constraints on properly operating cars. However, fully pulling down the gas pedal and abruptly turning the steering wheel may destroy a car. These transformations are permitted by generative constraints, but not by stability constraints. Generative and stability constraints differ for different types of systems, depending on their organization. How these notions can be applied to the evolution of organisms is detailed below (section 2).

Constraints and Opportunities

Constraints on and potentials (or opportunities) for transformations are two sides of the same coin (Alberch 1982; Maynard Smith et al. 1985; Gould 1989a; Hall 1996): all physically possible transformations that are not prohibited are permitted. Accordingly, generative constraints define a generative potential and stability constraints define a stability potential for any stable system, as illustrated in figure 4.1 schematically. It is evident that after perturbation of a stable system only some of the states that are causally reachable (permitted by generative constraints) may be stable states of

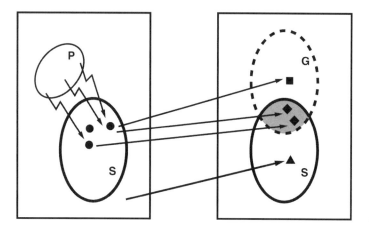

Figure 4.1
Constraints on transformations of a stable system. A stable system of type S is represented as a set of stable states, which is a subset of all physically possible states (rectangle). Only a subset G of all physically possible states is causally reachable from S given a certain set of perturbations P. G is the generative potential of S. Other transformations of S are impossible due to generative constraints. However, the system will remain stable only after transformations from one member of S to another member of S. The set S, therefore, defines the stability potential of the system. Other transformations disrupt S and are not permitted by stability constraints. Note that only some of the causally reachable states may be stable states of a system of type S, while only some of the stable states may be causally reachable. Together they define the set of permissible states (gray area).

the same kind of system (permitted by stability constraints). Conversely, only some of the stable states may be causally reachable.

The notion of constraint is usually reserved for transformations of stable systems that are subject to both kind of constraints: stability constraints prohibiting some states although they could be generated, and generative constraints prohibiting some states although they would be stable. Among all physically possible states, only the subset of states permitted by both generative and stability constraints is available for nondisruptive transformations of a particular kind of stable system. The more generative and stability potential overlap, the more will such nondisruptive transformations actually be facilitated by generative constraints (Wagner 1988; Hall 1996). Only states that are within that doubly constrained space of possibilities—let us call them *permissible*—can be realized without disrupting the system. Because constraints do not allow us to predict which of several permissible transformations will take place, this is the realm of contingency, the unpredictable (Gould 1989a, 1989b; Wake and Roth 1989; Schwenk 1994). Which of all the possibilities will be realized depends on which perturbations will take place, and this may depend, for instance, on parameters like mutation rates, population sizes, and so on (Barton and Partridge 2000).

The picture painted in figure 4.1 is, of course, much too simple, because there may be several nonoverlapping subsets of the set of stable states that can be reached by permissible transformations of a given stable state, and these may differ with respect to which other stable states are causally reachable from them (i.e., they may differ in their generative potential). As a consequence, the space of the permissible will be a very complicated, reticulating web (Figure 4.2). Contingent choices of the past may, therefore, be crucial in determining which constraints the system is subjected to in the present. In that sense, it could be claimed that present constraints grow out of "historical contingency" (Gould 1989a, 1989b), but one should not forget that contingency acted within the boundaries set by the constraints of the past.

Furthermore, the picture will be made more complicated but at the same time more realistic by construing constraints and potentials more broadly, namely, as conditions that define probabilities (propensities) rather than mere possibilities of transformations. Generative constraints/potentials may refer to conditions that bias transformations in the sense that they make the generation of some causally reachable stable states—so-called generic states (e.g., Newman 1992; Kauffman 1993; Webster and Goodwin 1996)—much more probable than others. And stability constraints/potentials may refer to conditions that bias transformations in the sense

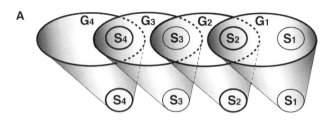

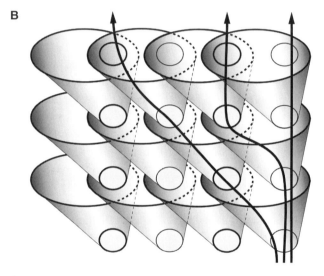

Figure 4.2
The reticulated space of permissible transformations. *A*: Stable systems of type S may exist in several distinct stable states (nonoverlapping subsets S_1, S_2, S_3, S_4). Only some of these stable states (e.g., S_1, S_2) may lie within the set of states that is causally reachable (e.g., G_1) from a particular state (e.g., S_1). The thin and thick lines bordering sets S_1, S_3 and S_2, S_4, respectively, indicate that these sets may differ in their probabilities for long-term stability—that is, may have different fitness values. *B*: Three possible trajectories from S_1 through the space of permissible transformations defined in *A*. Note that trajectories will tend to diverge in the long run, because any local choice will result in different choices being available subsequently and therefore increase the probability of divergence.

Functional and Developmental Constraints on Life-Cycle Evolution

that some states have a higher propensity than others for being stable, for persisting for a prolonged period of time, or for recurring with a certain frequency (figure 4.2). Together, generative and stability constraints will bias the decision probabilities at bifurcations in the reticulated space of the permissible.

To apply these ideas to the evolution of life cycles, we have to flesh out a bit more, how living systems are organized and what kinds of generative and stability constraints this kind of organization implies.

4.2 Constraints on the Evolution of Living Systems

The particular organization of living systems implies particular generative and stability constraints. Organisms are characterized by a special kind of dynamic stability that is realized by complex patterns of recurrent interactions among their components. All components that need to be present under at least some circumstances to ensure the maintenance (recurrence) of the process that we call "organism" will be referred to as constitutive components or *constituents*. These may be identified at different hierarchical levels. As necessary components of a process, constituents should be conceptualized as processes themselves—that is, patterns of interactions or activities (input-output transformations) at certain points in space and time. Structures (e.g., molecules like DNA, proteins or nutrients, or organs like heart, liver) are not the constituents of a dynamic system, but their states or activities at certain points in space and time are. These activities depend on the capacities of particular structures to interact (e.g., capacities for gene replication, gene transcription, protein binding or heart beating, glycolysis, and so on) as well as on the availability of defined inputs at a given point in space and time (for details see Schlosser 2002). Whenever the following text refers to structures (e.g., genes, proteins) as constituents of organisms, this is only to avoid unnecessary circumlocution; it always means that certain states or activities of these structures at certain points in space and time are constituents.

The Organization of Living Systems: Self-Maintenance and Reproduction

Many constituents of organisms are generated in a complex network of collectively autocatalytic interactions involving each of these constituents (e.g., Maturana and Varela 1975; Kauffman 1993; Schlosser 1993, 2002; Fontana and Buss 1994a; Szathmáry 1995). Such autocatalytic networks involve not only the recurrence of patterns of interactions, but also the multiplication of constituents—that is, the construction of multiple constituents (A′, A″, A‴), which are realized by structures that share the same capacities but may potentially engage in different interactions. Besides these proper, autocatalytically constructed or "internal" constituents ("autoconstituents" of Schlosser 1993), organisms require some input from biotic or abiotic environmental factors. While the latter are not reconstructed in the complex cycle of self-construction, they are required for the maintenance of this cycle. Therefore, they may be referred to as "external" constituents ("alloconstituents" of Schlosser 1993). Organisms are only stable in those environments where these external constituents are recurrently available at the right place at the right time. Systems with this kind of organization have been termed autopoietic (Varela, Maturana, and Uribe 1974; Maturana and Varela 1975), self-maintaining (Fontana and Buss 1994a, 1994b; Fontana, Wagner, and Buss 1995), or complex self-re-producing systems[2] (Schlosser 1998; see also Schlosser 1993).

In addition to being self-maintaining systems, organisms are also capable of multiplicative reproduction via "progeneration" (genealogical continuity via material overlap; Griesemer 2000), involving not only the collectively autocatalytic reconstruction of their constituents but also the recurrent production of other tokens of themselves (Szathmáry 1995; Maynard Smith and Szathmáry 1995; Szathmáry and Maynard Smith 1995, 1997; Griesemer 2000; Schlosser 2002): the entire "life cycle" (Bonner 1974) of an organism is reproduced—possibly multiple times—in its progeny (figure 4.3). This reproduction of the life cycle implies the reliable reproduction ("inheritance") of all patterns of interactions not only among internal but also between internal and external constituents in subsequent generations (Uexküll 1928; Lewontin 1983; Oyama 1985; Wimsatt 1986; Gray 1992; Schlosser 1993, 1996, 2002; Griffiths and Gray 1994).[3] Faithful multiplication of variants of life cycles ("heredity"; see Maynard Smith and Szathmáry 1995; Szathmáry 1995; Szathmáry and Maynard Smith 1995, 1997; Schlosser 2002) is possible due to the combination of mechanisms that guarantee the faithful multiplication of variants of some of their constituents (e.g., gene replication) as well as their reliable separation (as in mitosis and meiosis).

Functional and Developmental (Generative) Constraints on Life-Cycle Evolution

Because self-maintaining and reproducing systems are dynamically stable via the quasi-cyclical recurrence of states and processes, they can be described in functional terminology. All interactions that are necessary (at least under some conditions) for their own reconstruction and reproduc-

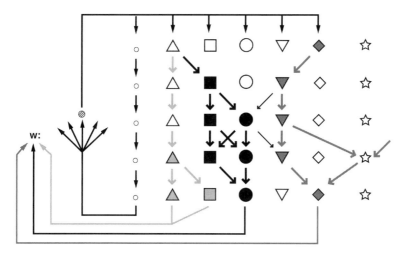

Figure 4.3
Reproduction of life cycles. The figure schematically depicts sequences of interactions (arrows) between various kinds of constituents (e.g., cells, molecules; symbols). Differences in activities of constituents at various points in time are indicated by filled versus nonfilled symbols. The small circle may represent cells of the germ line and the hatched circle the propagule (e.g., zygote in case of sexual reproduction). Interactions between all internal (triangles, circles, squares) and external (star) constituents contribute to the reproduction of the life cycle and jointly determine its fitness w (expected frequency of reproduction). For a more detailed scheme (e.g., including multiplication of constituents), see figure 4.7.

tion and thus for the self-maintenance and reproduction of the entire system may be termed *functional*. Their *functions* are those of their effects that ensure their reconstruction or reproduction—that is, that contribute to self-maintenance and reproduction of the containing system (see Schlosser 1998 for a more detailed account). Systems capable of self-maintenance or reproduction are therefore functional systems. Permissible, nondisruptive transformations of a functional system are those that again result in a state of the functional system. They are like engineering a machine, while it is running. Thus, the stability constraints for functional systems are *functional constraints*, constraints on the maintenance of a functional organization.

Unless there is functional redundancy, the functional organization of the system will be disrupted when existing functions of constituents are compromised. Therefore, functional constraints on evolutionary transformations of life cycles (e.g., Wake 1982; Hall 1996; Wagner and Schwenk 2000; Strathmann 2000; Schwenk 2001; often different terms are used such as *constructional constraints* in Dullemeijer 1980 or *selective constraints* in Maynard Smith et al. 1985; Arthur 1997; Barton and Partridge 2000) are mainly imposed by the existing functional organization of the life cycle in its environment.[4] The existing functional organization also determines what the propensity of a newly introduced variation for long-term persistence (iterated reproduction) or for reproduction with certain frequency will be—in other words, what its *fitness* will be.[5]

In addition to functional constraints, there are also generative constraints on evolutionary transformations of life cycles. The latter are often referred to as *developmental constraints* (e.g., Gould and Lewontin 1979; Alberch 1980, 1982, 1989; Wake 1982; Cheverud 1984; Maynard Smith et al. 1985; Gould 1989a; McKitrick 1993; Schwenk 1994; Raff 1996; Arthur 1997; Hodin 2000; Wagner and Schwenk 2000). This term emphasizes that existing developmental patterns of life cycles determine which kind of variants can be or are likely to be generated in progeny after the introduction of heritable variation. However, development is often regarded as synonymous with embryonic development excluding postembyronic changes during later phases of the life cycle as well as adult behavior. But evolutionary transformations of postembryonic development and behavior will of course also be subject to generative constraints. On the other hand, evolutionary transformations of development will also be subject to functional constraints (see below). Although the term *developmental constraints* is sometimes used in a sense that excludes constraints on postembryonic stages of the life cycle, or that includes functional constraints on development (e.g., Roth and Wake 1985; Zelditch and Carmichael 1989; Cheverud 1996; Hall 1996; Hodin 2000), I will use it here in accordance with most authors to denote generative constraints on evolutionary transformations of life cycles—that is, biases on the generation of heritable variation.

Internal and External Constraints on Life-Cycle Evolution

Heritable variation of life cycles is introduced in a localized manner—for example, by alteration of gene activity after mutation. However, whether a variation of a single constituent will again be functional and what developmental consequences it has depend on its interactions with other constituents of the life cycle, "internal" factors (autoconstituents) as well as those "external" environmental factors (alloconstituents) that are of critical importance for maintaining functionality. Or putting it differently, variability of constituents is functionally and developmentally constrained by other constituents of the life cycle. However, the constituents of the life cycle will not be equally important in determining whether some particular variation of a constituent will be functional or what developmental consequences it has. Instead there will be degrees of dependence. Constituents that interact with each other more frequently, strongly, or directly will exert stronger functional and developmental constraints on each other's variation than constituents that interact only rarely, weakly, or indirectly. Therefore, constraints should be predictable from the patterns of interactions among constituents during the life cycle (section 3). It should even be in principle possible (although not usually practicable) to quantify the degree to which a constituent A (e.g., expression of gene A) is developmentally or functionally constrained by some other constituent B (e.g., expression of gene B) of a containing system S (see section 4 for details).

Reproduction of the life cycle requires that all internal constituents (autoconstituents) interact strongly and directly with at least some other internal constituents, whereas only some internal constituents interact strongly and directly with external constituents (environmental factors). Therefore, it can be expected that all internal constituents of the life cycle will be subject to some strong functional and developmental constraints by other internal constituents ("internal constraints"), but only some of them will be additionally subject to strong functional and developmental constraints by external environmental factors ("external constraints"). Focusing on functional constraints, this simple argument implies that the role of the environment in determining the fitness of heritable variations is less prominent than often thought for two reasons. First, the fitness of heritable variations of some constituents may almost exclusively depend on other internal constituents and not on the environment (i.e., it may be the same in a large range of environments; see also Arthur 1997; Wagner and Schwenk 2000; Schlosser 2002). In recognition of this, some authors apply the expression "functional constraints" only to internal functional constraints (resulting in "internal selection"; e.g., Whyte 1965; Arthur 1997; Wagner and Schwenk 2000; Schwenk 2001). Second, even in cases where the fitness of heritable variations of some constituent does clearly depend on the environment, it will still additionally and perhaps more strongly depend on other internal constituents.

Constraints Are Boundary Conditions for the Dynamics of Life-Cycle Evolution

Because the two notions of external and functional constraints are not coincident, commitment to functionalism does not imply commitment to externalism (contra Alberch 1989). Furthermore, neither commitment to functionalism nor commitment to externalism implies commitment to selectionism (again contra Alberch 1989), because the question of whether there are functional or developmental, internal or external, constraints on variation of some constituent of a life cycle is completely distinct from the question concerning the prevalence of particular types of evolutionary dynamics (e.g., selection versus drift) by which variations spread in a population.

The most productive view of the relation between selection and constraints is to consider constraints as boundary conditions to all evolutionarily dynamic processes (figure 4.4): various kinds of constraints together define the space of the permissible or probable transformations for living systems. The dynamics of these processes, however, may either be algorithmic (Dennett 1995) and can be predicted by deterministic models such as the selection equations of population genetics (e.g., Hofbauer and Sigmund 1984; Maynard Smith 1989), or they may require probabilistic explanations (subsumed under the term *drift*).

It is uncontroversial that deterministic descriptions of evolutionary dynamics are appropriate only under certain boundary conditions, in particular large population sizes and low rates of introduction of new variants. Moreover, selection equations will only be applicable if additional conditions are met. In particular, selection requires that different variants are present in the population, that each variant reliably reproduces variants of the same type, and that the reproduced variants have relative fitness values similar to their parent variants[6] (Schlosser 2002). In other words (Lewontin 1970), selection requires the presence of different variants that differ in fitness and these fitness differences have to be heritable (for different versions of similar ideas, see also Eigen et al. 1981; Dawkins 1982; Bernstein et al. 1983; Maynard Smith 1987; Fontana, Wagner, and Buss 1995; Michod 1999). Only when these conditions hold will selection take place—that is, differential and cumulative changes in the relative frequencies of different variants that can be predicted by their initial frequencies and the fitness differences between them.

So the application of selection equations always *presupposes* the presence of different variants and of fitness differences between them; it cannot explain them. The presence of variants with differential fitness is a boundary conditions for the applicability of selection equations. The facts that only certain variants can be generated by a particular type of life cycle and that these have certain fitness values can only be explained by considering the developmental and functional constraints implied by the particular organization of this life cycle (sections 3 and 4). Consequently, terms like *internal selection* have to be understood as abbreviations for "selection between vari-

Functional and Developmental Constraints on Life-Cycle Evolution

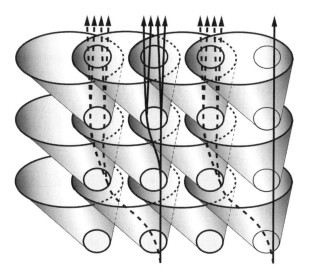

Figure 4.4
Constraints are boundary conditions for the dynamics of life-cycle evolution. Constraints define the space of permissible transformations (see figure 4.2). Arrows illustrate the dynamics of evolutionary frequency changes of different variants (broken versus solid arrows). Branching of arrows indicates reproduction. Evolutionary dynamics may involve selection (right half of figure), when there are fitness differences between the variants (sets bordered by thick and thin lines, respectively), when there is reliable reproduction of variants and fitness values, and when certain other conditions are met (e.g., large population size, low mutation rate) that favor a close match of actual reproductive rates to fitness values. Other cases of evolutionary dynamics do not involve selection—for instance, when there are no fitness differences among variants under conditions that favor a close match of actual reproductive rates to fitness values (left half of figure) or when actual reproductive rates differ from fitness values, as in small populations (drift; not shown).

ants differing in fitness (mainly) due to internal constraints" (for similar views see Arthur 1997; Wagner and Schwenk 2000).[7] Claims (e.g., in Whyte 1965) that internal selection is fundamentally different from external selection because it supposedly acts "prior to" the latter, is based on cooperation instead of competition, and does act in the individual rather than in the population, are therefore misleading and confused.

In conclusion, constraints are invoked to explain which variants are possible and what their fitness is, whereas selection or drift are invoked to explain by which type of process variants spread in a population. The two types of explanations have different explananda and they are in no way incompatible with each other. From this perspective, the question regarding the importance or preponderance of selection (versus drift) in evolution becomes a purely empirical question concerning how frequently certain boundary conditions (presence of different variants with heritable fitness differences in large populations) were realized during the evolutionary diversification of life on earth. This question is completely unrelated to other empirical questions regarding the relative importance of internal versus external or developmental versus functional constraints in evolution.[8] These latter questions—questions concerning the architecture of constraints—are the topic of the remainder of this chapter.

4.3 From Couplings to Constraints

How can the pattern of functional and developmental constraints on evolutionary changes be predicted from the organization of a given functional system? Remember that constraints on transformations of a stable system are relative to (1) a set of conditions specifying the initial organization of the system and (2) a set of possible perturbations. Let us focus on the former here. They can be empirically determined by an analysis of the self-maintaining and reproducing organization of the system under consideration. The pattern of interactions (couplings) between constituents underlying the current organization predicts constraints on evolutionary transformations of these constituents because it suggests stronger developmental (generative) and functional interdependence for those constituents that are more closely coupled. According to this process-oriented perspective, constraints on the transformation of structures arise from the processes or interactions in which the structures participate.

Interestingly, developmental and functional constraints on evolutionary transformations depend on different kinds of coupling (see also Roth and Wake 1989). Strong developmental constraints will be expected between constituents that are closely generatively (developmentally) coupled—that is, that have a recent common causal input in the life cycle (figure 4.5A). Strong functional constraints, in contrast, will be expected between constituents that are closely functionally coupled—in other words, that jointly contribute to the realization of a common functional output (figure 4.5B) and therefore are likely to make nonindependent fitness contributions. These notions of functional and generative coupling are also applicable to constituents that are nested within each other in a hierarchical fashion (see figure 4.5A–B). The constituents to which a given constituent is most strongly generatively coupled may differ from the constituents to which it

Functional and Developmental Constraints on Life-Cycle Evolution

is most strongly functionally coupled, so different constituents may be the prevalent determinants of developmental and functional constraints (figure 4.5C–D).

Examples for Generative (Developmental) and Functional Couplings and the Resulting Constraints

Let me just briefly discuss a few examples for generative and functional couplings at different levels and how they generate constraints. Vertebrate limb development has always served as a favorite model for the illustration of generative (developmental) couplings. Different limb bones do not develop independently from each other because they depend on common patterning mechanisms (reviewed in Shubin, Tabin, and Carroll 1997; Johnson and Tabin 1997; Ng et al. 1999; Gilbert 2006). This generative coupling of limb bones imposes developmental constraints biasing their response to various perturbations and their evolutionary modification. For example, loss of phalanges or digits can be induced by various kinds of experimental or genetic perturbations (Alberch and Gale 1983; Davis and Capecchi 1996). In such cases the phalanges or digits that form latest are usually lost first. Preferential loss of late-forming bones is also the typical pattern observed in most tetrapod taxa that have phylogenetically lost digits (Holder 1983; Alberch and Gale 1985; Shubin, Tabin, and Carroll 1997).

Many other examples of generative couplings are revealed by our increasing knowledge of molecular networks that operate during development, such as signaling cascades or networks of transcription factors (reviewed in Gerhart and Kirschner 1997; Gilbert 2006). For instance, various genes coding for transcription factors of the basic helix loop helix class (e.g., *NeuroD*) have been identified as so-called selector genes that regulate the transcription of a large battery of downstream genes involved in the specification and differentiation of various cell types (e.g., neurons; reviewed in Lee 1997; Guillemot 1999; Brunet and Ghysen 1999). The generative coupling of transcription of these downstream genes imposes developmental constraints/opportunities because it greatly facilitates coexpression of these downstream genes in novel domains. This becomes evident after experimentally induced ectopic expression of cell-type-specific selector genes (e.g., Lee et al. 1995; Ma, Kintner, and Anderson 1996; Blader et al. 1997; Perron et al. 1999) and may underlie the repeated evolution of new domains of coexpression of the same battery of cell-type-specific genes (e.g., the generation of neurons from neural crest cells and placodes in vertebrates; Northcutt and Gans 1983; Baker and Bronner-Fraser 1997; Schlosser 2005a).

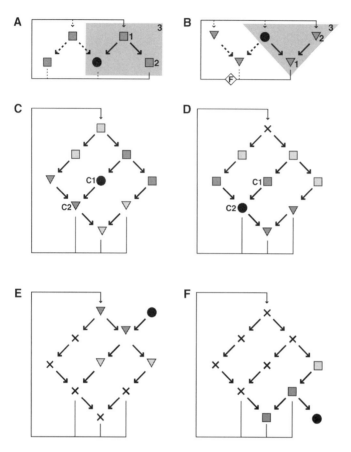

Figure 4.5

Generative and functional couplings among components (i.e., among states or patterns of activity of component structures). *A*: Roughly speaking, a component C (black circle) is generatively coupled to those constituents (constitutive components) of a functional system S (thin solid looping arrow), with which it shares a common causal input. More precisely, a component C is generatively coupled to some constituent of a functional system S, when some of the immediate determinants (i.e., immediate causal inputs or constitutive components) of the activity of the latter are also among the direct or indirect determinants of C's activity. Thus, constituents to which C is generatively coupled include (1) constituents causally required for its realization (activation), (2) other constituents that are causally dependent on constituents of type (1), (3) higher-level constituents that include C as a constitutive component, and (4) lower-level constituents (not shown), which are constitutive components of C. Note that C itself may or may not be necessary for the reproduction of the functional system (thin broken line), and therefore may or may not be a

Functional and Developmental Constraints on Life-Cycle Evolution   131

Functional couplings and the constraints they produce have probably been most intensely discussed in functional morphology (e.g., see Bock and von Wahlert 1965; Gutmann 1977; Dullemeijer 1980; Lauder 1981; Roth and Wake 1989; Liem 1991; Wagner and Schwenk 2000; Schwenk 2001). The functional cooperation of various muscles and bones necessary for vertebrate jaw movements during feeding may serve as a simple example (e.g., Romer 1977; Young 1981). Firm adduction of the lower jaw is required to hold prey. This is one of the functions of jaw movements, because it is necessary for feeding and therefore for survival and reproduction.

Figure 4.5 (continued)
constituent of S. It may also be generatively coupled to other components that may or may not be constituents of S (broken lines and arrows). *B*: Roughly speaking, a component C (black circle) is functionally coupled to those constituents with which it shares a common function. More precisely, a component C is functionally coupled to some constituent of a functional system, when the realization (activation) of one of C's functions (i.e., of one of its effects that is instrumental in the reproduction of the functional system including C) by C's activity is directly or indirectly dependent on (requires) the activity of this other constituent. Thus, constituents to which C is functionally coupled include (1) those constituents causally required for a function F, for whose realization (activation) C is in turn causally required, (2) constituents that are causally required in addition to C for the realization (activation) of constituents of type (1), (3) higher-level constituents that include C as a component constitutive for performing F, and (4) lower-level constituents (not shown), which are components of C constitutive for performing F. Note that C here is necessarily itself a constituent of S, because it is required for its reproduction. However, C itself may or may not be reproduced by the functional system (thin broken arrow), and therefore may or may not be an internal constituent of S. It may also be functionally coupled to other constituents that may or may not be constituents of S (broken lines and arrows). *C–F*: Schematic illustration of the pattern of generative and functional couplings of some particular constituent C (black circle) of a functional system. Constituents that are generatively coupled to C are drawn as squares, constituents that are functionally coupled to C as triangles. Constituents to which C is directly coupled, by virtue of a recent common input or output, are colored dark gray, indirectly coupled constituents are colored in light gray, and more distantly coupled constituents depicted as crosses. *C*: Pattern of couplings for constituent C1 (black circle). *D*: Pattern of couplings for constituent C2 (now drawn as black circle) immediately downstream of C1. Note that functional couplings of constituents such as C1, which act earlier in the causal network (*C*), are reflected in generative couplings of constituents such as C2, which they affect. Environmental components may also be functionally (*E*) or generatively (*F*) coupled to the constituents of a functional system. See text for further explanation.

During the process of jaw adduction, the jaw levator muscles create the necessary force, while the lower jaw transmits the force. This requires that the bony element forming the articulation of the lower jaw with the skull and the bony elements bearing teeth are firmly connected with each other and move in a coordinated fashion during muscle contraction. In nonmammalian tetrapods the articulation of the lower jaw with the skull is formed by the articular bone, while other bones such as the dentary bear the teeth. The functional coupling of these two bones during feeding in nonmammalian gnathostomes imposes functional constraints that prohibit the evolution of disconnected bones, although the development of the two bones does not appear to be tightly generatively coupled. Such lack of tight generative coupling may have permitted the formation of separate articular and dentary bones in mammals, where the articular bone becomes transformed into the hammer of the inner ear. This evolutionary change was presumably only possible functionally, after the dentary itself evolved a new articulation with the skull in ancestral mammals, liberating the articular bone from former functional constraints (e.g., Romer 1977; Young 1981).

Functional couplings are also evident in many molecular interactions. Signal transduction in cells, for instance, often depends on interactions of an extracellular ligand with a transmembrane receptor (e.g., Gerhart and Kirschner 1997; Gilbert 2006). Similarly, regulation of gene transcription requires the binding of various transcription factors to the cis-regulatory regions of the gene (e.g., Ptashne and Gann 1997, 1998; Arnone and Davidson 1997; Davidson 2001). The functional couplings between ligand and receptor as well as between transcription factors and the genes they regulate, impose functional constraints on evolutionary changes in these coupled elements. They result in reduced fitness of those (generatively possible) mutations that compromise presently functional molecular interactions. This leads to coordinated evolutionary changes of ligand and receptor or transcription factors and the cis-regulatory regions of the genes they regulate (e.g., Ludwig et al. 2000; Goh et al. 2000; reviewed in Dover 2000; Stern 2000).

Functional and Generative Couplings Are Present during All Phases of the LIfe Cycle and Between Internal and External Constituents

Of course, the characterization of generative and functional couplings given here is applicable to each stage of the life cycle. So functional couplings among constituents are not confined to the adult stage but also exist during development. For instance, the activity of two transcription factors

Functional and Developmental Constraints on Life-Cycle Evolution 133

is functionally coupled, when they are jointly required for transcription of a particular gene that is essential for development of some functionally important organ. Similarly, during embryonic inductions of such organs (e.g., of the lens; see Gilbert 2006) the development of the inducing tissue (eye-cup) is functionally coupled to the development of some other tissue competent to respond to such inducers (head ectoderm).

Conversely, generative couplings among constituents are not confined to the period of embryonic development, but also exist during the adult stage. For instance, glucose uptake and glycogen synthesis are generatively coupled to the inhibition of fat and glycogen breakdown because both depend on the binding of insulin to its receptor. Similarly, generative coupling exists among the movements of all bones in which the same muscle inserts. Because generative and functional couplings can be identified during each stage of a life cycle, the same will be true for generative (or *developmental* in the sense used here) and functional constraints. Unfortunately, the term *developmental constraints* is sometimes applied to all kinds of (i.e., generative and functional) constraints on development, while the notion of functional constraints is reserved for constraints on the organization of the adult stage (e.g., Roth and Wake 1985; Zelditch and Carmichael 1989; Cheverud 1996; Hall 1996; Hodin 2000). This use of the two terms should be abandoned because it misleadingly suggests that generative and functional considerations are relevant for understanding evolutionary changes of only the embryonic or adult phase of the life cycle, respectively.

It should be emphasized, too, that functional and generative couplings will not only bind together internal constituents of an organism, but will also link some environmental factors to the organization of the organism. This will generate constraints on the coevolution of organisms and their abiotic and biotic environment (Dawkins 1982; Lewontin 1983; Oyama 1985; Wimsatt 1986; Gray 1992; Schlosser 1993, 2002; Griffiths and Gray 1994). On the one hand organisms not only depend for the maintenance and reproduction of their organization on specific interactions among their internal constituents; they also require certain interactions with environmental factors. As a consequence of this functional coupling of environmental factors with internal constituents, the former may functionally constrain modifications of the latter (figure 4.5E). This is the case in many classical examples of environmental adaptations, like mimesis and mimicry, involving functional coupling to abiotic and biotic environmental factors, respectively, that underlie the production of color patterns in the environment. On the other hand, the generative coupling of certain environmental components (that may but do not need to be external constituents) to internal constituents may developmentally constrain the transformation of these environmental components (figure 4.5F). Developmentally constrained components of the environment include for instance abiotic factors constructed by organisms such as birds' nests or beavers' dams, and of course also developmental, physiological, or behavioral effects on other organisms—as in predator-prey, parasite-host, or symbiont-host relationships.

Functional and Developmental Constraints May Act in Concert
The distinction between developmental (generative) and functional couplings and constraints is conceptually clearcut. However, in reality both are often difficult to tear apart for several reasons. First, functional constraints on a constituent acting earlier in the causal network imply developmental constraints on constituents that they affect further down in the causal chain (figure 4.5C–D; see also Wagner and Misof 1993). For example, early developmental processes may become drastically modified in some mutant, but the modified processes may not be able to perform their normal function in subsequent development. This may even result in embryonic lethality and limits the kind of modifications of later developmental processes that can be generated. Consequently, heritable variations of constituents that are embryonically lethal (e.g., Galis 1999) can be modeled in two ways. Considering their early developmental effects, variants are produced but their fitness is zero, because they fail to perform their function, resulting in subsequent lethality (fitness decrease due to functional constraints). With respect to evolutionary dynamics, the loss of such variants from a population may be regarded as a degenerate example of selection between the lethal and the original nonlethal variant of early embryogenesis (see also Hall 1996; Arthur 1997). Considering their effects on later stages of development, however, no alternative variants are ever generated because embryos die already at earlier stages (absence of variants due to developmental constraints). Because a new variant of these late stages is never generated, questions of evolutionary dynamics do not even arise.

A second reason why functional and developmental constraints may act in concert is that constituents in organisms often interact with each other in a reciprocal fashion—that is, they are coupled to each other both generatively and functionally. Consequently, they will exert both developmental and functional constraints on each other's evolution. To give only one example, the hedgehog protein serves as a ligand for the transmembrane receptor patched. This interaction is required for signal transduction; hence, patched activity is functionally coupled to hedgehog activity. But the cascade of events initiated by binding of hedgehog to patched also increases transcription of the patched gene and thereby the availability of patched protein in the membrane; hence, patched activity is also generatively coupled to hedgehog activity (reviewed in Hammerschmidt, Brook, and McMahon 1997; Murone, Rosenthal, and deSauvage 1999). It can be predicted that the coevolution of patched and hedgehog activity will be guided by functional as well as by developmental (generative) constraints.[9]

4.4 Analyzing the Architecture of Constraints

It has been suggested above that the degree to which some constituent A of a functional system is functionally or developmentally constrained by another constituent B should in principle be quantifiable. I want to briefly sketch how this might be done, without providing a detailed and formalized analysis here.

To develop this account, it is useful to distinguish two kinds of variables: *capacity variables* a describing the capacities of a constituent A to interact with other constituents, that is, its logical properties (how it maps inputs to outputs), and *state or activity variables* α describing the actual state, behavior, or pattern of interactions of A, that is, its actual input-output transformations at a certain point in space and time (in order to describe the behavior of A at multiple points in space and time, of course, multiple activity variables α^1, α^2, etc. are needed). Variables α are a function of variables a as well as of the inputs available at a certain point in space and time (which are described by the activity variables β, γ, etc. of other constituents). For example, the capacities of a certain gene to be activated by certain "upstream" transcription factors (due to the presence of specific binding sites in the regulatory regions) and to turn on other "downstream" genes (due to the ability of the encoded protein to bind to specific regions of DNA) would be described by variable a, while the expression of the gene in a particular cell at a particular time of development would be described by variable α.

It will be further assumed that modifications of capacity variables (e.g., due to gene mutation) are the primary source of all heritable variation. However, heritable variations in the capacity variables of some constituents may be strictly and asymmetrically dependent on heritable variations in capacity variables of other constituents. Heritable modifications of genes or proteins in somatic cells, for example, reflect heritable gene mutations in germ-line cells but not vice versa. Moreover, heritable variations of an

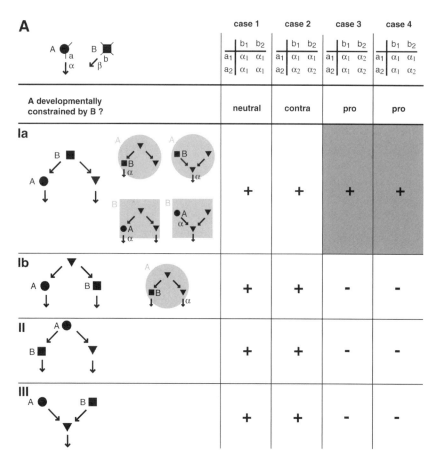

Figure 4.6
Analyzing the degree to which variability of A is constrained by B. Constituents A and B can be described by variables a and b, describing their respective interaction capacities and by (one or several) variables α and β describing their respective activities at certain points in space and time. Interaction capacities for higher-level constituents (gray squares and circles) are determined by the interaction capacities of their constitutive components (i.e., they may vary with variation of any of their constitutive components). A: The hypothesis that A is developmentally constrained by B predicts that α or the effect of variations of a on α depends on variations in b. This hypothesis may be tested by pairwise comparison of the effects of different variants of a and b on α. This may yield four different types of results. In case 1, α is neither affected by a nor by b; this neither supports nor contradicts the hypothesis. In case 2, the effect of a on α does not depend on variations of b; this contradicts the hypothesis. In case 3, α depends on variations of b; this supports the hypothesis. In case 4, the effect of a on α depends on variations of b; this likewise supports the hypothesis.

Functional and Developmental Constraints on Life-Cycle Evolution

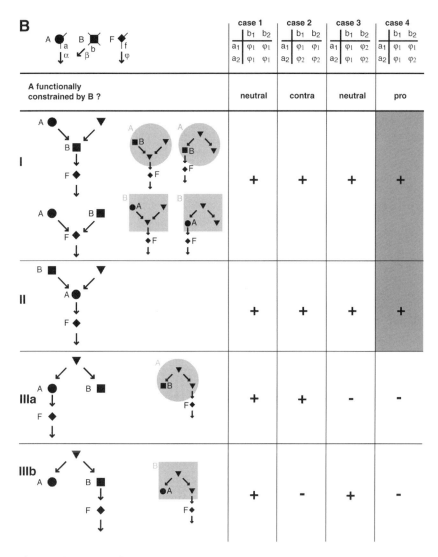

Figure 4.6 (continued)
The figure also indicates with − and + signs which of the four cases can be plausibly expected for the four patterns of couplings indicated in the left column of the table under the assumption that activity variables (e.g., α) of a constituent can only be affected by capacity variables (e.g., b) of another constituent, if the latter acts causally prior to the former. This demonstrates that support for the hypothesis can only be expected (gray area) in a subset (pattern Ia) of the cases where A is generatively coupled to B (pattern I), namely, when A is causally dependent on B, when A is component of B, or when B is a component of A with the potential of affecting α.

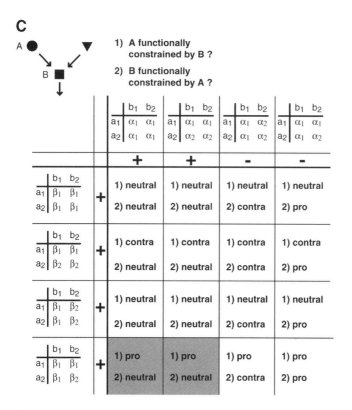

Figure 4.6 (continued)
No support is expected in cases where B is generatively coupled to A but not vice versa (pattern II) or where A and B are not generatively coupled (pattern III). The degree to which A is developmentally constrained by B can then be measured as the relative frequency of cases supporting the hypothesis among all possible pairwise constellations of variants of a and b from some predefined set. This kind of analysis is also applicable to constituents at different hierarchical levels (i.e., when A includes B or vice versa) as indicated. B: The hypothesis that A is functionally constrained by B predicts that the effect of variations in α resulting from heritable capacity variations of a on the activity φ of one of its functions F (or on the fitness w_A) depends on variations of b (epistasis). This hypothesis can be similarly tested. The table illustrates that support for the hypothesis can only be expected (gray area) when A is functionally coupled to B (pattern I) or vice versa (pattern II), but not when A and B are not functionally coupled (pattern III). The degree to which A is functionally constrained by B can then again be measured as the relative frequency of cases supporting the hypothesis among all possible pairwise constellations of variants of a and b from some predefined set. Again, this kind of analysis is also applicable to constituents at different hierarchical levels as shown. C: Epistatic relationships between two constit-

Functional and Developmental Constraints on Life-Cycle Evolution

activity variable α do not necessarily or exclusively reflect heritable variations of capacity variable a, but may reflect heritable variations in other capacity variables such as b or c, because these may alter inputs (e.g., β, γ) to A. For example, expression of gene A may be altered after mutation of this gene itself or of other genes B or C encoding transcription factors necessary for its activation.

In the following paragraphs, I will concentrate on this dependence of activities on capacities to illustrate how constraints may, in principle, be measured. The main idea is that the degree to which variability of A is constrained by B can be conceptualized as the probability (propensity) with which variation in α is restricted or biased by variation in b.

Determining the Degree to Which A Is Developmentally Constrained by B

Given these assumptions, the degree to which evolutionary transformations of a constituent A are *developmentally constrained* by another constituent B may be determined as the probability (propensity) that heritable variations of α or the effects of heritable variations of a on α depend on heritable variations of b (figure 4.6A). For instance, the level of expression of gene A may increase from α_1 to α_2 after mutation of b_1 into b_2 (e.g., because b_2 results in increased expression of a transcription factor that activates A). Or a particular mutation a_1 into a_2 may result in an increased level of expression α_2 (compared to the original expression level α_1) of gene A only

Figure 4.6 (continued)
uents A and B may also be analyzed by directly assessing the effects on α and β rather than some activity φ of another constituent F testing the two hypotheses, that A is functionally constrained by B or vice versa. Pairwise comparison of the effects of variation of a and b on α and β may yield sixteen types of results as indicated. Given the pattern of interactions depicted (A being causally prior to B), only eight of those can be plausibly expected (+/+). These include only cases supporting hypothesis 1 (gray area: cases where the effect of a on β depends on b, i.e., where b is epistatic to a) but no cases supporting hypothesis 2 (cases where the effect of b on α depends on a, i.e., where a is epistatic to b). Note, however, that this asymmetry of epistasis will disappear and both hypotheses will be equally supported when fitness values w_A and w_B rather than activities α and β are compared assuming that the fitness of A equals the fitness of B (this is likely to be the case whenever the reproduction of both A and B depends on the function of B—for instance, when both act in sequence as illustrated). Under this assumption, there will be only four cases (corresponding to the diagonal, but all causally possible), and every case supporting hypothesis 1 will also support hypothesis 2.

in the presence of variant b_1 of B, having no effect on gene expression in the presence of variant b_2 (e.g., because it increases binding capacity for the transcription factor encoded by b_1 but not for the one encoded by b_2). However, it is important to note that different variations of the same capacity variable may have different effects, and some—for instance, the mutation from b_1 into b_3—may have no such effects on α.

The probability that heritable variations of α depend on heritable variations of b becomes a well-defined notion, when some additional assumptions are made. First, we need to assume which kinds of heritable capacity variants of a and b are generatively possible to begin with, so the probability will be determined relative to some defined sets of heritable variants $a_1, \ldots, a_i$ and $b_1 \ldots b_j$ with known distributions of probabilities for occurrence and combinability (e.g., equiprobability and free combinability). Such a set may not always be straightforward to specify, but in case A and B represent genes or proteins, it may be definable as the set of all one-mutant neighbors of some reference variant, for instance the presently realized variant. When only certain types of mutations are allowed (e.g., point mutations), this set is finite and well defined.

Second, we need to assume that every particular combination of capacity variants of a and b will map in an unequivocal (but possibly many-to-one) fashion to a particular activity variant of the set $\alpha_1, \ldots, \alpha_m$ of A (assuming all other conditions being equal). Under the assumptions of equiprobability and free combinability of all heritable capacity variants, the degree to which A is developmentally constrained by B may be determined as the relative frequency of those pairwise constellations of heritable capacity variants among all possible constellations of variants of the predefined set, in which α or the effect of variations of a on α depends on variations in b—that is, in which α differs between b_v and b_w for at least one of two variants a_x and a_y (figure 4.6A).

As figure 4.6A illustrates, developmental constraints are likely to reflect the pattern of generative couplings among constituents. A is most likely to be developmentally constrained by B if β is one of the causes of α; by the same token, A and B are most likely to be each developmentally constrained by the same constituent C, in case both α and β depend causally on its activity γ (are generatively coupled by it). The degree to which A is developmentally constrained by B or C will probably depend on how direct or strong the interactions between A and B or C are. Unless there are reciprocal interactions between A and B, the degree of developmental constraint will be often asymmetrical—that is, A may be developmentally constrained by B but not vice versa.

This kind of analysis is also applicable to quantify constraints on variation between constituents at different hierarchical levels (figure 4.6A), but it would exceed the scope of this chapter to show this in detail. Briefly sketched, the account presented here can be applied to determine, for instance, to what degree evolutionary transformations of some organ A are developmentally constrained by any of the genes B, C, and so on that influence its development or activity (i.e., to what extent heritable variations of the organ's performance α or the effects of heritable variations of the organ's interaction capacity a on α depend on heritable variations of interaction capacities of genes B, C, and so forth due to mutation). Vice versa this account can also be applied to determine to what extent evolutionary changes of genes are developmentally constrained by the cells and organs whose development or operation they modulate.

Determining the Degree to Which A Is Functionally Constrained by B

The degree to which A is *functionally constrained* by B depends on the probability that variations in B are epistatic to variations in A regarding their effects on some other constituent F (i.e., on the performance of one of the functions of A) of the life cycle (see also Cheverud and Routman 1995; Wagner, Laubichler, and Bagheri-Chaichian 1998; Wagner and Mezey 2000; Hansen and Wagner 2001). There will be epistasis when the effect of variations in α resulting from heritable capacity variations of a on the activity φ of F depends on heritable capacity variations of b (figure 4.6B). As in the case of developmental constraints, this kind of analysis is applicable to constituents A and B that either reside at the same or at different hierarchical levels (figure 4.6B). To analyze functional constraints on evolutionary transformations, the fitness of A (effects on the probability of reproduction of A) serves as the most adequate choice of φ because it depends on the effects of all functions of A. Hence, the degree to which evolutionary transformations of A are functionally constrained by B reflects the probability that the effects of heritable capacity variations of a on the fitness of A (w_A) depend on heritable capacity variations of b. For internal constituents, this probability can be expressed as the probability for epistatic fitness interactions between A and B (figure 4.6B). Considering, for instance, variants a_1 and a_2 of A and variants b_1 and b_2 of B, there will be fitness epistasis when the relative fitness of a_1 versus a_2 is different, depending on whether these variants are combined with b_1 or with b_2. a_1 and a_2 may, for instance, be variants of a transmembrane receptor that differ in their ligand-binding domain. a_1 may have higher binding affinity and consequently higher fitness than a_2 in the presence of ligand b_1, while the reverse may be true in the presence of ligand b_2. Testing another constellation of variants of A and B—for example, a_1, a_3, b_1, and b_2—we may, however, not find any fitness epistasis. For instance, a_3 may differ from a_1 in its intracellular domain and may be more efficient in signal transduction, resulting in higher fitness regardless of whether it is combined with b_1 or b_2.

Again, the probability of fitness epistasis between heritable variations of a and b becomes a well-defined notion when several additional assumptions are made. First, whether a and b tend to show fitness epistasis will depend on which heritable variants of a and b can be generated. Therefore, the probability for fitness epistasis between a and b can be defined only relative to given sets of heritable variants with known distributions of probabilities for occurrence and combinability. These sets may be determined as specified above.

Second, it needs to be assumed that it can be unequivocally determined whether there is fitness epistasis for any particular combination of variants of a and b (with multiplicative fitness interaction as the null model). Under the assumptions of equiprobability and free combinability of all heritable capacity variants, the degree of functional constraint between A and B (the probability for fitness epistasis between a and b) may then be determined as the relative frequency of those pairwise constellations of heritable capacity variants among all possible constellations of variants of the predefined set that exhibit fitness epistasis.

As figure 4.6B illustrates, functional constraints are likely to reflect the pattern of functional couplings among constituents. A is most likely to be functionally constrained by B if the cooperation (either in sequence or in parallel) of A and B is necessary for the realization of a common function—for example, for the activity φ of some other constituent F affecting the fitness of A. The degree to which A is functionally constrained by B will again depend on how direct or strong the interactions between A and B are, epistatic interactions being most likely for the capacity variants of a and b that affect the capacities of A and B to directly interact with each other (e.g., variations affecting the receptor-binding domain of a ligand and the ligand-binding domain of its receptor).

In contrast to the asymmetry of developmental constraints, epistatic effects of variations in a and b on the fitness of A and B will typically be symmetrical (i.e., A will be functionally constrained by B as well as vice versa) because the fitness of both A and B (the probability of reproduction of A and B in the next generation) will equally depend on the performance of their common function. This symmetry may hold even when the epistatic effects of variations in a and b on activities α and β, respectively,

Functional and Developmental Constraints on Life-Cycle Evolution 143

may be strongly asymmetrical (downstream constituents being epistatic to upstream constituents but not vice versa; see figure 4.6C for further explanation). The higher the number of functional interactions of a constituent with other constituents, the higher the number of constituents by which it will tend to be functionally constrained. Such multiply constrained constituents have been said to carry a burden (Riedl 1975) or to be generatively entrenched (Wimsatt 1986; Schank and Wimsatt 1986; Wimsatt and Schank 1988), because due to potential conflict among different functional constraints, they have fewer options for permissible evolutionary transformation.

Coevolution Probabilities and Units of Evolution

It has been argued so far that the degree to which some constituent A of a functional system is developmentally or functionally constrained by another constituent B of the system is in principle determinable, as long as a finite and well-defined set of capacity variants can be specified. It will then also be possible to determine for each pair of constituents A and B a single probability for A and B being constrained by each other either developmentally or functionally relative to the defined sets of capacity variants. This will be referred to as the *constraint probability* $P_{constraint(AB)}$ of A and B. While the degree of developmental constraint of A by B may differ from the degree of developmental constraint of B by A, this asymmetry will be disregarded here to simplify the following argument. The constraint probabilities for pairs of constituents can form the basis for analyzing which constituents of a functional system will tend to evolve as a unit in a coordinated fashion. For instance, if there are high constraint probabilities exclusively among AB, AC, and BC, but not among AD, BD, or CD, A, B, and C are likely to change in a coordinated fashion but largely independent of D during evolution.

In quantitative genetics, estimations of such constraint probabilities are often based on genetic variance/covariance matrices (e.g., Lande 1980; Cheverud 1984). It should be emphasized, however, that such matrices will not necessarily give a good approximation of constraint probabilities as defined here, because they only compare variations actually present in a given population, rather than considering all capacity variations that are generatively possible (see also Mezey et al. 2000).

Assuming that there are no higher-order interactions, the pairwise constraint probabilities determined for each pair of constituents of a functional system (e.g., AB, AC, AD, BC, BD, CD) may be used to calculate for each combination of constituents (e.g., ABC, ABD, BCD, ABCD, etc.) a *coevolution probability*. The coevolution probability of each constellation (e.g., ABC of system ABCD) may be defined as the probability that there will be constraints on evolutionary transformations among all members of the constellation (AB, AC, BC) but not between any members of the constellation and any other constituent (AD, BD, CD) of the system (e.g., the coevolution probability of AB within a system ABCD may be calculated as the product of $P_{constraint(AB)}$, and $1 - P_{constraint(AC)}$, $1 - P_{constraint(AD)}$, $1 - P_{constraint(BC)}$, $1 - P_{constraint(BD)}$).

A unit ABC may be termed a *unit of evolution* if its coevolution probability is higher than its disruption probability (the latter is given by the combined coevolution probabilities of all constellations containing some but not all of the constituents of the unit ABC plus at least one other constituent of the system, i.e., it can be calculated from the coevolution probabilities of AD, BD, CD, ABD, BCD, and ACD). Units of evolution so defined are units of constituents that tend to coevolve because they reciprocally constrain each other's evolution.

These units of constraints persist as long as coevolution probabilities do not change—that is, as long as the pattern of couplings between constituents remains conserved even after iterated variation. Because functional constraints allow only evolutionary transformations that maintain functionality, a certain perseverance of coevolution probabilities is to be expected. Nonetheless, the architecture of coevolution probabilities can itself evolve (Cheverud 1984, 1996, 2001; Müller and Wagner 1991; Wagner and Mezey 2000; Wagner, Chiu, and Laubichler 2000; for empirical studies see also Atchley et al. 1992; Fink and Zelditch 1995). This is because heritable variations can occasionally alter constraint and coevolution probabilities (i.e., can introduce a reference variant with an altered distribution of constraint probabilities relative to the set of its one mutant neighbors), due to the evolution of new patterns of generative or functional couplings. Changes in the architecture of constraints are often observed under conditions of functional redundancy or neutrality—for example, after duplication of constituents (e.g., due to duplication of gene expression domains or genes; see Ohno 1970; Jacob 1977; Zuckerkandl 1994, 1997; Müller and Wagner 1996; Sidow 1996; Duboule and Wilkins 1998; Force et al. 1999; Galis 2001; Schank and Wimsatt 2001; Schlosser 2002, 2004). In consequence, novel units of evolution may be established, and old ones either disrupted or reorganized by partial replacement of some of their lower-level constituents.

Of course, the sheer number of possible combinations that would have to be tested practically precludes a complete analysis of the kind discussed

Functional and Developmental Constraints on Life-Cycle Evolution

here. However, with the advent of functional genomics, combining high-throughput analysis of gene expression patterns after combinatorial targeted mutagenesis with high-speed comparative analysis, the determination of constraints may become more feasible for gene and protein interactions in the future (e.g., Wen et al. 1998; Eisen et al. 1998; Niehrs and Pollet 1999; Lockhart and Winzeler 2000). In the meantime, the concepts of constraint probabilities or coevolution probabilities, although difficult to determine empirically, can serve as useful conceptual tools that allow us to give a well-defined meaning to the concept of a unit of constraints on evolutionary changes (unit of evolution).

4.5 Units of Evolution and Units of Selection

As units of elements that codetermine a context-insensitive fitness value, units of evolution may also be considered excellent candidates for "units of selection." Indeed, such context insensitivity for fitness has often been considered to be the hallmark of a nondecomposable unit of selection (Lewontin 1974; Wimsatt 1980, 1981; Sober 1981, 1984; Sober and Lewontin 1982; Lloyd 1988; but for different views see Dawkins 1982; Maynard Smith 1987, 1998; Sterelny and Kitcher 1988; Waters 1991). Consequently, whenever there are epistatic fitness interactions between constituents, units comprising multiple constituents (e.g., gene complexes or even entire organisms) seem better candidates for units of selection than single constituents (e.g., genes), contrary to the view of "gene selectionists" that genes should always be considered the units of selection (Williams 1966; Dawkins 1976, 1982; Maynard Smith 1987). An extensive discussion of the complex debate on the unit of selection problem is beyond the scope of this chapter (see e.g., Lewontin 1970, 1974; Wimsatt 1981; Sober 1981, 1984, 1987; Sober and Lewontin 1982; Dawkins 1982; Gould 1982; Brandon and Burian 1984; Maynard Smith 1987; Lloyd 1988; Sterelny and Kitcher 1988; Brandon 1990, 1999; Waters 1991; Sober and Wilson 1998; Schlosser 2002). However, in order to avoid misunderstandings, I want to explain briefly why context insensitivity of fitness is insufficient for defining units of selection, and why units of evolution as defined here are neither equivalent to nor necessarily coincident with units of selection.

Selection Processes and Units of Selection

To characterize a unit of selection, we need to remind ourselves (see section 2) that selection processes only take place among entities when different variants of these entities are present in the population, when each variant reliably reproduces variants of the same type in subsequent generations (variant heritability), and when the reproduced variants have similar fitness values relative to other variants (fitness heritability). Therefore, a set of constituents (e.g., genes) A,B of a reproducing system will only belong to a single unit of selection AB when each constituent is present in different variants (e.g., alleles a_1, a_2, b_1, b_2), when each variant of AB (i.e., each constellation a_1b_1, a_1b_2, a_2b_1, a_2b_2 of variants a_1, a_2, b_1, b_2) will be reliably reproduced, and when the fitness of each variant of AB is also relatively reliably reproduced.[10]

Such units of selection may be recognized at different levels, each of which corresponds to a particular way of partitioning a certain set of reproducing systems into different subsets (e.g., a unit defined by A partitions the set into two variants a_1 and a_2, while a unit defined by AB partitions the same set into four variants a_1b_1, a_1b_2, a_2b_1, a_2b_2; Wagner and Laubichler 2000; Laubichler and Wagner 2000; Schlosser 2002). A particular unit of selection may be called nondecomposable when either the reproduction of all of its constituents is strictly linked (e.g., due to asexual reproduction or due to low recombination rates of closely linked genes in sexually reproducing organisms) or when there is fitness epistasis between the variants of all of its constituents.

Given that units of selection can be recognized at many different levels, it may be asked for which of these units selection is most effective. To answer this question it has to be realized that the requirements for reliable variant reproduction and reliable fitness reproduction can be in conflict. While the variants of a relatively high-level unit of selection ABC may have more context-insensitive fitness values than a lower-level unit A (due to fitness epistasis between A, B, and C), they may at the same time be reproduced much less reliably than the lower-level unit A (e.g., because particular combinations $a_1b_1c_2$ tend to be broken up due to sexual recombination between A, B, and C). Selection will be most effective for intermediate units that guarantee an optimal compromise between the requirements for reliability of variant reproduction and reliability of fitness reproduction (Schlosser 2002).[11]

Units of Selection Do Not Delimit Units of Constraints in Evolution (Units of Evolution)

This concept of a unit of selection is important to understand the population dynamics of particular selection processes. However, attempts to define units of evolution—units of constituents that reciprocally constrain each other's evolution—in terms of nondecomposable units of selection would seriously underestimate the number of constituents that exert constraints on each other for two reasons. First, in the case of high recombination rates between constituents in sexually reproducing organisms and in large populations, reliable reproduction of a variant is less likely the more independent-variable constituents it comprises (e.g., Lewontin 1970; Maynard Smith 1987; Schlosser 2002). Under these circumstances high-level units (e.g., complexes ABC of many polymorphic genes) are unlikely to act as effective units of selection in a particular selection process, even when there will be fitness epistasis between them, because particular variants are extremely unlikely to be reproduced. Nonetheless, all constituents of the higher-level unit may constrain each other's evolution in sequences of selection processes—for example, because patterns of polymorphisms will differ for each selection process and the boundaries of units of selection will shift. (A unit of selection during a particular selection process may include only a subset of the constituents comprising a unit of evolution, but it may be a different subset in a subsequent selection process.)

Second, a unit of selection comprises by definition only constituents that exist in different variants during a particular selection process. Nonetheless, constituents that do not vary in a particular selection process may again impose constraints in sequences of selection processes. For instance, the fitness differences between variants of a constituent A in a particular selection process, in which only A but not B is polymorphic, may differ depending on the outcome of a *previous* selection processes among variants of constituent B. Therefore, B constrains the evolutionary modifications of A and both may belong to a single unit of evolution, although B does not belong to the unit of selection in a particular selection process among variants of A.

In conclusion, units of selection are conceptually distinct from the units of evolution introduced in the previous section, and we have to focus on the latter in order to obtain further insights into constraints on evolutionary transformations.

4.6 A Modular Architecture of Constraints?

Units of evolution are units of constituents that tend to coevolve because they reciprocally constrain each other's evolution due to their non-independent variability and reciprocal fitness dependence. As the other side of the same coin, constituents belonging to different units of evolution will be variable largely independently of each other and will make quasi-independent fitness contributions. To what extent organisms are decomposable into different, mutually quasi-independent units of evolution is, of course, an empirical question. However, it has long been argued that evolutionary changes in complex systems would be virtually impossible, if they were so tightly integrated that evolutionary changes in each constituent would be equally strongly constrained by every other constituent of the system (Lewontin 1978, 2001; see also Brandon 1999; Schank and Wimsatt 2001). Indeed, there is strong empirical support for mosaic evolution of characters at many different hierarchical levels (e.g., Wake and Roth 1989; Atchley et al. 1992; Raff 1996; Gerhart and Kirschner 1997; Shubin 1998; Barton and Harvey 2000; Schlosser 2001, 2004, 2005b). It has even be proposed that characters can be individuated and identified in spite of modifications throughout a lineage by virtue of their being "building blocks" or modules of evolutionary transformations (Roth 1991; Wagner 1995, 2001; Wagner and Altenberg 1996). Units of evolution may delimit such modules of evolutionary transformations.

The term *module* is, however, not only applied to units of constituents that *evolve* independently from each other in the chain of generations, but is also used to characterize units of constituents that *develop or operate* relatively independently from each other during the life cycle (in a single generation). It is often assumed that such modules of development or behavior are largely coincident with the modules of evolutionary transformations (e.g., Wagner 1996; Wagner and Altenberg 1996; Gilbert, Opitz, and Raff 1996; Raff 1996; Gerhart and Kirschner 1997; Hartwell et al. 1999; Bolker 2000; Gilbert 2006). Recognizing developmental or behavioral modules would then allow us to predict which constituents constrain each other's evolutionary modification. However, the validity of this assumption has been questioned (e.g., von Dassow and Munro 1999; Sterelny 2000). The purpose of this section is to show that it is likely to be true only under certain conditions. To elaborate on this argument, the notion of a module first needs to be briefly defined somewhat more precisely (for a detailed account see Schlosser 2002, 2004).

Modules Are Integrated and Relatively Context-Insensitive Subprocesses
Modules can be recognized at various levels of the biological hierarchy ranging from simple gene and protein networks to complex organs consisting of many interacting cells (e.g., Riedl 1975; Bonner 1988; Wagner 1995; Wagner and Altenberg 1996; Raff 1996; Gilbert, Opitz, and Raff 1996; García-Bellido 1996; Gerhart and Kirschner 1997; von Dassow and Munro 1999; Hartwell et al. 1999; Niehrs and Pollet 1999; Mendoza, Thieffry, and Alvarez-Buylla 1999; Thieffry and Romero 1999; von Dassow et al.

Functional and Developmental Constraints on Life-Cycle Evolution

2000; Schlosser and Thieffry 2000; Ancel and Fontana 2000; Bolker 2000; Schlosser 2001, 2002, 2004, 2005b; Schlosser and Wagner 2004; von Dassow and Meir 2004). From a process-oriented perspective, modules may be generally defined as those types of subprocesses of some containing process that are likely to operate in an integrated but relatively autonomous manner (Schlosser 2002, 2004). A subprocess is any pattern of interactions that is constitutive but not sufficient for the realization of a containing process, because it involves interactions among only a subset of all lower-level constituents of the process. A subprocess is *integrated* when the constituents of the subprocess do not merely additively contribute to its activity—that is, its input-output transformations (IOT)—but rather all cooperate in the generation of the IOT (due to their generative or functional coupling to at least some other constituents of the subprocess). A subprocess is *relatively autonomous* when its IOT is relatively insensitive to the context in which it operates, because constituents of the context do not contribute to the IOT (are not functionally coupled to constituents of the subprocess for the realization of the IOT; see figure 4.7).

A subprocess qualifies as a module only when it operates in an integrated and context-insensitive manner with high probability. And this probability depends on how robustly its defining constituents contribute to an integrated and context-insensitive IOT in the face of various perturbations. Hence, whether several constituents of any given process or system belong to a single module can only be analyzed (1) relative to some specified contribution of a subprocess to a containing process (i.e., some specified IOT), and (2) relative to a specified class of permitted perturbations of the interaction capacities of constituents. This is important because these specifications will differ, between modules of the development or behavior of life cycles and modules of their evolution.

In the former case, we look for the units of constituents that (1) make an integrated and context-insensitive contribution to some developmental process or some behavior in the face of (2) variations that do not need to be heritable (e.g., environmental perturbations or mutations in genes of somatic cells). However, in the case of modules of evolution, we look for the units of constituents that (1) make an integrated and context-insensitive contribution to their own reproduction in the next generation (i.e., an independent contribution to the stable perpetuation of a functional system across generation boundaries) in the face of (2) such variations that are heritable (e.g., mutations in genes of germ-line cells). Units of evolution define modules of this latter kind. The identity of a module or unit of evolution throughout lineages may be defined via its conserved contribution to its

Gerhard Schlosser

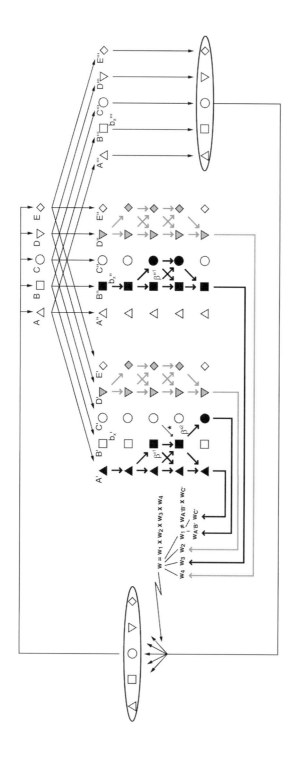

own reproduction even though there may be evolutionary replacements of some of its lower-level constituents (Roth 1991, 2001; Wagner 1995, 2001; Wagner and Altenberg 1996; von Dassow and Munro 1999; Wagner, Chiu, and Laubichler 2000; Schlosser 2002). There are many examples for this in cases where the genetic or cellular basis for homologous structures has changed (e.g., Striedter and Northcutt 1991; Dickinson 1995; Bolker and Raff 1996; Abouheif 1997).

Figure 4.7
Modules of development and modules of evolution. The figure schematically depicts a reproducing life cycle, as generally explained in figure 4.3 (environmental interactions are not shown here for simplicity). A–E (symbols) indicate various types of constituents of the life cycle that are replicated (involving the faithful multiplication of capacity variants) and therefore multiply instantiated during the life cycle (A′–E′, A″–E″, A‴–E‴). For example, A–E may represent different genes, which after cell divison are present in multiple cell types of a multicellular organism, one of which may represent a germ-line cell (incorporating A‴–E‴, also indicated by long ellipse), the others somatic cells (incorporating A′–E′ and A″–E″, respectively). Gene activity patterns in different cell types may differ as indicated. The initial differences in gene activity patterns among various cells may either be established via some stochastic or environmentally induced symmetry break or may be due to asymmetries in the activity of other constituents reproduced via germ cells (not shown).

Constituents that behave in an integrated but context-insensitive manner during development (i.e., are characterized by integrated and context-insensitive input-output transformations) form developmental modules. Three types of such developmental modules are illustrated: the interactions between D and E (instantiated in two domains D′–E′ and D″–E″), the interactions between A′ and B′, and the interactions between B″ and C″. Note that A′–B′ will only qualify as a module if it is relatively context insensitive—that is, when the input from C′ (asterisk) is weak. However, it may have strong outputs (β'^2) on C′ that may be indispensable for the activity of the latter. Developmental modules with such indispensable effects on constituents of the context may not act as modules of evolutionary transformations, because there is likely to be epistasis between the fitness w of A′B′ and C′. Furthermore, A′–B′ and B″–C″ may act as developmental modules, but not as modules of evolutionary transformations, because they are independently perturbable during development (e.g., interactions β'^1 and β''^1 may be separately affected by perturbation of b_x' and b_x'', respectively), but such independent perturbations may be impossible or less likely in the case of heritable variations (e.g., β'^1 and β''^1 may tend to be jointly affected by perturbation of b_x''') in germ-line cells. For the same reason, the two domains D′–E′ and D″–E″ will act as a single module of evolution, although they make independent (multiplicative) fitness contributions relative to developmental perturbations.

Modules of Development May Coincide with Modules (Units) of Evolution

Because developmental and evolutionary modules will both depend on the pattern of couplings between the constituents of a life cycle, modules of development and modules (units) of evolution will often coincide. However, due to the different IOTs and different classes of permitted variations for each type of module, this will not be the case under certain conditions, as illustrated in figure 4.7. First, some constituents A and B may act as an integrated and relatively context-insensitive unit with a robust IOT during development, even though they have strong and indispensable effects on other constituents C. Such modules may not act as modules with respect to evolutionary transformations (units of evolution) because their fitness may depend strongly on C and will, therefore, not be context insensitive (figure 4.7).

Second, a constituent A may have pleiotropic roles that may be independently perturbable during development (i.e., each role may be perturbed without pleiotropic effects on another role). For example, gene A may be expressed in several somatic cells, and gene activities α' and α'' in each of these expression domains A' and A'' may be perturbable independently from one another (e.g., by a mutation that alters capacity variable a' of A' in one somatic cell, without affecting a'' of A'' in another somatic cell). It is, therefore, possible that A' and A'' belong to two different developmental modules A'B' and A''C'', respectively. However, these two modules may not act as separate modules with respect to evolutionary transformations, because only heritable variations are relevant for the latter. But the class of possible heritable variations may be less inclusive than the class of possible developmental perturbations and may, for instance, either not allow for the independent variation of α' and α'' or make it less likely (e.g., heritable variation of α' may require a mutation altering capacity variable α''' of A''' in a germ-line cell, which may inadvertently affect α'' as well). Therefore, heritable variations of A' or A'' are more likely than developmental perturbations to have pleiotropic effects on B' as well as C''. By the same token, however, multiple domains (e.g., D'E' and D''E'') of the same developmental module will be tied together as a single unit of evolution, despite their independent developmental perturbability (figure 4.7).

Therefore, modules of development or behavior of some functional system will tend to coincide with the units of its evolution only to the extent that on the one hand the behavior of these modules is not only insensitive to, but also largely dispensable for the developmental context in which they are embedded, and on the other hand the effects of heritable variations tend to be congruent with the effects of developmental perturbations (similar to the "plastogenetic congruence" of Ancel and Fontana 2000) and do not tend to have a much wider scope for pleiotropic effects. Taken together with integration and context insensitivity of developmental/behavioral modules, these two conditions guarantee the modularity of the "genotype-phenotype map" (Riedl 1975; Cheverud 1984, 1996; Wagner 1996; Wagner and Altenberg 1996; Mezey, Cheverud, and Wagner 2000).

Although the available empirical data are clearly insufficient for a final evaluation, the following observations suggest that these conditions often seem to be met for life cycles, in particular for multicellular organisms. First, the modular architecture of these systems tends to be pervasive (e.g., Gerhart and Kirschner 1997). Therefore, modules may often be relatively dispensable for the context, because the latter is itself organized in a modular fashion. Second, different interactions of constituents subject to heritable variation will often act as lower-level modules themselves. Therefore, the interaction of a constituent A with one constituent B tends to be to some degree variable independently from its interactions with another constituent C, allowing for heritable variations without pleiotropic effects. For instance, the binding of a transcription factor B to the cis-regulatory region of gene A typically proceeds largely independently from the binding of another factor C to the cis-regulatory region of A, and mutations in different sections of the cis-regulatory regions of A (capacity variants a_B and a_C) may, therefore, differentially affect these interactions (e.g., Arnone and Davidson 1997; Ptashne and Gann 1998; Stern 2000).

There is still no consensus on how the prevalence of architectures with such a modular genotype-phenotype map can be explained, but several models for their evolutionary origin have been proposed, involving either selection on adaptation rate (Wagner 1981), clade selection (e.g., Gerhart and Kirschner 1997; Kirschner and Gerhart 1998), or the much more efficient mechanisms of stabilizing selection on functionally correlated constituents (Riedl 1975; Cheverud 1984, 1996; Wagner 1996; Wagner and Altenberg 1996) and selection for developmental robustness and flexibility (Conrad 1990; Gerhart and Kirschner 1997; Kirschner and Gerhart 1998; Barton and Partridge 2000).

The modular organization of living systems facilitates particular evolutionary transformations and hence contributes to their "evolvability" (e.g., Wagner and Altenberg 1996; Dawkins 1996; Gerhart and Kirschner 1997; Kirschner and Gerhart 1998; Brandon 1999) for two reasons (Schlosser 2002, 2004). First, it facilitates mosaic evolution, because it greatly reduces the number of constituents that constrain each other's evolutionary transformations. Constraints on evolutionary transformations of a constituent are mainly exerted by those (possibly few) other constituents with which it is strongly and multiply coupled, forming a module. Second, it facilitates the evolution of complexity, because it allows the combinatorial use of modules and the redeployment of modules in novel contexts. The reciprocal constraints among the constituents of a module ensure that they will behave in a similarly coordinated fashion in novel contexts. This can, for instance, explain why the same gene networks and signaling cascades have been redeployed over and over again for the development of a diverse array of tissues during metazoan evolution (for overviews see Raff 1996; Gerhart and Kirschner 1997; Schlosser 2004).

4.7 Testing Hypotheses about Constraints and Units of Evolution

I have argued in the last section that a better understanding of the developmental and behavioral modularity of living systems may frequently allow us to predict which constituents will form a unit of evolution with strong reciprocal constraints on evolutionary transformations. Fortunately, such predictions do not have to remain speculative, but can be tested by analyzing patterns of correlated changes in comparative studies. Conversely, such comparative studies are increasingly used as heuristic tools to formulate hypotheses about developmental modules, which can then be tested experimentally (Eisenberg et al. 2000; Vukmirovic and Tilghman 2000). Only the former approach will be sketched here briefly (see also Schlosser 2001, 2004, 2005b).

Repeated Dissociated Coevolution Indicates Units of Evolution in Comparative Studies

In the first step of such an analysis, experimental evidence from detailed developmental or physiological studies in a few model organisms is used to predict which constituents of a life cycle are likely to reciprocally constrain their evolutionary changes forming a unit of evolution. This may be a small network of genes or proteins or a complex organ consisting of many interacting cells like the vertebrate limb bud.

In a second step, certain assumptions about the time window of perseverance of this unit of evolution in phylogeny need to be made (McKitrick 1993; Schwenk 1994). This allows us to delimit the lineage in which we expect to find the same unit of evolution intact. These assumptions may be informed by our findings in other model organisms or by general biological knowledge. For example, detailed experimental evidence from studies in chicken and mouse embryos, together with a wealth of anatomical and embryological data on other species, suggests that similar patterns of interactions among genes and cells underlie limb development in all amniotes, and some of these patterns may even be shared with other tetrapods or gnathostomes (e.g., Shubin, Tabin, and Carroll 1997; Johnson and Tabin 1997; Ng et al. 1999).

In the third step, the hypothesis that a certain set of constituents acts as a unit of evolution in a certain taxon (e.g., certain gene networks or cell interactions in the limb bud in amniotes or in vertebrates) can be tested using comparative methods. This requires that an explicit phylogenetic tree of the taxon in question has been previously established by independent methods (e.g., Eldredge and Cracraft 1980; Wiley 1981; Raff 1996). For lineages in which a certain unit of evolution remains intact, we expect that coordinated evolutionary transformations (coevolution) of all constituents comprising a unit of evolution (including the special case of coordinated stasis discussed below) should be associated with uncoordinated evolutionary transformations (dissociation) in other constituents with a frequency that is significantly higher than would be expected by chance (the null hypothesis). The latter may be estimated based on assumptions on rates of evolutionary change in the lineage tested (e.g., Antonovics and van Tienderen 1991; Harvey and Pagel 1991; McKitrick 1993; Sanderson and Hufford 1996; Donoghue and Ree 2000). Such *dissociated coevolution* may involve internal as well as external constituents (environmental factors) and may take many forms. These forms include correlated/uncorrelated losses, redeployments in novel contexts, or shifts in relative timing (heterochrony), location (heterotopy), size, shape, or activity (Schlosser 2004, 2005b). A variety of comparative methods (e.g., Felsenstein 1985; Maddison 1990; Harvey and Pagel 1991; Losos and Miles 1994; Pagel 1999), such as outgroup comparison (Fink 1982; Kluge and Strauss 1985; Northcutt 1990; McKitrick 1993), are available to detect such independently recurring—that is, homoplasious—patterns of dissociated coevolution (figure 4.8).

Homoplasies, in particular the repeated evolution of similar traits in different environments, are widely recognized as evidence for constraints (e.g., Alberch and Gale 1985; Maynard Smith et al. 1985; Wake and Larson 1987; Wake 1991, 1996; Shubin and Wake 1996; Brooks 1996; Hufford 1996; Donoghue and Rhee 2000; Hodin 2000). It deserves to be emphasized that such evidence for constraints cannot be counted as evidence

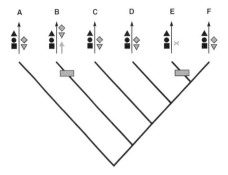

Figure 4.8
Detecting units of evolution using comparative methods. Repeated dissociated coevolution of constituents is revealed by mapping sequences of development of characters (symbols next to long arrows) on a phylogenetic tree of species A–F. Outgroup comparison among the species depicted allows us to infer two independent events (gray bars) of dissociated coevolution of the gray characters from the black characters: a coordinated heterochronic shift (short arrow) relative to other characters in B, and a coordinated loss (cross) in E.

against selection. Repeated dissociated coevolution of a unit in a lineage only suggests that a certain pattern of developmental and/or functional constraints persisted for a certain period of time, influencing which variants could be generated and what their fitness was, regardless of whether evolutionary changes involved selection or drift. With the exception of cases where units of evolution are tied together exclusively by developmental constraints, the observation of repeated dissociated coevolution of some unit even *supports* an important role of selection (which may, however, be due to fitness differences predominantly affected by internal constituents), because frequent homoplasies would not be expected if drift prevailed.

Evolutionary stasis (i.e., absence of change in some complex of constituents) is often cited as a different phenomenon that also indicates constraints (e.g., Wake, Roth, and Wake 1983; Maynard Smith et al. 1985; Schwenk 1994; Wagner and Schwenk 2000). However, stasis per se is uninformative, because just attesting the absence of change in some constituents does not allow us to delimit units of constraints. Only when relations between a set of constituents remain static or conserved despite repeated changes of their context do we have evidence that there are constraints among the constituents of the set but not between these constituents and their context. Therefore, stasis provides evidence for constraints

Functional and Developmental Constraints on Life-Cycle Evolution

only when it can be subsumed under the general pattern of dissociated coevolution.

Summary and Conclusion

Constraints on evolutionary transformations of a given life cycle arise from the facts (1) that due to heritable variations of its constituents only a limited number of modifications can be generated and that some modifications are more likely to be generated than others (developmental constraints), and (2) that these modifications differ in their probabilities (propensities) to maintain a functional organization and to be reproduced in another life cycle—that is, that they differ in fitness (functional constraints). Constraints therefore define boundary conditions (e.g., a probability distribution) for evolutionary transformations, while the populational dynamics of such events are either described by deterministic models (which assume that fitness differences are translated into actual reproductive rates) like selection or by probabilistic explanations (which accept drift). Explanations invoking constraints should, therefore, not be seen as opposed in any way to explanations invoking selection. Both types of explanations are truly complementary.

Consequently, the question whether constraints or selection make a larger contribution to channeling the direction of evolutionary transformations becomes obsolete. This allows us to focus attention on the more interesting question concerning the architecture of constraints for the evolutionary modifications of particular life cycles. While cooperation of all constituents of a functional life cycle is required for its maintenance and reproduction, not all constituents are coupled equally strongly with each other during that process. Therefore, variations of a constituent will not be equally strongly constrained by all other constituents. Rather there will be units of constituents (units of evolution) that are tied together by strong and reciprocal developmental constraints (high probabilities for interdependent variability) and functional constraints (high probability for epistatic fitness interaction among variants).

Only the constituents belonging to the same unit of evolution strongly constrain each other's evolutionary transformation. Hence, units of evolution act as modules of evolutionary transformations—that is, they reproduce in an integrated but relatively context-insensitive manner relative to a set of heritable variations. Under certain conditions, such modules of evolutionary transformations will coincide with modules of development or behavior (developmental or behavioral processes that operate in an integrated and context-insensitive fashion relative to a set of perturbations). Hypotheses about units of evolution may, therefore, be derived from experimental studies of developmental modules and can then be tested in comparative phylogenetic studies.

Despite iterated variation, the architecture of constraints (the units of evolution) of a lineage may remain stable for a certain period of time. During this time period the evolutionary transformations of the lineage are subject to these same constraints. However, constraints are clearly not immutable entities but can themselves evolve. Which constraints will be broken and which new constraints may evolve is—within the constrained space of permissible transformations—dependent on contingent events. Therefore, elucidating the architecture of constraints on variation in recent organisms may allow us to explain patterns of evolutionary transformations in the phylogenetic history of their lineages, but allows only local and short range prediction of future evolutionary trajectories. Such insights cannot be used to vindicate claims about global or insurmountable "constraints on adaptation" (Amundson 1994).

Acknowledgments

I am grateful to Günter Wagner and Roger Sansom for many helpful comments on this chapter.

Notes

1. Generally, transformations will be nondisruptive, when they are transformations between stable states, where the minimum requirement for any stable state is that a state of the same type is causally reachable from it in a finite number of transformations (i.e., the stable state is part of an attractor). Note that the notion of stability here also applies to cases of dynamic stability (i.e., to states that participate in a cycle of state transformations).

2. Re-production here refers to the general phenomenon of recurrent production of some state rather than to the multiplicative reproduction of progeny.

3. In multicellular organisms in particular, life-cycle reproduction often involves the reproduction of development—that is, of a more or less invariant sequence of states preceding reproduction in each progeny (Bonner 1974, 1988; Buss 1987; Maynard Smith and Szathmáry 1995; Griesemer 2000).

4. Two broad classes of functionally permissible transformations (e.g., of gene or protein interaction patterns) can be distinguished: (1) transformations due to nonherit-

Functional and Developmental Constraints on Life-Cycle Evolution

able perturbations (e.g., changes of protein activities due to environmental factors or changes of protein interaction capacities due to gene mutations in somatic cells), the effects of which are restricted to a single generation, and (2) transformations due to heritable variations (e.g., changes of protein interaction capacities due to gene mutations in germ cells) that have effects across generation boundaries. Regarding the first, only those perturbations are permitted that do not disrupt self-maintenance. This implies for every self-maintaining system a set of permissible environments. Regarding the second, only those heritable variations are permitted that do not disrupt transgenerational reproduction (implying that they also do not disrupt self-maintenance). Because the focus of this chapter is on evolutionary changes of life cycles, and these require heritable variations, only this second category will be considered further here.

5. Note that this is true regardless of whether the new variant spreads by selection or not, as will be discussed in more detail later.

6. Frequency-dependent selection is compatible with the last condition but complicates its formalization (see Schlosser 2002).

7. Because constraints may be imposed by external in addition to internal constituents, "internal selection" may in fact often not be *purely* internal, but may differ from "external selection" only regarding the *predominant* source of functional constraints.

8. Of course, this does not preclude that in explanations of why particular variants are present in certain frequencies or absent from a given population, the explanans may have to quote both constraints (to account for different fitness values of variants or to account for the fact that particular variants are not generated) and evolutionary dynamics (to account for the fate of variants with different fitness in a population of particular size and structure).

9. It has been argued that this kind of overlap of functional and generative couplings will tend to increase during evolution (e.g., Riedl 1975; Cheverud 1984, 1996; Wagner 1996; Wagner and Altenberg 1996): because functional coupling of two constituents will result in relative fitness advantages of variants that reduce the probability of disruptive noncoordinated perturbations, the evolution of particular kinds of developmental couplings (ensuring coordinated perturbation) between these constituents will be favored. Functional constraints therefore may favor the evolution of developmental constraints in a process known as stabilizing or canalizing selection (Schmalhausen 1949; Waddington 1957; Gibson and Wagner 2000).

10. Gradual shifts of fitness, however, may occur in case of frequency-dependent selection (see, e.g., Sober 1984; Maynard Smith 1989; Ridley 1993; Michod 1999).

11. Nondecomposable units of selection, as defined here, will exhibit nondecomposable additive variance for fitness, in accordance with Wimsatt's (1980, 1981; see also

Lloyd 1988) definition. However, Wimsatt and other adherents of this variance approach tend to link additive variance closely with the requirement for reliable fitness reproduction (context insensitivity for fitness), underemphasizing that reliability of variant reproduction is equally required and that both requirements tend to conflict (for detailed discussion see Schlosser 2002).

References

Abouheif, E. 1997. Developmental genetics and homology: A hierarchical approach. *Trends Ecol. Evolut.* 12: 405–408.

Alberch, P. 1980. Ontogenesis and morphological diversification. *Am. Zool.* 20: 653–667.

Alberch, P. 1982. The generative and regulatory roles of development in evolution. In D. Mossakowski and G. Roth, eds., *Environmental Adaptation and Evolution*, 19–36. Stuttgart: Fischer.

Alberch, P. 1989. The logic of monsters: Evidence for internal constraint in development and evolution. *Geobios* (*mem. spec.*) 12: 21–57.

Alberch, P., and E. A. Gale. 1983. Size dependence during the development of the amphibian foot: Colchicine-induced digital loss and reduction. *J. Embryol. Exp. Morph.* 76: 177–197.

Alberch, P., and E. A. Gale. 1985. A developmental analysis of an evolutionary trend: Digital reduction in amphibians. *Evolution* 39: 8–23.

Alexander, R. M. 1985. The ideal and the feasible: Physical constraints on evolution. *Biol. J. Linn. Soc.* 26: 345–358.

Amundson, R. 1994. The concept of constraint: Adaptationism and the challenge from developmental biology. *Phil. Sci.* 61: 556–578.

Ancel, L. W., and W. Fontana. 2000. Plasticity, evolvability and modularity in RNA. *J. Exp. Zool.* (*Mol. Dev. Evol.*) 288: 242–283.

Antonovics, J., and P. H. van Tienderen. 1991. Ontoecogenophyloconstraints? The chaos of constraint terminology. *Trends Ecol. Evol.* 6: 166–168.

Arnone, M. I., and E. H. Davidson. 1997. The hardwiring of development: Organization and function of genomic regulatory systems. *Development* 124: 1851–1864.

Arthur, W. 1997. *The Origin of Animal Body Plans*. Cambridge: Cambridge University Press.

Atchley, W. R., D. E. Cowley, C. Vogl, and T. McLellan. 1992. Evolutionary divergence, shape change and genetic correlation structure in the rodent mandible. *Syst. Biol.* 41: 196–221.

Baker, C. V. H., and M. Bronner-Fraser. 1997. The origins of the neural crest, Part II: An evolutionary perspective. *Mech. Dev.* 69: 13–29.

Barton, N., and L. Partridge. 2000. Limits to natural selection. *Bioessays* 22: 1075–1084.

Barton, R. A., and P. H. Harvey. 2000. Mosaic evolution of brain structure in mammals. *Nature* 405: 1055–1058.

Bernstein, H., H. C. Byerly, F. A. Hopf, R. E. Michod, and G. K. Vemulapalli. 1983. The Darwinian dynamic. *Quart. Rev. Biol.* 58: 185–207.

Blader, P., N. Fischer, G. Gradwohl, F. Guillemot, and U. Stähle. 1997. The activity of neurogenin1 is controlled by local cues in the zebrafish embryo. *Development* 124: 4557–4569.

Bock, W. J. and G. von Wahlert. 1965. Adaptation and the form-function complex. *Evolution* 19: 269–299.

Bolker, J. A. 2000. Modularity in development and why it matters to evo-devo. *Am. Zool.* 40: 770–776.

Bolker, J. A., and R. A. Raff. 1996. Developmental genetics and traditional homology. *Bioessays* 18: 489–494.

Bonner, J. T. 1974. *On Development.* Cambridge: Harvard University Press.

Bonner, J. T. 1988. *The Evolution of Complexity.* Princeton, NJ: Princeton University Press.

Brandon, R. N. 1990. *Adaptation and Environment.* Princeton, NJ: Princeton University Press.

Brandon, R. N. 1999. The units of selection revisited: The modules of selection. *Biology and Philosophy* 14: 167–180.

Brandon, R. N., and R. M. Burian. 1984. *Genes, Organisms, Populations: Controversies over the Units of Selection.* Cambridge: MIT Press.

Brooks, D. R. 1996. Explanations of homoplasy at different levels of biological organization. In M. J. Sanderson and L. Hufford, eds., *Homoplasy: The recurrence of similarity in evolution*, 3–36. San Diego: Academic Press.

Brunet, J. F., and A. Ghysen. 1999. Deconstructing cell determination: Proneural genes and neuronal identity. *Bioessays* 21: 313–318.

Buss, L. W. 1987. *The Evolution of Individuality.* Princeton, NJ: Princeton University Press.

Cheverud, J. M. 1984. Quantitative genetics and developmental constraints on evolution by selection. *J. Theor. Biol.* 110: 155–171.

Cheverud, J. M. 1996. Developmental integration and the evolution of pleiotropy. *Am. Zool.* 36: 44–50.

Cheverud, J. M. 2001. The genetic architecture of pleiotropic relations and differential epistasis. In G. P. Wagner, ed., *The Character Concept in Evolutionary Biology*, 411–433. San Diego: Academic Press.

Cheverud, J. M., and E. J. Routman. 1995. Epistasis and its contribution to genetic variance components. *Genetics* 139: 1455–1461.

Conrad, M. 1990. The geometry of evolution. *Biosystems* 24: 61–81.

Davidson, E. H. 2001. *Genomic Regulatory Systems*. San Diego: Academic Press.

Davis, A. P., and M. R. Capecchi. 1996. A mutational analysis of the 5′ HoxD genes: Dissection of genetic interactions during limb development in the mouse. *Development* 122: 1175–1185.

Dawkins, R. 1976. *The Selfish Gene*. Oxford: Oxford University Press.

Dawkins, R. 1982. *The Extended Phenotype*. Oxford: Oxford University Press.

Dawkins, R. 1996. *Climbing Mount Improbable*. New York: Norton.

Dennett, D. 1995. *Darwin's Dangerous Idea*. New York: Touchstone.

Dickinson, W. J. 1995. Molecules and morphology: Where's the homology. *Trends Genet.* 11: 119–121.

Donoghue, M. J., and R. H. Ree. 2000. Homoplasy and developmental constraint: A model and an example from plants. *Am. Zool.* 40: 759–769.

Dover, G. 2000. How genomic and developmental dynamics affect evolutionary processes. *Bioessays* 22: 1153–1159.

Duboule, D., and A. S. Wilkins. 1998. The evolution of bricolage. *Trends Genet.* 14: 54–59.

Dullemeijer, P. 1980. Functional morphology and evolutionary biology. *Acta Biotheor.* 29: 151–250.

Eigen, M., W. Gardiner, P. Schuster, and R. Winkler-Oswatitsch. 1981. The origin of genetic information. *Sci. Am.* 244(4): 88–118.

Eisen, M. B., P. T. Spellman, P. O. Brown, and D. Botstein. 1998. Cluster analysis and display of genome-wide expression patterns. *Proc. Natl. Acad. Sci. USA* 95: 14863–14868.

Eisenberg, D., E. M. Marcotte, I. Xenarios, and T. O. Yeates. 2000. Protein function in the postgenomic area. *Nature* 405: 823–826.

Eldredge, N., and J. Cracraft. 1980. *Phylogenetic Patterns and the Evolutionary Process*. New York: Columbia University Press.

Felsenstein, J. 1985. Phylogenies and the comparative method. *Am. Nat.* 125: 1–15.

Fink, W. L. 1982. The conceptual relationship between ontogeny and phylogeny. *Paleobiol.* 8: 254–264.

Fink, W. L., and M. L. Zelditch. 1995. Phylogenetic analysis of ontogenetic shape transformations: A reassessment of the Piranha genus *Pygocentrus* (Teleostei). *Syst. Biol.* 44: 343–360.

Fontana, W., and L. Buss. 1994a. "The arrival of the fittest": Toward a theory of biological organization. *Bull. Math. Biol.* 56: 1–64.

Fontana, W., and L. Buss. 1994b. What would be conserved if "the tape were played twice"? *Proc. Natl. Acad. Sci. USA* 91: 757–761.

Fontana, W., G. Wagner, and L. W. Buss. 1995. Beyond digital naturalism. In C. G. Langton, ed., *Artificial Life*. Cambridge, MA: MIT Press.

Force, A., M. Lynch, F. B. Pickett, A. Amores, Y.-L. Yan, and J. Postlethwait. 1999. Preservation of duplicate genes by complementary degenerative mutations. *Genetics* 151: 1531–1545.

Galis, F. 1999. Why do almost all mammals have seven cervical vertebrae? Developmental constraints, Hox genes, and cancer. *J. Exp. Zool. (Mol. Dev. Evol.)* 285: 19–26.

Galis, F. 2001. Key innovations and radiations. In G. P. Wagner, ed., *The Character Concept in Evolutionary Biology*, 581–605. San Diego: Academic Press.

García-Bellido, A. 1996. Symmetries throughout organic evolution. *Proc. Natl. Acad. Sci. USA* 93: 14229–14232.

Gerhart, J., and J. Kirschner. 1997. *Cells, Embryos, and Evolution*. Malden, MA: Blackwell Science.

Gibson, G., and G. Wagner. 2000. Canalization in evolutionary genetics: A stabilizing theory? *Bioessays* 22: 372–380.

Gilbert, S. F. 2006. *Developmental Biology*. 8th ed. Sunderland, MA: Sinauer.

Gilbert, S. F., J. M. Opitz, and R. A. Raff. 1996. Resynthesizing evolutionary and developmental biology. *Dev. Biol.* 173: 357–372.

Goh, C. S., A. A. Bogan, M. Joachimiak, D. Walther, and F. E. Cohen. 2000. Co-evolution of proteins with their interaction partners. *J. Mol. Biol.* 299: 283–293.

Gould, S. J. 1982. Darwinism and the expansion of the evolutionary theory. *Science* 216: 380–387.

Gould, S. J. 1989a. A developmental constraint in *Cerion*, with comments on the definition and integration of constraint in evolution. *Evolution* 43: 516–539.

Gould, S. J. 1989b. *Wonderful Life*. London: Hutchinson.

Gould, S. J., and R. C. Lewontin. 1979. The spandrels of San Marco and the Panglossian paradigm: A critique of the adaptationist programme. *Proc. R. Soc. Lond.* B 205: 581–598.

Gray, R. 1992. Death of the gene: Developmental systems strike back. In P. Griffiths, ed., *Trees of Life*, 165–209. Dordrecht: Kluwer.

Griesemer, J. 2000. Development, culture, and the units of inheritance. *Phil. Sci. (Suppl.)* 67: S348–S368.

Griffiths, P. E., and R. D. Gray. 1994. Developmental systems and evolutionary explanation. *J. Philos.* 91: 277–304.

Guillemot, F. 1999. Vertebrate bHLH genes and the determination of neuronal fates. *Exp. Cell Res.* 253: 357–364.

Gutmann, W. F. 1977. Phylogenetic reconstruction: Theory, methodology, and application to chordate evolution. In M. K. Hecht, P. C. Goody, and B. M. Hecht, eds., *Major Patterns in Vertebrate Evolution*, 645–669. New York: Plenum Press.

Hall, B. K. 1996. Baupläne, phylotypic stages, and constraint: Why are there so few types of animals? *Evol. Biol.* 29: 215–261.

Hall, B. K. 2000. Evo-devo or devo-evo—does it matter ? *Evol. Dev.* 2: 177–178.

Hammerschmidt, M., A. Brook, and A. P. McMahon. 1997. The world according to hedgehog. *Trends Genet.* 13: 14–21.

Hansen, T. F., and G. P. Wagner. 2001. Modeling genetic architecture: A multilinear theory of gene interaction. *Theor. Pop. Biol.* 59: 61–86.

Hartwell, L. H., J. J. Hopfield, S. Leibler, and A. W. Murray. 1999. From molecular to modular cell biology. *Nature* 402 Suppl.: C47–C52.

Harvey, P. H., and M. D. Pagel. 1991. *The Comparative Method in Evolutionary Biology*. Oxford: Oxford University Press.

Hodin, J. 2000. Plasticity and constraints in development and evolution. *J. Exp. Zool. (Mol. Dev. Evol.)* 288: 1–20.

Hofbauer, J., and K. Sigmund. 1984. *Evolutionstheorie und dynamische Systeme*. Berlin: Parey.

Holder, N. 1983. Developmental constraints and the evolution of vertebrate digit patterns. *J. Theor. Biol.* 104: 451–471.

Hufford, L. 1996. Ontogenetic evolution, clade diversification, and homoplasy. In M. J. Sanderson and L. Hufford, eds., *Homoplasy: The Recurrence of Similarity in Evolution*, 271–301. San Diego: Academic Press.

Jacob, F. 1977. Evolution and tinkering. *Science* 196: 1161–1166.

Johnson, R. L., and C. J. Tabin. 1997. Molecular models for vertebrate limb development. *Cell* 90: 979–990.

Kauffman, S. A. 1993. *The Origins of Order.* New York: Oxford University Press.

Kirschner, M., and J. Gerhart. 1998. Evolvability. *Proc. Natl. Acad. Sci. USA* 95: 8420–8427.

Kluge, A. G., and R. E. Strauss. 1985. Ontogeny and systematics. *Annu. Rev. Ecol. Syst.* 16: 247–268.

Lande, R. 1980. The genetic covariance between characters maintained by pleiotropic mutations. *Genetics* 94: 203–215.

Laubichler, M. D., and G. P. Wagner. 2000. Organism and character decomposition: Steps towards an integrative theory of biology. *Phil. Sci. (Suppl.)* 67: S289–S300.

Lauder, G. C. 1981. Form and function: Structural analysis in evolutionary morphology. *Paleobiol.* 7: 430–442.

Lee, J. E. 1997. Basic helix-loop-helix genes in neural development. *Curr. Opin. Neurobiol.* 7: 13–20.

Lee, J. E., S. M. Hollenberg, L. Snider, D. L. Turner, N. Lipnick, and H. Weintraub. 1995. Conversion of *Xenopus* ectoderm into neurons by NeuroD, a basic helix-loop-helix protein. *Science* 268: 836–844.

Lewontin, R. C. 1970. The units of selection. *Annu. Rev. Ecol. Syst.* 1: 1–18.

Lewontin, R. C. 1974. *The Genetic Basis of Evolutionary Change.* New York: Columbia University Press.

Lewontin, R. C. 1978. Adaptation. *Sci. Am.* 239/3: 156–169.

Lewontin, R. C. 1983. The organism as the subject and object of evolution. *Scientia* 118: 65–82.

Lewontin, R. 2001. Foreword. In G. P. Wagner, ed., *The Character Concept in Evolutionary Biology*, xvii–xxiii. San Diego: Academic Press.

Liem, K. F. 1991. A functional approach to the development of the head of teleosts: Implications on constructional morphology and constraints. In N. Schmidt-Kittler and K. Vogler, eds., *Constructional Morphology and Evolution*, 231–249. Berlin: Springer.

Lloyd, E. 1988. *The Structure and Confirmation of Evolutionary Theory.* Princeton, NJ: Princeton University Press.

Lockhart, D. J., and E. A. Winzeler. 2000. Genomics, gene expression and DNA arrays. *Nature* 405: 827–836.

Losos, J. B., and D. B. Miles. 1994. Adaptation, constraints, and the comparative method: Phylogenetic issues and methods. In P. C. Wainwright and S. M. Reilly, eds., *Ecological Morphology*, 60–98. Chicago: University of Chicago Press.

Ludwig, M. Z., C. Bergman, N. H. Patel, and M. Kreitman. 2000. Evidence for stabilizing selection in a eukaryotic enhancer element. *Nature* 403: 564–567.

Ma, Q. F., C. Kintner, and D. J. Anderson. 1996. Identification of neurogenin, a vertebrate neuronal determination gene. *Cell* 87: 43–52.

Maddison, W. P. 1990. A method for testing the correlated evolution of two binary characters: Are gains or losses concentrated on certain branches of a phylogenetic tree? *Evolution* 44: 539–557.

Maturana, H. R., and F. J. Varela. 1975. *Autopoietic Systems: A Characterization of the Living Organization*. Report 9.4. Urbana: Biological Computer Laboratory, University of Illinois.

Maynard Smith, J. 1987. How to model evolution. In J. Dupré, ed., *The Latest on the Best*, 119–131. Cambridge, MA: MIT Press.

Maynard Smith, J. S. 1989. *Evolutionary Genetics*. New York: Oxford University Press.

Maynard Smith, J. 1998. The units of selection. *Novartis Foundation Symp.* 213: 203–217.

Maynard Smith, J., R. Burian, S. Kauffman, P. Alberch, J. Campbell, B. Goodwin, R. Lande, D. Raup, and L. Wolpert. 1985. Developmental constraints and evolution. *Quart. Rev. Biol.* 60: 265–287.

Maynard Smith, J., and E. Szathmáry. 1995. *The Major Transitions in Evolution*. Oxford: Freeman.

McKitrick, M. C. 1993. Phylogenetic constraint in evolutionary theory: Has it any explanatory power? *Annu. Rev. Evol. Syst.* 24: 307–330.

Mendoza, L., D. Thieffry, and E. R. Alvarez-Buylla, 1999. Genetic control of flower morphogenesis in *Arabidopsis thaliana:* A logical analysis. *Bioinformatics* 15: 593–606.

Mezey, J. G., J. M. Cheverud, and G. P. Wagner. 2000. Is the genotype-phenotype map modular?: A statistical approach using mouse quantitative trait loci data. *Genetics* 156: 305–311.

Michod, R. E. 1999. *Darwinian Dynamics*. Princeton, NJ: Princeton University Press.

Müller, G. B., and G. P. Wagner. 1991. Novelty in evolution: restructuring the concept. *Annu. Rev. Ecol. Syst.* 22: 229–256.

Müller, G. B., and G. P. Wagner. 1996. Homology, Hox genes, and developmental integration. *Am. Zool.* 36: 4–13.

Murone, M., A. Rosenthal, and F. J. deSauvage. 1999. Hedgehog signal transduction: From flies to vertebrates. *Exp. Cell Res.* 253: 25–33.

Newman, S. A. 1992. Generic physical mechanisms of morphogenesis and pattern formation as determinants in the evolution of multicellular organization. *J. Biosci.* 17: 193–215.

Ng, J. K., K. Tamura, D. Buscher, and J. C. Izpisua-Belmonte. 1999. Molecular and cellular basis of pattern formation during vertebrate limb development. *Curr. Topics Dev. Biol.* 41: 37–66.

Niehrs, C., and N. Pollet. 1999. Synexpression groups in eukaryotes. *Nature* 402: 483–487.

Northcutt, R. G. 1990. Ontogeny and phylogeny: A re-evaluation of conceptual relationships and some applications. *Brain Behav. Evol.* 36: 116–140.

Northcutt, R. G., and C. Gans. 1983. The genesis of neural crest and epidermal placodes: A reinterpretation of vertebrate origins. *Q. Rev. Biol.* 58: 1–28.

Ohno, S. 1970. *Evolution by Gene Duplication.* New York: Springer.

Oyama, S. 1985. *The Ontogeny of Information.* Cambridge: Cambridge University Press.

Pagel, M. 1999. Inferring the historical patterns of biological evolution. *Nature* 401: 877–884.

Perron, M., K. Opdecamp, K. Butler, W. A. Harris, and E. J. Bellefroid. 1999. X-ngnr-1 and Xath3 promote ectopic expression of sensory neuron markers in the neurula ectoderm and have distinct inducing properties in the retina. *Proc. Natl. Acad. Sci. USA* 96: 14996–15001.

Ptashne, M., and A. Gann. 1997. Transcriptional activation by recruitment. *Nature* 386: 569–577.

Ptashne, M., and A. Gann. 1998. Imposing specificity by localization: Mechanism and evolvability. *Curr. Biol.* 8: R812–R822.

Purugganan, M. D. 2000. The molecular population genetics of regulatory genes. *Mol. Ecol.* 9: 1451–1461.

Raff, R. A. 1996. *The Shape of Life.* Chicago: University of Chicago Press.

Rasskin-Gutman, D. 2001. Boundary conditions for the emergence of form. In G. Müller and S. Newman, eds., *Origins of Organismal Form.* Cambridge, MA: MIT Press.

Raup, D. M. 1966. Geometric analysis of shell coiling: General problems. *J. Paleontol.* 40: 1178–1190.

Resnik, D. 1996. Developmental constraints and patterns: Some pertinent distinctions. *J. Theor. Biol.* 173: 231–240.

Ridley, M. 1993. *Evolution*. Cambridge, MA: Blackwell.

Riedl, R. 1975. *Die Ordnung des Lebendigen*. Hamburg: Parey.

Romer, A. S. 1977. *The Vertebrate Body*. Philadelphia: Saunders.

Roth, G., and D. B. Wake. 1985. Trends in the functional morphology and sensorimotor control of feeding behavior in salamanders: An example of the role of internal dynamics in evolution. *Acta Biotheor.* 34: 175–192.

Roth, G., and D. B. Wake. 1989. Conservatism and innovation in the evolution of feeding in vertebrates. In D. B. Wake and G. Roth, eds., *Complex Organismal Functions: Integration and Evolution in Vertebrates*, 7–21. Chichester: Wiley.

Roth, V. L. 1991. Homology and hierarchies: Problems solved and unresolved. *J. Evol. Biol.* 4: 167–194.

Roth, V. L. 2001. Character replication. In G. P. Wagner, ed., *The Character Concept in Evolutionary Biology*, 81–107. San Diego: Academic Press.

Sanderson, M. J., and L. Hufford. 1996. Homoplasy and the evolutionary process. In M. J. Sanderson and L. Hufford, eds., *Homoplasy: The Recurrence of Similarity in Evolution*, 327–330. San Diego: Academic Press.

Schank, J. C., and W. C. Wimsatt. 1986. Generative entrenchment and evolution. *PSA* 2: 33–60.

Schank, J. C., and W. C. Wimsatt. 2001. Evolvability: Adaptation and modularity. In R. S. Singh, C. B. Krimbas, D. Paul, and J. Beatty, eds., *Thinking about Evolution*, 322–335. Cambridge: Cambridge University Press.

Schlosser, G. 1993. *Einheit der Welt und Einheitswissenschaft: Grundlegung einer Allgemeinen Systemtheorie*. Braunschweig: Vieweg.

Schlosser, G. 1996. Der Organismus—eine Fiktion? In H. J. Rheinberger and M. Weingarten, eds., *Jahrbuch für Geschichte und Theorie der Biologie III*, 75–92. Berlin: Verlag für Wissenschaft und Bildung.

Schlosser, G. 1998. Self-re-production and functionality: A systems-theoretical approach to teleological explanation. *Synthese* 116: 303–354.

Schlosser, G. 2001. Using heterochrony plots to detect the dissociated coevolution of characters. *J. Exp. Zool. (Mol. Dev. Evol.)* 291: 282–304.

Schlosser, G. 2002. Modularity and the units of evolution. *Theory. Biosci.* 121: 1–79.

Schlosser, G. 2004. Modules—Developmental units as units of evolution? In G. Schlosser and G. P. Wagner, eds., *Modularity in Development and Evolution*, 519–582. Chicago: University of Chicago Press.

Schlosser, G. 2005a. Evolutionary origins of vertebrate placodes: Insights from developmental studies and from comparisons with other deuterostomes. *J. Exp. Zool. Part B Mol. Dev. Evol.* 304: 347–399.

Schlosser, G. 2005b. Amphibian variations—the role of modules in mosaic evolution. In D. Rasskin-Gutman and W. Callebaut, eds., *Modularity: Understanding the Development and Evolution of Complex Natural Systems.* Cambridge, MA: MIT Press.

Schlosser, G., and D. Thieffry. 2000. Modularity in development and evolution. *Bioessays* 22: 1043–1045.

Schlosser, G., and G. P. Wagner. 2004. *Modularity in Development and Evolution.* Chicago: University of Chicago Press.

Schmalhausen, I. I. 1949. *Factors of Evolution: The Theory of Stabilizing Selection.* Philadelphia: Blakiston.

Schwenk, K. 1994. A utilitarian approach to evolutionary constraint. *Zoology* 98: 251–262.

Schwenk, K. 2001. Functional units and their evolution. In G. P. Wagner, ed., *The Character Concept in Evolutionary Biology*, 165–198. San Diego: Academic Press.

Shubin, N. 1998. Evolutionary cut and paste. *Nature* 394: 12–13.

Shubin, N., C. Tabin, and S. Carroll. 1997. Fossils, genes and the evolution of animal limbs. *Nature* 388: 639–648.

Shubin, N., and D. Wake. 1996. Phylogeny, variation, and morphological integration. *Am. Zool.* 36: 51–60.

Sidow, A. 1996. Gen(om)e duplications in the evolution of early vertebrates. *Curr. Opin. Genet. Develop.* 6: 715–722.

Sober, E. 1981. Holism, individualism and the units of selection. *PSA* 93–121.

Sober, E. 1984. *The Nature of Selection.* Chicago: University of Chicago Press.

Sober, E. 1987. What is adaptationism? In J. Dupré, ed., *The Latest on the Best*, 105–118. Cambridge, MA: MIT Press.

Sober, E., and R. C. Lewontin. 1982. Artifact, cause and genic selection. *Philos. Science* 49: 157–180.

Sober, E., and D. S. Wilson. 1998. *Unto Others.* Cambridge, MA: Harvard University Press.

Stadler, B. M., P. F. Stadler, G. P. Wagner, and W. Fontana. 2001. The topology of the possible: Formal spaces underlying patterns of evolutionary change. *J. Theor. Biol.* 213: 241–274.

Sterelny, K. 2000. Development, evolution, and adaptation. *Phil. Sci. (Suppl.)* 67: S369–S387.

Sterelny, K., and P. Kitcher. 1988. The return of the gene. *J. Philos.* 85: 339–361.

Stern, D. L. 2000. Evolutionary developmental biology and the problem of variation. *Evolution* 54: 1079–1091.

Strathmann, R. R. 2000. Functional design in the evolution of embryos and larvae. *Semin. Cell Dev. Biol.* 11: 395–402.

Striedter, G. F., and R. G. Northcutt. 1991. Biological hierarchies and the concept of homology. *Brain Behav. Evol.* 38: 177–189.

Szathmáry, E. 1995. A classification of replicators and lambda-calculus models of biological organization. *Proc. R. Soc. Lond.* B 260: 279–286.

Szathmáry, E., and J. Maynard Smith. 1995. The major evolutionary transitions. *Nature* 374: 227–232.

Szathmáry, E., and J. Maynard Smith. 1997. From replicators to reproducers: The first major transitions leading to life. *J. Theor. Biol.* 187: 555–571.

Thieffry, D., and D. Romero. 1999. The modularity of biological regulatory networks. *Biosystems* 50: 49–59.

Thomas, R. D. K., and W. E. Reif. 1993. The skeleton space—A finite set of organic designs. *Evolution* 47: 341–360.

Uexküll, J. v. 1928. *Theoretische Biologie.* Berlin: Springer.

Varela, F., H. R. Maturana, and R. B. Uribe. 1974. Autopoiesis: The organization of living systems, its characterization and a model. *Biosystems* 5: 187–196.

Von Dassow, G., and E. Meir. 2004. Exploring modularity with dynamical models of gene networks. In G. Schlosser and G. P. Wagner, eds., *Modularity in Development and Evolution,* 244–287. Chicago: University of Chicago Press.

Von Dassow, G., E. Meir, E. M. Munro, and G. M. Odell. 2000. The segment polarity network is a robust development module. *Nature* 406: 188–192.

Von Dassow, G., and E. Munro. 1999. Modularity in animal development and evolution: Elements of a conceptual framework for EvoDevo. *J. Exp. Zool. (Mol. Dev. Evol.)* 285: 307–325.

Vukmirovic, O. G., and S. M. Tilghman. 2000. Exploring genome space. *Nature* 405: 820–822.

Waddington, C. H. 1957. *The Strategy of the Genes.* London: George Allen and Unwin.

Wagner, G. P. 1981. Feedback selection and the evolution of modifiers. *Acta Biotheor.* 30: 79–102.

Wagner, G. P. 1988. The influence of variation and of developmental constraints on the rate of multivariate phenotypic evolution. *J. Evol. Biol.* 1: 45–66.

Wagner, G. P. 1995. The biological role of homologues: A building block hypothesis. *N. Jb. Geol. Paläont. Abh.* 19: 36–43.

Wagner, G. P. 1996. Homologues, natural kinds and the evolution of modularity. *Am. Zool.* 36: 36–43.

Wagner, G. P. 2001. Characters, units and natural kinds: An introduction. In G. P. Wagner, ed., *The Character Concept in Evolutionary Biology*, 1–10. San Diego: Academic Press.

Wagner, G. P., and L. Altenberg. 1996. Perspective—complex adaptations and the evolution of evolvability. *Evolution* 50: 967–976.

Wagner, G. P., C. H. Chiu, and M. Laubichler. 2000. Developmental evolution as a mechanistic science: The inference from developmental mechanisms to evolutionary processes. *Am. Zool.* 40: 819–831.

Wagner, G. P., and M. D. Laubichler. 2000. Character identification in evolutionary biology: The role of the organism. *Theory Biosci.* 119: 20–40.

Wagner, G. P., M. D. Laubichler, and H. Bagheri-Chaichian. 1998. Genetic measurement theory of epistatic effects. *Genetica* 102/103: 569–580.

Wagner, G. P., and A. Mezey. 2000. Modeling the evolution of genetic architecture: A continuum of alleles model with pairwise A × A epistasis. *J. Theor. Biol.* 203: 163–175.

Wagner, G. P., and B. Y. Misof. 1993. How can a character be developmentally constrained despite variation in developmental pathways? *J. Evol. Biol.* 6: 449–455.

Wagner, G. P., and K. Schwenk. 2000. Evolutionarily stable configurations: Functional integration and the evolution of phenotype stability. *Evol. Biol.* 31: 155–217.

Wagner, G. P., and P. F. Stadler. 2003. Quasi-independence, homology and the unity of type: A topological theory of characters. *J. Theor. Biol.* 220: 505–527.

Wake, D. B. 1982. Functional and developmental constraints and opportunities in the evolution of feeding systems in urodeles. In D. Mossakowski and G. Roth, eds., *Environmental Adaptation and Evolution*, 51–66. Stuttgart: Fischer.

Wake, D. B. 1991. Homoplasy: The result of natural selection, or evidence of design limitations? *Am. Nat.* 138: 543–567.

Wake, D. B. 1996. Introduction. In M. J. Sanderson and L. Hufford, eds., *Homoplasy: The Recurrence of Similarity in Evolution*, xvii–xxv. San Diego: Academic Press.

Wake, D. B., and A. Larson. 1987. Multidimensional analysis of an evolving lineage. *Science* 238: 42–48.

Wake, D. B., and G. Roth. 1989. The linkage between ontogeny and phylogeny in the evolution of complex systems. In D. B. Wake and G. Roth, eds., *Complex Organismal Functions: Integration and Evolution in Vertebrates*, 361–377. Chichester: Wiley.

Wake, D. B., G. Roth, and M. H. Wake. 1983. On the problem of stasis in organismal evolution. *J. Theor. Biol.* 101: 211–224.

Waters, K. 1991. Tempered realism about the force of selection. *Philos. Science* 58: 553–573.

Webster, G., and B. Goodwin. 1996. *Form and Transformation.* Cambridge: Cambridge University Press.

Wen, X., S. Fuhrman, G. S. Michaels, D. B. Carr, S. Smith, J. L. Barker, and R. Somogyi. 1998. Large-scale temporal expression mapping of central nervous system development. *Proc. Natl. Acad. Sci. USA* 95: 334–339.

Whyte, L. L. 1965. *Internal Factors in Evolution.* London: Tavistock.

Wiley, E. O. 1981. *Phylogenetics: The Theory and Practice of Phylogenetic Systematics.* New York: Wiley.

Wilkins, A. S. 1998. Evolutionary developmental biology: Where is it going? *Bioessays* 20: 783–784.

Williams, G. C. 1966. *Adaptation and Natural Selection.* Princeton, NJ: Princeton University Press.

Wimsatt, W. C. 1980. Reductionistic research strategies and their biases in the unit of selection controversy. In T. Nickles, ed., *Scientific Discovery: Case Studies*, 213–259. Dordrecht: Reidel.

Wimsatt, W. C. 1981. Units of selection and the structure of the multilevel genome. *PSA* 2: 122–183.

Wimsatt, W. C. 1986. Developmental constraints, generative entrenchment, and the innate-acquired distinction. In W. Bechtel, ed., *Integrating Scientific Disciplines*, 185–208. Dordrecht: Nijhoff.

Wimsatt, W. C., and J. C. Schank. 1988. Two constraints on the evolution of complex adaptations and the means for their avoidance. In M. Nitecki and K. Nitecki, eds., *Evolutionary Progress*, 231–273. Chicago: University of Chicago Press.

Young, J. Z. 1981. *The Life of Vertebrates.* New York: Oxford University Press.

Zelditch, M. L., and A. C. Carmichael. 1989. Ontogenetic variation in patterns of developmental and functional integration in skulls of *Sigmodon fulviventer. Evolution* 43: 814–824.

Zuckerkandl, E. 1994. Molecular pathways to parallel evolution, 1: Gene nexuses and their morphological correlates. *J. Mol. Evol.* 39: 661–678.

Zuckerkandl, E. 1997. Neutral and nonneutral mutations: The creative mix—evolution of complexity in gene interaction systems. *J. Mol. Evol.* 44, Suppl. 1: S2–S8.

5 Legacies of Adaptive Development

Roger Sansom

The Adaptedness of Gradual Mutation

Richard Dawkins (1989) coined the term *evolvability* in an investigation involving simulated life. Wagner and Altenberg (1996, 967) define the same term as "the ability of random mutations to sometimes produce improvement." Natural selection can only increase the fitness of organisms that can have mutations that increase their fitness. What allows evolvability? Most generally, accurate reproduction improves evolvability by reducing the severity of mutation.

Mutation is a generic feature of reproduction; any reproduction system will have a mutation rate of greater than zero. While random changes to well-designed systems tend to be deleterious, some fraction of random mutations are adaptive and the smaller the effect, or the fewer the effects, of a mutation, the better chance it has. It is always unlikely that a mutation will be adaptive, but a smaller mutation still has a greater chance than a larger mutation. Thus, there is generally selective pressure in favor of limiting mutation and mutating only gradually.[1]

One example of the results of this general selective force is the widespread presence of gene proofreading mechanisms, which limit mutation (Drake et al. 1998). These mechanisms act at the gene level. Evolutionary biology has focused intensely on the genes. This chapter is a contribution to the developmental synthesis, which is supposed to widen our focus from just genes to include phenomena to do with development. I will offer some lessons about development later in this chapter, but at the moment it is important that mutations are not just changes in genotype, but also phenotype and ecotype (i.e., the relationship between an organism and its environment) as well.

To make the claim that there is selective pressure for gradual mutation, one must have a way to measure the size of a mutation. One could try to measure the size of a mutation in strictly genetic terms, such as the number of nucleotides that vary across generations. The shortcoming of such an approach is that some mutations are more significant to the process of evolution than others.[2] Some mutations are fatal to an organism and others make no phenotypic difference at all. Therefore, I measure the size of a mutation in terms of its effects on fitness. I will concentrate on fitness at the organism level and so measure the size of a mutation in terms of how it changes the relationship between an organism and its environment (i.e., the way it functions in its environment, or its ecotype). Most heritable mutations to an organism do involve mutation to genes, but organismal fitness change is due to the developmental and ecological consequences.[3] Organisms are selected to produce offspring with similar ecotype to themselves.[4] There are two general ways to mutate more gradually: have fewer mutations, and have smaller mutations.[5] While mutation is a generic feature of replication, gradual mutation is an adaptation. Indeed, I think that gradual mutation is the most general adaptation for evolving life.

Selection for Evolvability

If gradual mutation is an adaptive trait that allows offspring to be more functionally similar to their parents than to other members of the population, what is it a trait *of*? What is the unit of evolution of such a trait? Gradual mutation is a trait of metamutation. It is a trait about how organisms mutate, rather than a simple mutation, which is about how organisms are. Therefore, it cannot be a trait of an organism, because this trait is only expressed through mutation across generations. Clearly it is due to the reproductive capacities of organisms. However, selection works on realizations, not capacities. Therefore, the unit of selection for gradual mutation must be multigenerational. Just how many generations is enough is an interesting conceptual and empirical issue. It may be that one could never point to a complete multigenerational unit, because, while some individuals that are parts of the unit may be present at the same time, others are not.

   I distinguish two types of mutations, metamutations, which are mutations to how mutation occurs across generations, and simple mutations, which are mutations of traits fully realized by an individual organism (such as the length of a wing). By definition, all and only metamutations happen to traits of multigenerational units and are selected for at that level of selection. In contrast, all and only simple mutations happen to traits of organisms that make up the multigenerational unit and are selected for at the organism level of selection.[6] I will say more about the multigenerational units of selection for evolvability in a later section.

Modularity

Lewontin's (1978) discussion of quasi-independence may have been a precursor to the current interest in modularity (e.g., Gilbert et al. 1996). There are two ways to think about modularity—evolutionarily and developmentally. Evolutionary modularity is a concept about the way organisms evolve. It is the view that the parts of organisms evolve quasi-independently of each other. For example, at some stage in our ancestry, our hind limbs changed in significantly different ways from our front limbs. Developmental modularity is a concept about the way an organism develops.[7] If an organism's development is modular, its various parts develop relatively independently of each other (i.e., there is less causal integration between parts than within parts). This relative independence extends from the morphological level down to the genetic level. Developmental modularity is a trait of an organism, while evolutionary modularity is a trait of a multigenerational unit. Both notions are relative and analogue. Parts that might be viewed as parts of the same module at a broader level of analysis, may make up different modules in a more detailed analysis. One should not use the notion of modularity to count the number of modules in an organism. Rather, one should prescribe a number of modules (i.e., a level of analysis) and look to where the boundaries between that number of modules are and how independent they are, compared to some other organism analyzed at the same level. It is not so much that some species are modular and some are not, but some species are more modular than others. Nevertheless, both notions are determined by the objective causal interactions between parts of an organism. It is objective that parts A and B are more independent than parts B and C and it is objective that parts A and B of species X are more independent than parts A and B of species Y.

In what follows, I will show how developmental modularity can be a mechanism for evolutionary modularity and gradual mutation and how selection for gradual mutation will ultimately result in selection for organisms with developmental modules that have unitary function. Many have assumed that there is a relationship between evolutionary and developmental modularity. Even before the theory of evolution by natural selection, physiology had revealed that parts of organisms appear to be designed for particular functions. More recently, Brandon (2005) analyzed a module as a character complex with two features: the character complex has

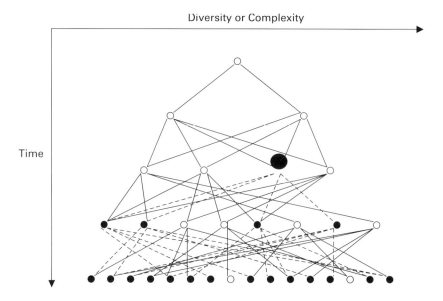

Figure 5.1
Generic development.

unitary function and the genes responsible for the character complex have their effects largely limited to that character complex. My work will show the selective pressure on multigenerational units that are responsible for this relationship between evolutionary and developmental modularity and, with it, why developmental modules tend to have unitary function.

Development

Figure 5.1 is a generic view of development (developed from Waddington 1966). Each stage of development causes the next, as the organism gains complexity or diversity. In this figure, each node is the developmental ancestor of four nodes in the next stage of development, selected at random. This way of developing does not allow for traits to be very quasi-independent. For example, mutation of the one node (shown larger) will influence the development of nineteen nodes (shown with dotted lines and black nodes). If changing any node has a 1/10 chance of being adaptive, then the probability of an adaptive mutation of this developmental node is 10^{-19}, at first approximation.

In figure 5.1, each node is the developmental ancestor of a random collection of four nodes in the next stage of development. In figure 5.2, each

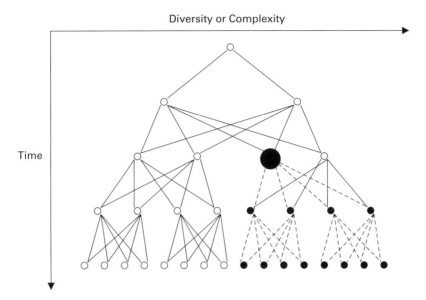

Figure 5.2
Modular development.

node still influences four nodes in the next stage, but there is a modular structure to which nodes are affected. Each node influences the four nodes closest to it. The result of this modular development is that mutation to the same node that was mutated in figure 5.1 influences only thirteen nodes, and the probability of an adaptive mutation is increased significantly to 10^{-13}.

In figure 5.3, modularity is increased further by reducing the number of developmental descendants of each node in the next stage of development to just two; the number of nodes affected by the same mutation is reduced to seven and the probability of an adaptive mutation is increased to 10^{-7}. Lewontin's view is that only quasi-independent traits have a sufficiently high probability of adaptive mutation for the evolution of organisms as we know them. We see here that developmental modularity is a way to achieve quasi-independence.

Generative Entrenchment

Schank and Wimsatt (1988) suggest a lesson that is contrary to Lewontin's about the evolution of complex organisms. In 1828, embryologist Karl Ernst von Baer observed that traits that appear early in development tend to be

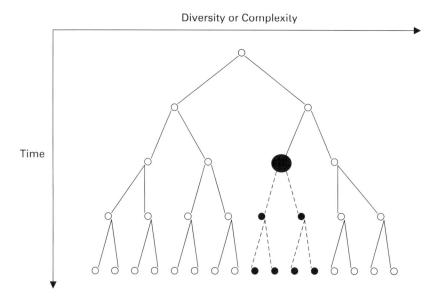

Figure 5.3
Highly modular development.

shared across more species than traits that appear later in development. Different species look much more alike early in their development than they do later. For example, many more species have gills early in development than later. Even human beings develop gills early in development only to lose them later. Schank and Wimsatt explain this phenomenon with Wimsatt's concept of generative entrenchment.

Generatively entrenched (GE'd) traits are those that have widespread developmental effects. There is an especially strong selective force to maintain GE'd traits, because any mutation to a GE'd trait will have many effects and random mutations tend to reduce fitness. Schank and Wimsatt argue that features that appear early in development tend to become GE'd and so are highly conserved by natural selection.

Figure 5.4 shows the number of effects of mutation of a developmental node that is deeply GE'd. Even though it is in a highly modular developmental system, mutation of the GE'd node affected fifteen nodes. Therefore, the probability of adaptive mutation of this GE'd trait is just 10^{-15}.

Figure 5.5 displays graphically Schank and Wimsatt's explanation for the general phenomenon of higher conservation of traits that appear early in development. The size of each node of development represents the number of effects of mutation of the node and the color of the node represents the

Legacies of Adaptive Development 179

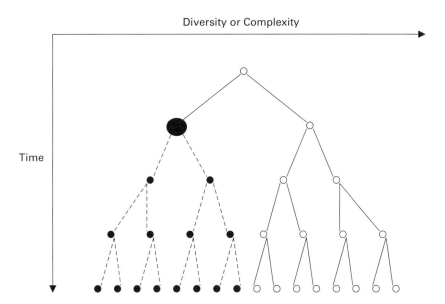

Figure 5.4
Generative entrenchment of highly modular development.

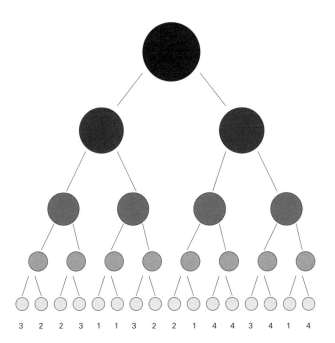

Figure 5.5
Generic generative entrenchment of highly modular development.

likelihood of adaptive mutation of the node (the darker the color, the lower the chance of adaptive mutation). Generically, mutations of earlier nodes have more effects, so earlier nodes are larger, and because there is an inverse relationship between the size of the effect and the chance of adaptive mutation, earlier nodes are darker too. The result of these generic phenomena is that there will be more stable selection on earlier nodes and more evolution by natural selection on later nodes.

The phenomenon of GE suggests interesting things about how natural selection may increase fitness. If early stages of development are so deeply GE'd that their evolution by natural selection becomes nearly impossible, what options does natural selection have for improving the chance of increasing the survival rates of embryos? The simplest solution might be to modify their early development in order to make them stronger, faster, or smarter. However, this would be extremely hard to evolve, because the traits of early development tend to be so GE'd. So, natural selection may come up with a more complex solution to the problem by evolving more evolvable traits, which typically appear later in development. For example, adult traits, such as the production of eggs or nesting practices, could evolve to improve the environment of the embryo. It has long been noted that more complex organisms tend to take better care of their young than simpler organisms. This can be partially explained by the fact that the options for more complex organisms evolving the traits of their young are more limited by GE and so the best resource available to natural selection for improving the survival rate of the young is to evolve adults so that they provide their young with a better environment.

The model of development shown above suggests that traits will evolve quasi-independently from each other to the extent that they are distantly related in the developmental ancestry. Conversely, two traits are evolutionarily integrated to the extent that they are developmentally integrated. Therefore, developmental modularity generally increases quasi-independence by reducing the number of developmental ancestors and descendants of each developmental node.

GE'd traits are not quasi-independent. Lewontin sees quasi-independence as necessary for the evolution of complex organisms. Schank and Wimsatt believe that the more complex an organism, the less quasi-independent some traits can be. We should understand generative entrenchment as a generic limit on quasi-independence. It is a generic feature of any developmental system that some traits will be more GE'd than others. The evolution of GE'd traits will be highly conserved by natural selection, because their mutation will have a greater effect on development than those that are not GE'd. In general, GE'd traits appear earlier in development.

Adaptive Integration

I have argued for a selective force in favor of quasi-independence. Evolution is the story of compromise between selective forces, so this investigation of the adaptive advantage and limits on quasi-independence would be incomplete without also examining any selective forces against quasi-independence and in favor of integration. Before doing this, I would like to point out that while the selective pressures for quasi-independence and integration are opposite, they still have a lot in common. They are both metamutation selective forces (i.e., forces on ways of mutating), which are traits of multigenerational units and selected at that level. Both traits are also supposed to increase evolvability (i.e., make mutations more likely to be adaptive), even though they do so in opposite ways.

There are a number of examples that defy the lesson of generative entrenchment. For example, some phyla, such as salamanders, have evolved neoteny. They have brought forward the developmental stage that organisms become sexually mature. Figure 5.6 offers the simplest way to model the evolution of neoteny as a huge change to quite early stages of development, which generative entrenchment suggests is very hard to do. So how is this possible? I have argued that quasi-independence of traits is generally adaptive. Now I want to supplement that with the claim that integration between functional parts can be adaptive too. Neoteny in some salamanders suggests that they have evolved links of adaptive developmental integration between the different components of sexual maturity and independence from other traits. Salamanders have evolved a sexual-maturity developmental module. This makes it easier for natural selection to tweak the variable of timing sexual maturity in species. Natural selection has discovered that this is a good variable to tweak and has selected for multigenerational units that are better able to mutate this suit of characters because they form a developmental module. Another way to think about the adaptive value of a sexual-maturity module in salamanders is to see that a neotenic species does not have to reinvent how to become sexually mature. This suggests that a mutation to neoteny in salamanders should actually be represented as it is in figure 5.7, where the sexual-maturity module remains unchanged in neotenic salamanders.

Early in this chapter I argued for the general adaptive value of quasi-independence. Schank and Wimsatt argued for a limit on quasi-independence.

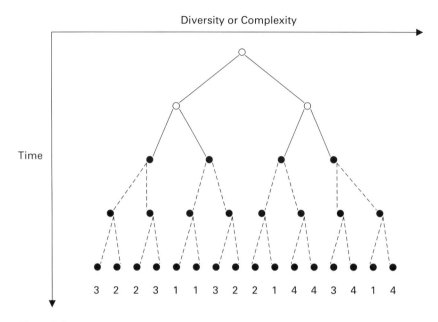

Figure 5.6
Neoteny of highly modular development.

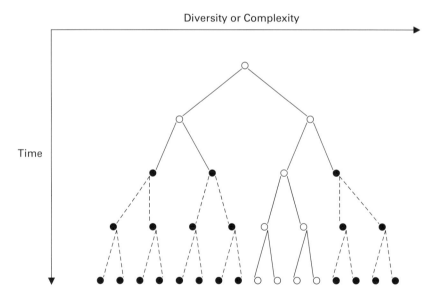

Figure 5.7
Neoteny of functional highly modular development.

Legacies of Adaptive Development

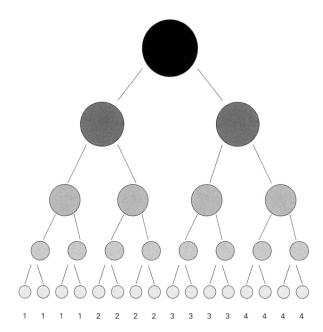

Figure 5.8
Generative entrenchment of highly functionally modular development.

Now I have argued for the adaptive value of integration between traits that are functionally related to each other. Together, these ideas suggest that evolution will evolve multigenerational units with individuals with developmental modules that have unitary function. Such multigenerational units will tend to have larger mutations change the relationships between functional modules and smaller mutations change the relationships within them. There are at least three possible explanations for this. One is that there is a fitness advantage in being able to significantly change the relationships between functional modules. The second is that any developmental structure has some nodes with many developmental effects and mutation of these nodes is somewhat less likely to be maladaptive if those effects are limited to one functional module. Finally, it is more adaptive to fine-tune functional parts within functional modules than the relationships across functional modules. These explanations are not exclusive and may well combine in various degrees in various cases.

Figure 5.8 represents both the number of effects as the size of the node and the likelihood of adaptive mutation as the color of the node. The parts of developmental modules share a function, which is represented by the numbers along the bottom of the figure. In contrast, figure 5.5 represents the parts of developmental modules involved in different functions. The intermediate nodes of development in figure 5.8 are lighter than the intermediate nodes of figure 5.5, representing a higher probability of adaptive mutation of these nodes. These diagrams prove nothing, but represent my conclusions based on the rationale above.

I have argued for general selective pressure in favor of a trait (i.e., the trait of having developmental modules with unitary function). Other general adaptive traits have been proposed. The main argument against them is that few traits are good in all environments. I have argued for selection for a developmental structure that allows natural selection to better tweak functional joints to better cope with a changing environment. The same type of counterargument still can be applied. If the ways that environments change are sufficiently variable, the best functional joints will change sufficiently to prevent consistent selection. However, both the trait and the argument against it have moved up a level of variation of the environment, from variation of the environment to variation of variation of the environment. With that step, the trait has a better chance of being consistently adaptive, because it will necessarily apply to a longer time frame and more dynamic environment.

The Evolvability of Hopeful Monsters

Gould (1977) attacked the view that evolution is completely dominated by gradual mutation. Instead, he revived Richard Goldschmidt's (1940) notion of "hopeful monsters," which boldly broke new morphological territory and struck it rich, in his explanation of some major evolutionary transitions. Both Goldsmidt and Gould proposed mutation early in development as likely origins of such drastic mutation. A multigenerational unit that adopted the "hopeful-monster strategy" would consistently produce hopeful monsters by consistently mutating drastically. My argument for a trend toward gradual mutation relies on the premise that there is selection against the hopeful-monster strategy and for quasi-independence.

I am sure that gradual mutation has adaptive value, but Gould's view suggests that while each offspring that results from drastic mutation is likely to have low fitness, because some will have very high fitness, there may be some adaptive value to drastic mutation too. So, do we evaluate the relative adaptedness of the hopeful monster strategy by multiplying the lower probability of an adaptive drastic mutation by the higher average increase in adaptivity of those mutations, and comparing the product to that of the gradual mutation strategy? If it turned out that the hopeful-monster strategy looks good, my project here runs into serious trouble, because it is founded on selection for gradual mutation. Fortunately for the project, there are two other variables that must be considered, which favor the gradual-mutation strategy.

Even if the hopeful-monster strategy does offer bigger payoffs, a multigenerational unit will most likely have to wait longer for it to pay off, because it offers a smaller chance of a bigger payoff. In the meantime, the selective pressure in favor of the gradual-mutation strategy will be brought to bear. More importantly, the hopeful-monster strategy might not offer such a payoff after all. While a hopeful monster may stumble across a very different and highly adaptive phenotype, to cash in it must be able to effectively pass on the same phenotype to its offspring. A multigenerational unit that adopts the hopeful-monster strategy will have difficulty doing this, because it will continue to create hopeful monsters instead of confident less monstrous offspring. So, gradual mutation is the best general strategy to adopt.

Note that my argument supports the conclusion that gradual mutation is the best *general* strategy. It may be adaptive to augment the gradual-mutation strategy in some way. For example, a multigenerational unit that produced one hopeful monster out of every ten offspring each generation would be able to cash in on any particularly adaptive phenotype that a hopeful monster happened on. This is the simplest form of a strategy that is hybrid between gradual and drastic mutation, but natural selection may have hit on a more subtle adaptive hybrid strategy. One example is that *Escherichia coli* increase mutation rates when under environmental stress (Loewe, Textor, and Scherer 2003). This is a smart hybrid strategy, because such stress is likely to reduce the fitness of the phenotype of the population, and a phenotype that is facing harder times faces a lower opportunity cost of mutating drastically than one facing good times (i.e., the fitness of gradually mutated offspring is lower in hard times). In contrast, the benefit of mutating drastically will be similar (i.e., the chance of drastically mutated offspring landing on an adaptive phenotype will be similar). Therefore, a hybrid strategy is improved, in general, if drastic mutation increases in times of environmental stress.

I have also argued for selection for integration in development of functional parts. Integration is the opposite of quasi-independence and allows bigger mutations. I do not think that my arguments for integration are arguments for the hopeful-monster strategy, for a rather subtle reason. While adaptive integration between functional parts does allow for bigger mutations, it is not selection for the creation of hopeful monsters, because it makes monsters that are the result of large mutations not merely "hopeful" after all. Their leap into new morphological territory is less of a blind leap of hope for being the result of evolved integration. That integration has an evolutionary history. It has been designed by natural selection to hit on a joint within development that is an adaptive one to tweak. In a sense, then, it is evolution for the monstrous results of larger mutations. But while this monster is no less a monster, it is filled less with hope and more with the confidence of being founded on a type of past evolutionary success.

Wagner and Altenberg (1996) provide support for my general conclusion about the adaptedness of quasi-independence with their acknowledgment that the evolution of differentiation was dominated by quasi-independence, not integration. However, they point out that integration does occur—for instance, in the symbiotic cells of different origin (mitochondria and plastids). This idea, together with the above discussion on hopeful monsters, suggests that organisms that hit on new symbiotic relationships are the true hopeful monsters. They are hopeful monsters because they take a huge leap of integration with little or no evolutionary history. We would expect mutations in a new symbiotic relationship to be rather drastic. Once symbionts become exclusive to each other, we can expect natural selection to start a trend toward gradual mutation of the new unit of both symbionts together. One of the strategies that natural selection adopts may be to break up the old modules within the symbionts to form new modules across them. But this would take a long time and until then you have a pretty hopeful monster. So, the most successful hopeful monster may have been the first eukaryote.

The Multigenerational Units of Selection for Evolvability

Earlier, I claimed that evolvability can only be selected for across generations, because evolvability is only realized across generations. Above, I discussed various strategies to increase evolvability. The general hopeful-monster strategy is one of inaccurate heredity. It may be modified such that at every generation there is a variety in the accuracy of heredity of offspring, or, perhaps more adaptively, such that there is variance in the accuracy of heredity across generations (most adaptively, there is less accuracy when under environmental stress). While all of these approaches can only be selected for in multigenerational units, they are not all expressed in the same multigenerational unit.

Legacies of Adaptive Development

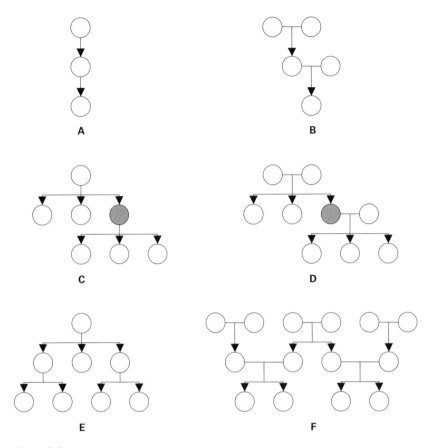

Figure 5.9
a. asexual legacy, *b.* sexual legacy, *c.* narrow asexual lineage, *d.* narrow sexual lineage, *e.* conventional asexual lineage, *f.* conventional sexual lineage.

General accuracy (and inaccuracy) of heredity are expressed in what I call a "legacy." With asexual reproduction, this is just an individual and its parent, grandparent, and so on (see figure 5.9a). Generally, I suggest that accurate legacies are selected over inaccurate legacies. In sexually reproducing organisms, if one were to look at such a narrow legacy, one would miss half of the picture, because it includes only one parent. Therefore, with sexual reproduction, the unit of selection for high heritability is a sexual legacy. A sexual legacy is made up of an individual, its two parents, one pair of grandparents, and so on (see figure 5.9b). For asexual organisms, accuracy (in terms of the relationship between an organism and its environment[8]) is the simple issue of producing identical clones. Accuracy for sexual species is more complicated, because novel combinations of genes must allow the development of novel, but functional, organisms. Genes must "play nice" with their new gene mates at each generation.

It is unclear how many generations make up a legacy that can be effectively selected for accurate heredity. The unit must be long enough to allow differences in accuracy to become apparent to natural selection. One generation may do it, but this would be contingent on the conceptual issue of how much difference in accuracy makes a difference and the empirical issue of how many generations are necessary to exhibit that difference. The number of generations necessary may vary with species.[9]

So far, I have only discussed selection for accurate heredity, which is one way to improve evolvability. Another way is to vary the accuracy of heredity, such that it is highly accurate when the environment does not put the organism under stress, but less accurate when it does. The units of selection for this trait will also be legacies, but they will be longer than the legacies that select for accurate heredity. This is because a legacy can only exhibit such variation in variance of accuracy of heredity over more generations than it takes to exhibit variation in accuracy of heredity.

Finally, selection for variance of accuracy of heredity among descendants of each generation can only be expressed by a unit that includes multiple descendants of a parent or pair of parents. It will most effectively be expressed in a unit that includes all of the descendants of a parent in asexual reproduction and all descendants of a pair of sexually reproducing parents. Again, the number of generations necessary is a conceptual and empirical issue. Let us call these units narrow asexual and sexual lineages (figures 5.9c and 5.9d).

Why is the unit of selection for variance of accuracy of heredity among descendants of each generation made up of all of the descendants of an individual (or breeding pair) generation x, but only the descendants of one of generation $x + 1$ (shown in gray)? Why not include all of the descendants of generation $x + 1$, and perhaps all of their breeding partners' ancestors too? I call such units conventional lineages (see figures 5.9e and 5.9f). I think that narrow lineages are more appropriate than conventional lineages, because one conventional lineage encompasses too many individuals to allow selection between them. If a gene, for example, were to increase the variance of accuracy of heredity among descendants of each generation, then that gene may only be shared by some of the narrow lineages, or some proportion of a conventional lineage. To reasonably talk about the level of that gene increasing within a breeding population, one may talk about selection in favor of narrow lineages, but there may be no selection in favor of conventional lineages within the breeding population, because the entire breeding population may make up only one conventional lineage. My explanation for the selection is improved by precisely identifying the (narrow) lineages that carry this gene.

It is true that a legacy falls within a lineage (if the lineage has at least as many generations the legacy). So, why don't we just say that all of the traits that may contribute to evolvability are selected for at the lineage level? The answer is that it is not good enough to simply consider the larger unit when smaller units express variants in a trait. Selection occurs at the smallest level where trait value differences are expressed. After much debate, centered around defeating gene selectionism and also concerning the issue of group selection (e.g., Sober and Wilson 1998), there is a growing consensus that selection is hierarchical. Junk DNA that has a higher tendency to be repeated may be selected for at the level of DNA, organisms that are especially fast or cunning may be selected for at the level of organisms, and groups that are especially altruistic may be selected for at the group level. All of this is so despite the fact that groups are made up of organisms and organisms are made of DNA (and other things). Similarly, accurate heredity may be selected for at the legacy level, variance in accuracy across generations may be selected for at the level of longer legacies, and variance in accuracy within a generation may be selected for at the narrow lineage level.

This section has assumed that there is selection for accuracy and both types of variance of accuracy in heredity. Just as there is no group selection if groups do not vary in fitness, so too if there is no selection for accuracy and its variations, then the corresponding unit of selection that would have expressed that difference in fitness does not exist either. It seems uncontroversial that there has been selection for accurate heredity. The evidence for variation in accuracy across generations is intriguing. The variation in accuracy of heredity across different offspring of the same parents in one generation has only been hypothesized here. Accordingly, selection between shorter legacies seems assured; there is evidence for selection between longer legacies, but selection between narrow lineages remains merely a hypothesis here.[10]

Evo-Devo Meets Eco-Devo

I have argued for consistent selection for separate developmental modules performing separate functions. In order for this claim to have any substance, the delineation of functions must be determined independently of the developmental modules performing them. It is circular to claim that wings are for one function (flight) if one counts flight as one function because one developmental module (wings) performs it. So I face the question, how do we distinguish functions? Before I answer this question, I will ask another question that leads to the same answer. I have suggested that the modular structure of development is under selective pressure to increase the probability of adaptive mutation (i.e., to increase evolvability). Given that the relationships between modules are more variable than the relationships within a module, which traits should be grouped together into the same module and which should be kept in different modules? I think that the answer to both of these questions lies in ecology.

Consider an environment that has a number of variables, such as temperature and salinity, which vary independently of each other. Some of those variables affect the fitness of a legacy more than others. Let us call these variables "important." My suggestion is that the species will evolve a modular developmental structure such that the function of developmental modules is linked to an important environmental variable. For example, because temperature is an important variable that varies somewhat independently of others, the adaptive legacy is one that has a thermoregulation module. Having such a module will allow both fine-tuning of the module and significant overhaul without affecting the development of other modules. The other functional modules will be linked to other ecological variables. The fact that an evolutionary overhaul of one module will not affect other modules is adaptive to the extent that the variation of the ecological variable that each module is functionally responsible for varies independently.

Combine the idea that modules are responsible for important independent ecological variables with the idea that modules have a particular function and you get an account of how to distinguish functions. Functions are to be distinguished by how independently the ecological variables that they have evolved to cope with vary. To the extent that temperature varies independently from the color that will best camouflage an organism, thermoregulation and camouflage are separate functions for that organism. To the extent that they covary (e.g., if the temperature drops to make snow likely and white the best color for camouflage), thermoregulation and camouflage are the same function. We can think of important environmental variables that covary as parts of an environmental module and those that vary independently making up separate environmental modules. A species' niche is determined by what environmental modules are important to the species.[11]

A Cautionary Note about All This Speculation

I have launched far into the world of adaptationist speculation about evo-devo and ultimately eco-devo. At the very least, responsibility requires a cautionary note. Gould and Lewontin (1979) have taught us that we ignore developmental constraints at our peril. One may think of the evolution of development, based on legacy selection, as the evolution of those constraints, but developmental constraints are probably more constrained than I have suggested here. I have talked about metamutations and meta-adaptations. These constraints on the evolution of developmental constraints are metaconstraints. For example, the fact that developmental modules tend to be physically separated from each other, which is loosely implied by the model of development used throughout this chapter, suggests that the division between functional modules is not nearly as free as my adaptationist speculation above acknowledges.

I do not apologize for speculating, but willingly acknowledge that it is not the full story. A better understanding will be gained when we even the scale by looking at evo-devo based not just on meta-adaptation of legacies, but on metaconstraints on legacies as well.

Notes

1. In this chapter, I will be focusing exclusively on mutations to the germ line, because these mutations are more significant to the evolutionary process. Darwinism has always acknowledged the power and importance of gradual mutation. Gould has argued that bold mutations have played a significant role in evolutionary history. I address the issue of Gould's "hopeful monsters" in section 7.

2. Most simply, the mutation of a single nucleotide might have no effect on the organism beyond that change because the nucleotide triplet that it is a part of might still code for the same amino acid, due to the redundancy in the coding system. Additionally, even if the gene now codes for a different sequence of amino acids, the protein it codes for might function in the same way if the characteristics of its active site are unchanged. Finally, even if the mutation does make a physiological difference, not all physiological differences are created equal, because some have larger effects on ecological relations than others.

3. One can use the notion of gradual mutation at levels below and above the organism if one also cashes out the size of a mutation in terms of fitness change at that level.

4. Richard Lewontin's (1978) discussion of "quasi-independence" and "continuity" inspired my notion of gradual mutation. I follow Lewontin in cashing out the similarity between a reproducer and its ancestors in ecological terms.

5. Here I am assuming that we have no difficulty individuating mutations so as to distinguish a case of many small mutations from a case of fewer larger mutations. However, my view about gradual mutation does not rely on this, because it treats both types of cases in the same way. They are different ways to produce a change in ecotypes of similar magnitude.

6. Here, I am ignoring other levels of selection, such as gene selection and some types of group selection, to distinguish legacy selection from individual selection.

7. Development is the process of growth and differentiation of an individual.

8. In section 1, I offer a justification of why I use similarity of the relationship between organism and environment to measure the degree of mutation. Accurate heredity is low degree of mutation.

9. The difference between asexual and sexual legacies suggests that species that can reproduce sexually and asexually may face intriguingly complex competing selective pressures at more units than those that reproduce only asexually or only sexually.

10. Some may object that what I have called a legacy is what they call a lineage, so I have not suggested that selection at the lineage level is more likely than selection at the legacy level after all. I have acted in good faith in my choice of terminology, but I make no claim here that such an objection is definitely to be rejected. However, I have never seen anything to suggest that a family tree that has descendants of multiple siblings is not a lineage, for example, and such an objector would have to be willing to make such a claim.

11. I have presented a view that functions are determined by environmental modularity and niches are determined by what environmental modules are important to the fitness of the species. This suggests an image of an organism only reacting to its environment rather than affecting its environment as well. Lewontin is rightly critical of this picture. He has argued that we should acknowledge that organisms play such a significant role in determining their environment that the notion of a niche without an organism occupying it risks incoherence. My analysis of function is not supposed to conflict with Lewontin's ideas. Rather, I present it in its simplest form, making it appear that the environment sets up the problem for the organism to solve. Having presented that idea, I acknowledge Lewontin's significant wrinkle. Which variables are important will change with the evolution of the species in question, and the values of the variables and their patterns of variation can also change as a direct result of evolution of the organism. This epicycle complicates matters and may make the empirical job of discovering the delineation between modules significantly harder, but I do not believe it undercuts the coherence of my view.

References

Brandon, R. 2005. Evolutionary modules: Conceptual analyses and empirical hypotheses. In W. Callebaut and D. Rasskin-Gutman, eds., *Modularity: Understanding the Development and Evolution of Natural Complex Systems*. Cambridge, MA: MIT Press.

Dawkins, Richard. 1989. The evolution of evolvability. In G. C. Langton, ed., *Artificial Life: Synthesis and Simulation of Living Systems*, 201–220. Redwood City, CA: Addison Wesley.

Drake, J. W., B. Charlesworth, D. Charlesworth, and J. F. Crow. 1998. Rates of spontaneous mutation. *Genetics* 148: 1667–1686.

Gilbert, S. F., J. M. Opitz, and R. A. Raff. 1996. Resynthesizing evolutionary and developmental biology. *Dev. Biol.* 173: 357–372.

Goldschmidt, R. 1940. *The Material Basis of Evolution*. New Haven, CT: Yale University Press.

Gould, S. 1977. The return of hopeful monsters. *Natural History* 86 (June/July): 22–30.

Gould, S., and R. Lewontin. 1979. The Spandrels of San Marco and the Panglossian paradigm: A critique of the adaptationist programme. *Proc. Roy. Soc. Lond.* B 205: 581–589.

Lewontin, R. 1978. Adaptation. *American Scientist* 239(3): 213–224.

Loewe, L., V. Textor, and S. Scherer. 2003. High deleterious mutation rate in stationary phase of Escherichia coli. *Science* 302: 1558–1560.

Schank, J., and W. Wimsatt. 1988. Generative entrenchment and evolution. In A. Fine and P. K. Machamer, eds., *PSA* 2: 33–60.

Sober, E., and D. Wilson. 1998. *Unto Others: The Evolution and Psychology of Unselfish Behavior*. Cambridge, MA: Harvard University Press.

Waddington, C. 1966. *Principles of Development and Differentiation*. New York: Macmillan.

Wagner, G., and L. Altenberg. 1996. Complex adaptations and the evolution of evolvability. *Evolution* 50(3): 967–976.

6 Evo-Devo Meets the Mind: Toward a Developmental Evolutionary Psychology

Paul E. Griffiths

What Is Evolutionary Developmental Biology?

In his influential introduction to the emerging discipline of evolutionary developmental biology, Brian Hall defines it as the study of "*how* development (proximate causation) impinges on evolution (ultimate causation) and *how* development has itself evolved" (Hall 1992, 2; original emphasis). However, to capture the commitments of most who endorse evo-devo as a scientific program it is necessary to add that the two projects Hall identifies are linked and that the first project is assumed to result in something other than a straightforward endorsement of neo-Darwinian orthodoxy. A maximally conservative evolutionary approach to developmental biology would merely apply contemporary neo-Darwinian theory to a new range of explananda, namely, development. The ways organisms develop would be explained the same way adult phenotypes are commonly explained. Populations of variants change over time so as to better fit the environment they occupy (the lock-and-key model of adaptation). These changes occur when genes are selected because their presence or absence causes a difference in some trait (the gene as unit of selection). This kind of evolutionary explanation of development, however, would not be evo-devo as we know it today.[1] Evo-devo is associated with the idea that paying attention to development problematizes both the idea that form is shaped in a one-sided manner by the demands of the environment and the idea that the unit of selection is the individual gene. Evo-devo problematizes the lock-and-key model of adaptation because the developmental biology of organisms is an input to the evolutionary process as well as an output. The particular developmental biology of an evolving lineage of organisms makes some phenotypes relatively accessible and others relatively inaccessible. Development thus affects the range of variation available for selection and partly determines the evolutionary trajectory of the lineage. Evo-devo also problematizes the idea that the unit of selection is the individual gene because it describes emergent levels of organization in the developing phenotype. Although characters at these levels of organization are constructed through the interaction of gene products, they retain their identity when they are constructed using different developmental resources. The selectionist narratives associated with at least some evo-devo work focus on selection for features at these levels rather than for traits uniquely associated with specific genes or other specific "atomic" inputs to development.

The premise behind this chapter is simple. If the ideas that make up evo-devo have been so productive in opening up new lines of investigation into morphological evolution, they may be equally productive for psychological evolution. In the following sections I explore how two of the core theoretical concepts in evo-devo—modularity and homology—apply to psychology. The next section examines how the "mental modules" at the heart of today's Evolutionary Psychology[2] relate to the "developmental modules" that play a prominent role in evo-devo. I distinguish three sorts of modules—developmental, functional, and mental modules. I argue that mental modules need only be "virtual" functional modules. Evolutionary Psychologists have argued that separate mental modules are solutions to separate evolutionary problems. I argue that the structure of developmental modules in an organism helps determine what counts as a separate evolutionary problem for that organism. In the third section, I suggest that homology as an organizing principle for research in evolutionary psychology has been severely neglected in favor of analogy (adaptive function). I consider some arguments suggesting that determining homology is less epistemically demanding than determining adaptive function. I argue more definitively that psychological categories defined by homology are more suitable objects of psychological—and particularly neuropsychological—investigation than categories defined by analogy. The extrapolation of experimental results in these fields to homologues of the experimental system is warranted, but similar extrapolations to analogues are not warranted by current models of evolution.

Modularity

Developmental Modules and Functional Modules
The fundamental notion of modularity in evolutionary developmental biology is that of a region of strong interaction in an interaction matrix.[3] A metazoan embryo is modularized to the extent that, at some specific stage in development, it consists of a number of spatial regions that are develop-

ing relatively independently of one another. For example, most events of gene transcription in one segment of a developing arthropod have relatively little effect on the immediate future state of other segments when compared to that of the segment in which the transcription occurs. Developmental modules are typically organized hierarchically, so that modules exist on a smaller physical scale within individual, larger-scale modules. The individual cell represents one prominent level of this spatial hierarchy. At a lower level than the cell are particular gene-control networks, for example, and at a higher level are classic embryological units like limb buds or arthropod segments. Although they evolve independently of one another, modules are not windowless monads. The increasing differentiation of various parts of the embryo over time owes a great deal to interactions between modules, as classically described using the concept of "induction." Contact between tissue composed of cells of one type and tissue composed of another type causes—or induces—further differentiation of one or both cell types. A classical example occurs in the development of the vertebrate eye, with the interaction between the incipient retina and what will later, as a result of this interaction, become the lens. The importance of such interactions between modules is entirely consistent with the basic picture of modules as regions that interact more strongly with themselves that with one another. The immediate effect of one module on the other is small. Its importance comes from the cascade of subsequent events that occurs because of causal connections within the module rather than because of the direct causal influence of other modules. For example, the activation of a regulatory gene as a consequence of an interaction between modules is significant not because of the transcribed product per se, but because of the resultant developmental cascade that transforms the affected module.

There are important similarities between the treatment of modularity in evo-devo and the way neuropsychology and cognitive science individuate their "subsystems" or modules. The claim that two functions are performed by separate neural subsystems has traditionally been established by presenting evidence of "double dissociation"—sets of clinical or experimental cases in which each function is impaired while the other is performed normally.[4] A commentary on the continuing dispute between modular and distributed accounts of brain function gives a characteristic example: "Lesions to temporal areas thought to encompass the FFA [fusiform face area] are associated with proposagnosia [deficits in face recognition]. Conversely, at least one patient with widespread damage to the visual cortex has shown severely impaired object recognition but selectively spared face recognition.... Such behavioral double-dissociations in response to brain damage provide intuitively appealing evidence of distinct neural mechanisms for processing each type of information" (Cohen and Tong 2001, 2406). Thus, like developmental modules, neural-functional modules are defined using the idea of semidecomposability (Simon 1969). A semidecomposable system can be divided into subsystems that are connected internally more strongly than they are connected to one another. The two disciplines have been attracted to the idea that their objects of study are semidecomposable systems for similar reasons. Semidecomposable systems can evolve more easily, because the effects of mutations are likely to be localized. They are robust when damaged, for the same reason. They are also relatively easy to study, because in such systems functions are structurally localized. Experimental interventions at different loci in the system have characteristically different effects, making the experimental elucidation of structure-function relationships tractable. Methodologically, the use of dissociation evidence in neuropsychology is paralleled by what is also called "dissociation" in developmental biology: if a region of the embryo can develop normally in the absence of another region, then it is not part of the same module.

Despite these striking similarities, there is an important difference between developmental and neural-functional modules. In development, the system whose dynamic properties are being studied is the matrix of genetic and other developmental resources that is required for the organism to develop. In neuropsychology, the system is the brain—a piece of morphology at some specific stage in development, traditionally an idealized "adult" stage. The difference between the two sciences comes out when we consider how they treat the very same organ—the brain. Neuropsychology regards brain activity as the dynamic expression of a more or less fixed neural architecture. Neuropsychologists aim to characterize that architecture. Developmental biology, in contrast, regards that neural architecture itself—extended over developmental time—as a dynamic expression of a developmental system. Developmental biologists aim to characterize the properties of that, quite different, system. Obviously, these two ways of looking at phenotypic structure are not unique to the brain, but exist for any structure whatever. Hence, from now on I will contrast developmental and functional modules in general, rather than developmental and neural-functional modules in particular. The fact that these two kinds of modules are actually very different needs to be kept in mind when importing results on modularity from one area of science into another. It is entirely possible, for example, that areas of the brain are parts of a single functional module, but derive from several, separate developmental modules. This relationship is entirely unproblematic—some single bones in the skull, for instance, correspond to more than one developmental module (Schlosser and Thieffry 2000, 1043). The reverse relationship—in which more than one functional module corresponds to a single developmental module—appears at first to be problematic. The two functional modules have to become separate at some stage, and this would presumably involve the emergence of two separate developmental modules. However, if the separation occurs late in development, then most of the work of understanding how these functional modules acquire their characteristic form would be done by studying a single developmental module. It would, therefore, be safest when comparing findings about modularity in development to findings about modularity in adult function to assume that that the two sets of modules may stand in a many-to-many relationship to one another.

Mental Modules

With the notions of a developmental module and a functional module in hand, I turn to the "mental modules" that play such a prominent role in contemporary Evolutionary Psychology. Evolutionary Psychologists often introduce the idea of modularity using dissociation evidence from neuropsychology (e.g., Gaulin and McBurney 2001, 24–26). However, a mental module is a very different thing from a neural-functional module. Evolutionary Psychologists themselves are quite clear that their mental modules need not be localized in single regions of the brain (Gaulin and McBurney 2001, 26) and I will argue that they need not be neural-functional modules in any standard sense. In neuropsychology, the double-dissociation experiment is a means for exploring structure-function relationships in the brain. But for the purposes of evolution, what matters is not how the brain is structured, but how it appears to be structured when "viewed" by natural selection. For Evolutionary Psychology, the fact that two functions are dissociated is significant in its own right, and not only as a clue to how those functions are instantiated in the brain. Thus, there are architectures that produce double dissociations but that neuropsychology regards as nonmodular, cases where apparent double dissociations are simply misleading. Evolutionary Psychology, in contrast, would regard these architectures as different ways to produce mental modularity. We might aptly term such mental modules "virtual modules."

The modularity concept of Evolutionary Psychology derives from that developed in the cognitive science of the early 1980s and synthesized by Jerry Fodor in *The Modularity of Mind* (1983). In Fodor's account, the definitive property of a module is informational encapsulation. A system is informationally encapsulated if there is information unavailable to that system but that is available to the mind for other purposes. For example, in a phobic response the emotional evaluation of a stimulus situation ignores much of what the subject explicitly believes about the situation, suggesting that the emotional evaluation is informationally encapsulated. Fodor lists several other properties of modules, including domain specificity and the possession of proprietary algorithms. A system is domain specific if it only processes information about certain stimuli. It has proprietary algorithms if it treats the same information differently from other cognitive subsystems, something that Evolutionary Psychology identifies with the older idea that the module has "innate knowledge." John Tooby and Leda Cosmides (1992) make it clear that it is these two properties, rather than informational encapsulation, that are the two definitive properties of mental modules. A mental mechanism is not a module if "it lacks any a priori knowledge about the recurrent structure of particular situations or problem domains, either in declarative or procedural form, that might guide the system to a solution quickly" (p. 104). In the Evolutionary Psychology literature the properties of being domain specific and of having proprietary algorithms are generally referred to simultaneously as "functional specialization." Modules are "complex structures that are functionally organised for processing information" (p. 33).

When Evolutionary Psychologists present experimental evidence of domain specificity in cognition, it is generally evidence suggesting that information about one class of stimuli is processed differently from information about another class of stimuli—that is, evidence of the use of different proprietary algorithms in the two domains. For example, Cosmides and Tooby (1992) show that how subjects reason when performing the Wason card-sorting task depends on how the task is described. They use this to argue that certain ways of describing the task activate a domain-specific device for social cognition. Similarly, David Buss (2000) has argued that people leap to conclusions about sexual infidelity more readily than about other subjects. He uses this to support the view that there is a domain-specific system for judging infidelity. Evidence for separate, domain-specific cognitive systems could, of course, be provided without postulating that the systems have proprietary algorithms. This could be done if the evidence for domain specificity came from dissociation studies. There is no conceptual difficulty in demonstrating double dissociation between deficits in performance on tasks in two domains while simultaneously showing that information about the two domains is processed using the same algorithms.

The brain might resemble a computer network with two identical mail servers, one used by the sales department and the other by the accounts department. Both run the same software, but when each goes down it causes a distinctive set of performance deficits. However, Evolutionary Psychologists have not tended to collect evidence of dissociation, relying instead on evidence that the brain has "innate knowledge" of certain domains, as in the two cases just described. In fact, if dissociation results were available but there was no evidence of "innate knowledge," Evolutionary Psychologists would not regard these functional modules as mental modules, as the quotation from Tooby and Cosmides makes clear. This is because the evolutionary rationale for the existence of domain-specific mental modules requires them to have proprietary algorithms. Separate mechanisms for reasoning about separate domains but reasoning about them in the same way would, from the perspective of Evolutionary Psychology, be merely bizarre. Evolutionary Psychologists argue that evolution would favor multiple mental modules over domain-general cognitive mechanisms because each module can be fine-tuned for a specific adaptive problem. From this perspective, separate mechanisms that deal with separate domains but have identical internal workings make no evolutionary sense.

The evolutionary rationale for mental modules also implies that mental modules must be developmentally dissociable. Domain-specific modules are superior to domain-general cognitive mechanisms, it is argued, because each module can be fine-tuned by natural selection to be good at performing tasks in a single cognitive domain. But independent evolutionary fine-tuning of mental modules requires that those modules are developmentally dissociable. If mutations affecting one mental module typically had effects on other mental modules, then there would be no difference, with respect to their ability to be fine-tuned to perform tasks in a single domain, between domain-specific modules and domain-general cognition. A similar argument suggests that mental modules will be functionally dissociated, since if they are functionally entangled, then changes to one are likely to impair performance in the other. So mental modules are expected to be both developmentally and functionally dissociable from one another, but this is a prediction of "adaptive thinking," not part of the core of what is meant by modularity in Evolutionary Psychology, which seems to be only functional specialization. In fact, as I suggested above, mental modules might sometimes be only "virtual modules" from the viewpoint of neuropsychology. Recall that a virtual module is a pattern of dissociability between aspects of the systems performance that does not correspond to the existence of separate neural systems. Tim Shallice (1988, 250) describes six kinds of neural architecture that can produce double dissociation without corresponding functional modules. It would take far too long to describe all these here, so I will give only the simplest one. If a range of inputs is processed by a continuum of processing space in the brain, as is the case for inputs across the visual field and areas of the visual cortex, then lesions to specific portions of that processing space will affect specific domains of input. It would not be illuminating, however, to divide such a continuous processing space into several discrete modules, or the corresponding input domain into several domains. This admittedly rather trivial example illustrates an important general point. Neuropsychologists use dissociation results to study structure, function, and structure-function relationships in the brain. They have become aware in the last fifty years of many difficulties in interpreting dissociation results and have come to treat dissociation as suggestive evidence in need of further interpretation. A modularity concept that simply identified a module with a neural system that produces double dissociations would not be useful for neuropsychology because it would serve only to blur distinctions between different neural architectures. Evolutionary Psychology, in contrast, is interested in what the brain must be like if it is a product of evolution. Central to the research program of Evolutionary Psychology is an argument that selection will favor many, functionally specialized modules rather than a few, domain-general cognitive mechanisms. The argument relies on a "thin" definition of a module that counts as modular any architecture that produces dissociations between performances in different domains. This is entirely in keeping with the fact, strongly emphasized by many Evolutionary Psychologists, that they offer a theory of function and not of structure: "When applied to behavior, natural selection theory is more closely allied with the cognitive level of explanation than with any other level of proximate causation. This is because the cognitive level seeks to specify a psychological mechanism's function, and natural selection theory is a theory of function" (Cosmides and Tooby 1987, 284).

Since the aims of the two disciplines are so distinct, it is unsurprising that they have different concepts of modularity.[5]

Modularity and Adaptation
The evolutionary rationale for modularity rests on the idea that the environment contains a series of separate adaptive problems. Since the best solution to one problem may not be the best solution to another, a suite of specialized mechanisms will be superior to a single, general-purpose mech-

anism. I am not concerned to assess the soundness of this argument here, only to explore its consequences. Closely related ideas can be found in evolutionary developmental biology. Günther Wagner and others have constructed population-genetic models in which developmental modularity at the level of gene-control networks is the result of selection for the ability to alter one trait of the organism without altering others (Wagner 1996; but see the more recent views in Schlosser and Wagner 2004). Modularity is selected for because it allows the organism to solve problems separately rather than settling for a single compromise solution. This selection scenario seems to presume that the environment contains a number of discrete problems (although I will suggest below that it does not, in the final analysis, rest on this presumption).

Elsewhere, Kim Sterelny and I have identified a fundamental difficulty for the idea that mental modules correspond to separate adaptive problems, a difficulty we called the "grain problem" for Evolutionary Psychology (Sterelny and Griffiths 1999, 328–332). We suggest that whether certain features of the environment of evolutionary adaptedness constitute one problem or many problems depends on the developmental structure of the mind. Problems whose solutions cannot be developmentally dissociated must be solved as a single problem and so are not separate problems from the standpoint of adaptive evolution. The grain problem is an aspect of a much better known conundrum in selection theory—the coconstructing and codefining nature of populations and their ecological niches.[6] It is not possible to take a region of space-time devoid of life and determine what niches it contains for life to evolve into. It contains vastly many overlapping possible niches, and which ones become actual will depend on the biota that evolves to occupy that region of space-time. Of course, there is a sense in which every possible niche that an evolving biota could forge in an area of space and time "exists." This sense becomes still more tenuous, however, once it is recognized that occupied landscapes owe many of their *abiotic* properties to the activities of the organisms that occupy them. In this tenuous sense there were niches for species requiring high rainfall in the Amazon Basin before the biota that make it a high-rainfall region had evolved. So a region of space and time contains not only all the niches that can be defined using its existing abiotic features, but also all those that can be defined using biotic and abiotic features that *could* be created by the action of all the species that *could* evolve so as to make a niche in that region. In a small-scale example of the creation of a niche in this sense, some eucalypt species can establish and sustain "islands" of dry, sclerophyll forest in rainforest regions by facilitating bushfires (Mount 1964). The existence of this niche in that region of space and time is a consequence of the evolution of the eucalypts that fill the niche as much as the reverse.

The grain problem for Evolutionary Psychology results from applying the insight that populations and niches coevolve with one another to the question of how many separate adaptive problems the niche contains. In one sense, the niche contains indefinitely many, overlapping problems and which of these problems the organism adapts to depends on the structure of the organism occupying the niche. An obvious example is the evolved basic emotion of fear (Ekman 1972). Cosmides and Tooby have consistently used the danger posed to our ancestors by predators as an example of the sort of recurring ecological problem that would shape a specific emotional adaptation—an emotion module (Tooby and Cosmides 1990; Cosmides and Tooby 2000). But the problem could be viewed in a more coarse-grained way as the problem of responding to danger, or in a more fine-grained way as the problem of responding to snakes on the one hand or to big cats on the other. The empirical evidence suggests that in humans the actual fear response—the output side of fear—is an outcome of very coarse-grained selection, since it responds in the same way to danger of all kinds. The emotional appraisal mechanism for fear—the input side—seems to have been shaped by a combination of very fine-grained selection, since it is primed to respond to a number of specific gestalts (Öhman 1993), and selection for developmental plasticity, since very few stimuli elicit fear without relevant experience.

There are many ways of parsing the environment into separate evolutionary problems. Although humans have a single fear response, many other animals have one fear response for aerial predators and another quite different fear response for terrestrial predators (e.g., the junglefowl *Gallus gallus* (Collias 1987)). More fine-grained systems can readily be imagined—a bird's unconstrained optimal response to snakes would no doubt be different from its unconstrained optimal response to small mammalian carnivores. So the idea that the environment itself sets a determinate number of problems that impose a structure on the mind is inadequate. Something must determine how finely an organism perceives its adaptive environment. One suggestion would be that grain size is optimized given constraints such as the costs of obtaining and processing information. Those costs, of course, will be a function of existing mental structure, among other things. Another obvious candidate is the capacity of the organism to developmentally disassociate the mechanisms that respond to the separate stimuli. I do not pretend to have a final solution to this problem, and it is in any case clear that empirical research is needed as well as conceptual clarification. However, I suggest that any adequate treatment of the grain problem will need to combine the idea that the emergence of separate mental modules is influenced by the selective advantages of a more fine-grained response to the environment with the recognition that the environment does not contain any determinate number of separate problems. One way to satisfy these two desiderata would be to postulate a coevolutionary process in which the psychological phenotype of an organism at some point in time imposes a structure on the environment (for example, dividing it into "aerial predation" and "terrestrial predation") and to postulate that intrinsic features of the environment then afford the possibility of subdividing that structure (for example, into "terrestrial–snake" and "terrestrial–cat or quoll"). A coevolutionary picture of this kind would be consistent with Robert Brandon's (1999) claim that the units of phenotypic evolution are simultaneously developmental modules and ecologically meaningful units that correspond to some set of features in the "selective environment" (sensu Brandon 1990). Why should each developmental module correspond to a single ecological problem? The coconstructing relationship between developmental modules and ecological "problems" would tend to ensure that, as a result of evolution rather than as a matter of definition, these two ways of "parsing the phenotype" would coincide on the same units. Ecological problems are individuated in terms of developmentally dissociable responses, and the advantages of being able to dissociate responses contribute to the evolution of developmental modularity.

Work on developmental modules has coped well with the coconstructing relationship between modules and evolutionary problems and has turned this relationship into an actual object of evolutionary investigation. This should be the model for work on mental modules, but it has the cost that the modular structure of the mind cannot be inferred from an independently derived taxonomy of problems in the environment of evolutionary adaptedness. The less we know about the human mind, the less we know about the structure (in the relevant sense) of the environment that shaped it. Psychological modules will have to be discovered through a reciprocal growth of knowledge based on simultaneously empirical investigation of the mind and its development and ecological/evolutionary modeling of mental evolution. Work of this kind in evolutionary psychology would resemble contemporary evo-devo research into the evolution of morphology.

Homology

The Unity of Type and the Conditions of Existence

Like Darwin, evolutionary developmental biologist Brian Hall defines evolution as "descent with modification" (Hall 1992, 10). There is a difference of emphasis here with at least some contemporary neo-Darwinists. I suspect that Richard Dawkins, if asked to define evolution in as few words as possible, might be happier with the formulation "adaptation by natural selection." This would better express his commitment to what Peter Godfrey Smith has called "explanatory adaptationism"—the view that the overwhelmingly important task of biology is to explain the adaptation of organisms to their local environment (Godfrey-Smith 1999, 2000). Hall's preferred formulation draws equal attention to the other major phenomenon that evolution explains, namely, comparative morphology. While both tuna and dolphin are superbly adapted as fast-swimming predators, one is built on a mammalian body plan and the other on the body plan of the teleostei (modern ray-finned fish). These two explanatory projects are, of course, not only compatible, but, as Darwin (1964, 206) famously recognized, they are two aspects of one process:

> It is generally acknowledged that all organic beings have been formed on two great laws—Unity of Type and the Conditions of Existence. By unity of type is meant that fundamental agreement in structure, which we see in organic beings of the same class, and which is quite independent of their habits of life. On my theory, unity of type is explained by unity of descent. The expression of conditions of existence, so often insisted upon by the illustrious Cuvier, is fully embraced by the principle of natural selection. For natural selection acts by either now adapting the varying parts of each being to its organic conditions of life; or by having adapted them in long-past periods of time.

In previous work I have argued that these two aspects of evolution give rise to two patterns in the distribution of biological forms. Evolution is a matter of genealogical actors playing ecological roles (Hull 1987), and the units of evolution need to be classified both genealogically and ecologically to capture this fact. Hence, organisms and parts of organisms are classified in terms of common descent, as we see in modern systematics and in classifications by homology in sciences such as anatomy, physiology, and comparative morphology. The same organisms and parts are classified in terms of their ecological role, as we see when organisms are classified into ecological categories like predator and prey or when parts are classified by their adaptive function (Griffiths 1994, 1996a, 1996b, 1997; Goode and Griffiths 1995).

In most areas of biology, the interaction of the "two great laws" is well understood. The application of that understanding is the comparative method, perhaps the single most powerful epistemological technique available to biology. The comparative method can be seen as the use of each of the two patterns in the distribution of biological forms to illuminate the other. Contemporary Evolutionary Psychology, however, attends predominantly to the effect of the conditions of existence rather than to the unity of type. Most work in Evolutionary Psychology focuses on explaining psychological traits as adaptations to the "environment of evolutionary adaptiveness." These explanations categorize psychological traits in terms of their adaptive function (biological analogy). There is relatively little work explaining psychological traits as the result of descent from a common ancestor, and thus explaining them as homologues. This emphasis is perhaps due to the fact that explanatory adaptationists like Dawkins have acted as the public face of evolutionary biology in recent decades.

The situation in contemporary Evolutionary Psychology is in stark contrast to the situation in the 1950s and 1960s, when Konrad Lorenz and Niko Tinbergen reintroduced a Darwinian approach to psychology to the English-speaking world (Burkhardt 1983). In his Nobel Prize acceptance speech in 1973, Lorenz identified the main contribution of ethology as the recognition that behavior as much as morphology could be treated as a topic in comparative biology. His "good old Darwinian procedures" (Lorenz 1966b, 274) require that behavior be homologized before adaptive explanations are advanced. Identifying where a behavior fits into the comparative pattern is a crucial step in constructing an evolutionary explanation for at least two reasons. First, it determines character polarity—the precursor state from which the current state evolved. Offering adaptive explanations without knowing character polarity is like setting out to explain the Revolutionary War without knowing if the United States seceded from the United Kingdom or vice versa (O'Hara 1988). The second reason why classifying by homology is important is because it allows the application of the comparative method. The best positive evidence for an adaptive explanation is a correlation between an ecological factor and the trait that is hypothesized to be an adaptation to that factor. A fundamental methodological principle in measuring such correlations is not to count the number of existing species that display the trait in association with the ecological factor, but to count the number of independent originations of the trait (homologies) in association with the original factor. Conversely, the easiest way to falsify adaptive hypotheses is usually to test predictions about the comparative pattern of the homologous traits that figure in those hypotheses.[7]

The emphasis on analogy in contemporary Evolutionary Psychology is shared by contemporary philosophy of psychology. With a very few exceptions (Griffiths 1997, 2001; Matthen 1998, 2000), philosophers of psychology seem to be simply unaware of the possibility of classifying behavioral, psychological, and neurological states by homology. Philosophers have hotly debated the value of thinking about the mind in terms of adaptive function, but both sides have assumed that if biology offers valuable explanations of psychological traits, then the traits it explains must be defined by their adaptive function. Just how deeply engrained this idea has become is clear from philosopher of psychology Valerie Hardcastle's response to my (1997) proposal that emotional states should be classified by homology, so that, for example, fear in the rat would be classified with fear in humans, since both are modified forms of a response to danger that existed in their common mammalian ancestor, while fear in squid would be excluded (the rationale for this proposal is given later). Rejecting this proposal, Hardcastle (1999, 244) writes that "they [scientists] want to know what various structures are doing now in an organism, not what led to their being there over evolutionary time.... Neurophysiologists and neuropsychologists want to know what roles and responsibilities isolated brain structures have, and not so much why we have them in the first place. Though an evolutionary perspective can certainly be useful in thinking about functions and individuating psychological categories, the buck doesn't stop there." Hardcastle has incorporated the idea that evolution equals adaptation so thoroughly that she apparently mistakes my proposal to classify by homology for a proposal to classify by analogy (adaptive purpose). Misunderstandings aside, I think Hardcastle's position and my own are actually very close. She is happy to classify brain structures in the traditional way using anatomical categories like "amygdala" and "cingulate cortex," which, like all traditional neuroanatomical categories, are categories of homology. Conversely, I agree with her that knowing the actual causal role of brain structures is the primary goal of neuroscience. On my account, homology is of interest to neuroscientists primarily because of the need to use animal models for research into those causal roles.

Homology and Evolutionary Psychology
Homology as an organizing principle for the study of the mind has enormous potential but has hardly been exploited in Evolutionary Psychology. I will not attempt here to convince those who are, on general grounds,

skeptical of the heuristic value of evolutionary thought for psychology. I will only suggest that if what is wanted is a heuristic organizing principle for psychology, and particularly neuropsychology, based on evolution, then there are a number of reasons why psychological traits defined by evolutionary homology may be better targets for scientific study than psychological categories defined by evolutionary analogy. The first reason is that assignments of homology may be less epistemically demanding than assignments of adaptive function. This is surely a good reason to use classifications based on homology wherever possible; working on categories that are hard to establish firmly and whose membership consequently tends to fluctuate with changes in theory is simply inconvenient. I am concerned not to be seen as defending the position, sometimes seen in the literature on "antiadaptationism," that almost nothing can be known about the adaptive, evolutionary function of traits. In many cases well-confirmed knowledge of adaptive function can be obtained, typically by the use of comparative methods. But philosophers of mind and Evolutionary Psychologists have erred in the other direction, supposing that reasonable estimates of adaptive function can be obtained with surprisingly little work and that knowledge of adaptive function is easier to come by than knowledge of homology.

One reason why assignments of adaptive function are epistemically demanding was extensively discussed in the second section of this chapter (see also Stotz and Griffiths 2002). The idea that the mind has been shaped by a set of independently defined problems in the "environment of evolutionary adaptedness" ignores the coevolutionary relationship between niche structure and phenotypic structure. The less we know about a phenotype, the less we know about the parameters of niche space that constitute the selective environment of that phenotype. Thus, our ability to infer the adaptive pressures that shaped the human mind will be proportional to our understanding of the human mind (and the primate mind more generally). This severely undermines the potential of "adaptive thinking" as a heuristic method for defining complex mental functions we do not yet understand.

This reservation about "adaptive thinking" does not speak directly to the question of the *relative* difficulty of classifying by adaptive function or by homology. A second argument, however, speaks very directly to this issue. As I discussed briefly above, it is extremely difficult to test adaptive hypotheses without knowing where the adapted trait and other traits to which it is functionally related fit into the comparative pattern—that is, without knowing to what they are homologous. Since homology is assessed either from background knowledge of relatedness (today often derived from molecular systematics) or by the traditional operational criteria described below, knowledge of homology does not depend on knowledge of adaptive function. Hence, while homology can typically be determined in ignorance of adaptive function,[8] assignments of adaptive function in ignorance of the relevant facts about homology are almost untestable.

In an influential paper Ronald Amundson and George V. Lauder (1994), a philosopher of science and a functional morphologist respectively, have offered a third argument that assignments of homology are less epistemically demanding than assignments of adaptive function. Amundson and Lauder claim that homology is a more "observational" and less "inferential" concept than adaptive function. This claim has been strongly disputed by Karen Neander, a long-standing advocate of the importance of classifications by adaptive purpose in the sciences of the mind. Neander (2002, 409) argues that "it would be hard to choose between function and homology as to which of them was more or less inferential or observational. In order to determine homologous relations, we have to make inferences about evolutionary history, just as we do when we determine normal *[adaptive]* functions."

What Neander says is quite correct, but Amundson and Lauder have not forgotten that the Darwinian homology concept is defined in terms of common descent. Their point as I understand it is that homology is a highly operational concept and that the operational criteria for judging homology carry a great deal of weight when compared to its theoretical definition. In fact, there is good reason to suppose that if future empirical findings place the operational criteria of homology in conflict with a theoretical definition of homology (e.g., Darwinian homology), then "homology" would continue to refer to whatever turns out to be the property picked out by the operational criteria.[9] For this reason, judgments of homology tend to be relatively stable in the light of evolving background knowledge.

The operational criteria of homology were developed by the great descriptive embryological tradition of the first half of the nineteenth century. Richard Owen's (1843, 374) definition of a homologue, "The same organ in different animals under every variety of form and function," was given at a point where the theoretical definition of homology was exceedingly obscure. Owen himself offered a theory of "archetypes," while his contemporary Karl Ernst von Baer explained homology in terms of shared developmental potentials in the egg. The inability to agree on a theory of homology did not prevent early nineteenth-century biologists from developing the criteria of homology to the point where strong scientific consensus could be established on even subtle, distant homologies. Thus, well before Darwin dared to announce his theory in public, some of the most powerful supporting evidence for that theory had already been established as biological orthodoxy. For example, M. H. Rathke had established the homology of the hyoid (hyomandibular) bone in tetrapods and the second gill arch (hyoid arch) in early fish, and von Baer had identified the notochord—the homologous structure that defines the chordates and from which the backbone develops in vertebrates. The criteria they used—such as the relative position of parts in an abstracted representation (body plan) and the embryological origin of the parts—have remained in continual use since then (e.g., Remane 1952).[10] The concept of "Darwinian homology" is therefore best conceived as a Darwinian explanation of the *phenomenon*[11] of homology: the existence of "corresponding parts" is explained by descent from common ancestors (Griffiths 2006). This interpretation is strongly reinforced by current theoretical debates over the definition of homology.[12] The evolutionary developmental biologist Günter Wagner has argued that the Darwinian definition of homology leaves unexplained what makes a part of a parent homologous to a part of an offspring and merely relies on the traditional criteria of homology to identify these relationships. Wagner calls for the development of a "biological homology concept" (Wagner 1989, 1994, 1999) that would define homology in terms of the developmental mechanisms that generate correspondences between parent and offspring. If successful, this would provide a mechanistic explanation of the success of the traditional criteria of homology in identifying stable phenomena. Another criticism of the Darwinian homology concept is that it does not apply straightforwardly to serial homologues, since these homologous parts of a single organism, such as vertebrae, need not share a common ancestor even in the extended, embryological sense of "ancestor." Proposals to capture the phenomenon of serial homology by defining homologues as characters caused by the same biological information are perhaps best regarded as promissory notes for some developmental homology concept, such as Wagner's, since the only operational definition that that could correspond to this proposal at present would define homology in terms of shared gene expression and this is clearly inadequate. Traits that are not homologous can be built from the same genes, and homologous structures can be identified even when the genes involved in the relevant developmental pathways have been substituted by evolutionary change.[13] But to return to the central point of this discussion, proposals to redefine homology have a striking feature: they take specific judgments of homology based on widely accepted operational criteria as firm ground to stand on while arguing for revision in the theoretical definition of homology.

I have been arguing that specific homologies constitute relatively theory-independent phenomena, and that scientific confidence in these categories derives from the track record of certain operational criteria developed during the long history of morphology and allied sciences. Hence, classifications in terms of homology will be relatively stable across changes in background theory. This is an obviously desirable property for categories that one science (psychology) is going to borrow from another (biology) in the hope of getting an external guarantee that nature is being carved at its joints. But does this establish that judgments of homology will be *more* stable than judgments of adaptive function? Enthusiasts for adaptive function can point to their own pretheoretical phenomenon—the apparent design of many traits in relation to the organism's way of life. The tradition of natural theology identified design as a manifest fact that cried out for explanation and the scientific biology of design—Cuvier's "law of the conditions of existence"—was well established before Darwin came along. It is arguable, however, whether the study of design before Darwin had generated a definitive methodology for assigning purposive functions to the parts of animals comparable to the methods of early nineteenth-century comparative morphology. It is still more arguable whether specific assignments of purposive function have exhibited since that time the same sort of stability as assignments of homology. Darwinians agree with natural theologians that organisms show evidence of design, but they frequently disagree about what they are designed for: Richard Dawkins and Pope Benedict agree that sex in *Homo sapiens* has one or more proper functions but disagree about which. More seriously, sociobiologists and others who reject naive group selectionist models of evolution thereby reject a wide range of claims about the adaptive function of specific behaviors made by the classical ethologists—Konrad Lorenz's work on the functions of aggression being a famous example (Lorenz 1966a; Wilson 1978). The persistent tendency of twentieth-century biology to redefine *adapted* to mean "the product of natural selection" and to internalize this redefinition to the point where the older concept is almost forgotten (Godfrey-Smith 1999) also suggests that, unlike homology, adaptive function is a concept for which current theory takes precedence over old operational criteria. Ultimately, however, a proper assessment of Amundson and Lauder's claim that assignments of adaptive function are more "inferential" than assignments of homology would require that we settle some complex debates about the meaning of function statements in biology, something that I cannot attempt here.[14]

The Argument from Causal Depth

I have considered three arguments for the view that assignments of adaptive function are more epistemically demanding than assignments of homology. But even if this were established, it would only be a subsidiary argument for the more general conclusion that homology is the best candidate for a heuristic classificatory principle for psychology, and particularly for neuropsychology. The main argument for this conclusion is what might be called the "argument from causal depth." Categories of analogy group together traits that resemble one another in the causal role they fulfill in a containing system or in the causal role they play in interaction with the environment. Categories of homology group together traits that resemble one another in the underlying mechanisms by which they fulfill one or more causal roles. Psychology is concerned to discover the mechanisms by which tasks are performed and there are typically several ways to perform a task. Hence homology groups like with like in the respects relevant to psychology. Neuropsychology is the discipline among the sciences of the mind in which the premises of this argument are most obviously correct, and hence in which it carries most weight. I have presented this argument at length in other publications (Griffiths 1994; 1997, chaps. 7–8; 1999), so I can be relatively brief here.

First, for the most basic forms of scientific inference to be valid, they must be applied to categories that correspond to some inherent structure in the subject matter and not merely to the whim of the person who creates the classification. Such categories have traditionally been termed "natural kinds" (but see Brigandt 2003b; Griffiths 2004). The very idea of repeating an experiment, for example, presumes that the new subjects are "of the same kind" as the earlier subjects. Extrapolation from samples presumes that there are other things "of the same kind" as the samples. The fundamental property that underlies these and other scientific inferences is "projectibility": the existence of some grounds for supposing that correlations between the different properties of the samples can be "projected" onto unexamined members of the category (Goodman 1954). Thus, for example, chemical elements are projectible with respect to their chemical properties. If a sample of gold dissolves in aqua regia, probably other samples will too. The grounds for projecting observations made on samples of chemical elements used to be merely empirical: chemistry conducted in this manner was highly successful. Today there are deeper reasons in physical theory for supposing classification by atomic number to define projectible categories for the purposes of chemistry.

Second, chemical elements—despite being the traditional flagship example of "natural kinds"—turn out to be a rather restrictive special case. In many sciences, extrapolations of even the most sensible kind from the best possible data are not absolutely reliable. No one supposes, for example, that even the best diagnostic categories in medicine will allow deterministic predictions of treatment outcomes. Hodgkin's lymphoma is nevertheless a very useful category in oncology. So this aspect of the traditional model of natural kinds needs to be relaxed. "Projectibility" is a matter of degree, and any category that allows better than chance predictions to be made from samples to unexamined instances has a degree of projectibility. Science seeks the strongest projectible categories that are available in the domain being investigated.

Third, another feature of chemical elements that turns out to be merely a special case of a wider phenomenon is that their being "of the same kind" is a matter of their sharing an intrinsic structure. This requirement needs to be relaxed in the life sciences. Species are projectible categories for the purposes of many sciences—anatomy, physiology, molecular biology, and psychology, among others. Findings about one rat can reliably be projected onto other rats in all these sciences. Since the modern synthesis, however, it has been accepted that natural populations are pools of variation, that they are continually evolving, and consequently that it is futile to try to define species in terms of a common intrinsic structure. Instead, we explain the projectibility of species categories by the fact that members of a species are part of a pattern of ancestry and descent. The members of a species inherit many shared features, and interbreeding and ecological forces ensure that species members at any one time are all very similar.

Fourth, different categories are projectible with respect to different domains of properties. Both analogies and homologies are projectible categories, but only with respect to the specific domains of properties that biological theory links to them. Analogous traits share a common function, so discoveries about what it takes to perform that function or to perform it optimally can be projected from a trait in one species to analogous traits in other species. Categories of analogy are thus suited to the study of the process of adaptation. Optimal foraging theory, for example, can be tested on any organism that forages, whether it is a bird, a mammal, or a snake. If the theory is confirmed, it can be applied to other foragers with increased confidence. In contrast, homologous traits need not share a function, but they do share a common ancestor: homologues are modified forms of a single, ancestral trait. Because parents resemble their offspring in the whole gamut of functional and structural properties, discoveries about a very wide range of properties can be projected from a trait in one species to homologous traits in other species. This is why anatomy and physiology are structured around categories of homology. We expect discoveries in morphology, biomechanics, physiology, biochemistry, and a host of other domains to be projectible from one snake to another snake with a reliability proportional to the taxonomic distance of the two species in a phylogeny of the group.

If these considerations are applied to the study of psychology and neuropsychology, the "argument from causal depth" emerges automatically. Suppose two animals have psychological traits that are analogous—perhaps they are both mechanisms for predator detection. Then we should expect both performances to look like a solution to a signal-detection problem (Godfrey-Smith 1991). For example, we might expect sensitivity (the probability of making type II errors) to be a positive function of the value of the tasks the organism is engaged in while it monitors for predators. All that follows, however, about the computational processes that evaluate the relevant noise/signal ratios or whatever surrogate measure the organism uses, is that their output will approximate the optimal phenotype predicted by signal-detection theory. Even less follows about how the relevant computations will be realized in the animal's brain. Some prominent Evolutionary Psychologists have arrived at these conclusions independently by reflecting on the relationship between David Marr's "levels of analysis" in cognitive science and the theory of natural selection: "When applied to behavior, natural selection theory is more closely allied with the cognitive level of explanation than with any other level of proximate causation. This is because the cognitive level seeks to specify a psychological mechanism's function, and natural selection theory is a theory of function" (Cosmides and Tooby 1987, 284).

In contrast to this case of analogy, suppose two animals have psychological traits that are homologous—the basic emotion of fear in humans and fear in chimpanzees, for example. We can predict that, even if the function of fear has been subtly altered by differences in the ecology of the two species (the different meaning of *danger* for humans and for chimps), the computational methods used to process danger-related information will be very similar and the neural structures that implement them will be very similar indeed. After all, Joseph LeDoux's (1996) widely accepted account of the implementation of fear in the human brain is largely, and legitimately, based on the study of far more distantly homologous processes in the rat.

Inferences to shared mechanism based on homology are not absolutely reliable, but they are reliable enough to build good science with, and, what matters in this context, they are more reliable than inferences to shared mechanism based on analogy.

The argument from causal depth seems to me hard to evade. Psychology is in the business of uncovering the mechanisms that produce behavior. This is even more evident in the case of neuropsychology. Hence these disciplines seek categories that are heuristically valuable for the study of underlying mechanisms. It is a truism in comparative biology that similarities due to analogy (shared adaptive function) are "shallow." The deeper you dig the more things diverge. Bat wings and bird wings have similar aerodynamic properties but their structure diverges radically, despite their deep homology as tetrapod limbs. In contrast, similarities due to homology (shared ancestry) are notoriously deep—even when function has been transformed, the deeper you dig the more similarity there is in mechanisms.[15] Threat displays in chimps look very different from anger in humans, but the more you understand about the facial musculature, the more similar they appear. The same is almost certainly true of the neural mechanisms that control them. The only reason to suppose that psychology is different from morphology is to suppose that, unlike anatomical functions, psychological functions can only be realized by one set of computations and those computations can only be realized by a single neural architecture. With a few prominent exceptions, that claim is implausible.

Conclusion

I have contrasted three concepts of modularity: developmental, functional, and mental modularity. Developmental modules are the parts of a semidecomposable developmental system. Functional modules are the parts of a semidecomposable phenotype, such as a neural architecture. Mental modules are "virtual modules"—aspects of an organism's psychological performance profile that can be developmentally and functionally dissociated from one another in such a way as to allow performance in one domain to be optimized independently of performance in the other. This may be because the mental module maps simply onto one or more functional modules, but it need not be. Hence neuropsychology and Evolutionary Psychology may not recognize the same list of modules. The same may be true of mental modules and developmental modules, but I have not attempted to establish this here. I have, however, argued that an adequate account of the relationship between modularity and adaptation must recognize a coevolutionary process in which the organism's capacity to developmentally dissociate performances is part of what makes separate adaptive problems "separate," while the existence of separate adaptive problems influences the evolution of developmental modules.

I have commented on the neglect of homology as a principle of categorization, both in Evolutionary Psychology and in the philosophy of psychology. I presented three arguments suggesting that the assignment of homology may be less epistemically demanding than assignments of adaptive function. More definitively, the "argument from causal depth" provides a powerful reason to prefer homology to analogy for the purposes of psychology, and particularly neuropsychology. Extrapolation of experimental results in neuropsychology to homologues of the experimental system are warranted, but similar extrapolations to analogues are not warranted. The most fundamental scientific inferences—such as induction—are thus warranted in one case and not in the other.

My aim in this chapter was to explore the potential value of the central themes of evolutionary developmental biology for the study of psychology from an evolutionary perspective. I hope I have succeeded in convincing at least some readers that these themes could prove as revolutionary in the study of psychology as they have already proved in the study of morphology.

Notes

This chapter was presented at a workshop at Duke University in July 2002. An earlier version was presented at the Pittsburgh-London Workshop in Philosophy of Biology and Neuroscience, September 2001, at Birkbeck College, London, and I benefited a great deal from subsequent discussion with two members of that audience, Jackie Sullivan and Jim Bogen. My ideas on homology have developed considerably since this chapter was completed and a fuller treatment can be found in (Griffiths 2006).

1. Key works in evo-devo include Hall 1992; Raff 1996; Arthur 1997. On the development of evo-devo as a research tradition, and its relation to neo-Darwinism, see Love 2001; Love and Raff 2003; Raff and Love 2004; Love, forthcoming. Kim Sterelny (2000) has argued that it is unlikely that evolutionary developmental biology will lead to findings that are inconsistent with contemporary neo-Darwinian orthodoxy. Sterelny takes a broad view of what is orthodox, but it is certainly true that no leading researchers in the field think of their work as inconsistent with fundamental tenets of Darwinism. In this respect, evo-devo differs from some other "developmental" approaches to evolution, such as process structuralism (Ho and Saunders 1984).

2. I capitalize Evolutionary Psychology/Psychologists when these are used as proper names referring to the views of Leda Cosmides, John Tooby, and their collaborators (Barkow, Cosmides, and Tooby 1992).

3. Since this chapter was completed a major new collection on developmental modularity has appeared (Schlosser and Wagner 2004). For a philosophical look at some of the issues raised in that volume, see Winther 2001.

4. For a brief history and critique, see Shallice 1988, 245–253.

5. I believe that the same is true of developmental biology, and that a mental module "virtual" with respect to developmental modularity is possible, but I will not argue for this conclusion here.

6. Richard Lewontin (1982, 1983) was an important early advocate of the idea that niches are shaped by populations as much as the reverse. A careful philosophical analysis of the notion of the environment in this context has been constructed by Robert Brandon (Brandon 1990; Brandon and Antonovics 1996) and discussed elsewhere (Sterelny and Griffiths 1999; Godfrey-Smith 1996). An important program of empirical and theoretical research into "niche construction" has been pioneered by John Odling Smee and his collaborators (Odling-Smee 1988; Laland, Odling-Smee, and Feldman 1996, 2001; Odling-Smee, Laland, and Feldman 2003). A discussion of these issues from the perspective of "developmental systems theory" can be found in Griffiths and Gray 2001.

7. For a philosophical overview of the use of the comparative method in the study of adaptation, see Griffiths 1996b; 1997, chap. 5; Sterelny and Griffiths 1999, chap. 10. For full treatment, see Harvey and Pagel 1991; Brooks and McLennan 1991.

8. This is denied by Karen Neander (2002). For a response, see Griffiths 2006.

9. Some philosophers would put this point by saying that Darwinian homology is the "secondary intention" of the homology concept, while the traditional criteria are its "primary intention."

10. Although they have been supplemented by other criteria and the weight placed on the various criteria has differed over time and between different fields that make use of the homology concept (see Brigandt 2002, 2003a).

11. In the sense popularized by "new experimentalists" in the philosophy of science: we have theory-independent arguments for supposing that the phenomenon is not an artifact and that its existence should be explained by our theories of the domain in question (Hacking 1981, 1983, [1982] 1991; Franklin 1986, 1990). See also Brigandt 2003b for a taxonomic application of these ideas.

12. Three major collections of papers on these issues are Hall 1994, 1999; Wagner 2001. For philosophical analysis of this literature, see Brigandt 2002, 2003a.

13. The need to distinguish "levels of homology" has become increasingly apparent in recent years. For example, the fact that the gene *bicoid* controls the formation of the anterior-posterior axis in *Drosophila* but not in other dipteran species does not undermine the claim that the elements that form along that axis in *Drosophila* (and indeed the axis itself) are homologous to those in other insects (Laubichler and Wagner 2001, 65–66). Likewise, the cascade of gene expression that induces masculinization of the fetus in *Ellobius* rodents and the sexual characteristics that result from that process are homologous to those seen in other mammalian species, despite the fact that some *Ellobius* species have lost the Y chromosome and the SRY gene (Just et al. 1995). The lesson of these examples is that evolution can preserve a structure at one level while transforming the underlying mechanism that produces it and, conversely, evolution can redeploy an existing mechanism to underpin the development of an evolutionary novelty. Arguably, behavior can also form an independent level of homology, with the anatomical structures that support the behavior being transformed over time while the behavior (e.g., the biomechanical profile of a movement) remains the same (Lauder 1990).

14. So different are their background assumptions about how to understand function statements that Amundson and Lauder on the one hand and Neander on the other often seem to be talking past one another. Amundson and Lauder distinguish between "causal-role" and "selected-effects" functions. They take it for granted that functional morphologists can describe the causal-role function of a trait (for example, the biomechanics of a behavior) without knowing the selection pressures that led to the evolution of that trait. In stark contrast, Neander (2002, 409–413) believes that all references to the function (and indeed structure) of traits in anatomy, physiology, comparative morphology, and so on presuppose claims about the selection pressures that brought those traits into existence. For a discussion, see Griffiths 2006.

15. Neander adds a useful qualification to this point, namely, that what she calls "functional homologues" will be even more reliable guides to mechanism. If I understand her correctly, these are homologies all of whose instances have continued to be selected for the same adaptive purpose. For similar reasons, analogies nested within a taxonomic category (such as convergent features of aquatic *mammals*) should be more reliable guides to mechanism that analogies not so nested (Griffiths 1994, 218–219).

References

Amundson, R., and G. V. Lauder. 1994. Function without purpose: The uses of causal role function in evolutionary biology. *Biology and Philosophy* 9(4): 443–470.

Arthur, Wallace. 1997. *The Origin of Animal Body Plans: A Study in Evolutionary Developmental Biology*. Cambridge: Cambridge University Press.

Barkow, Jerome H., Leda Cosmides, and John Tooby, eds. 1992. *The Adapted Mind: Evolutionary Psychology and the Generation of Culture*. Oxford: Oxford University Press.

Brandon, R. 1990. *Adaptation and Environment*. Princeton, NJ: Princeton University Press.

Brandon, Robert N. 1999. The units of selection revisited: The modules of selection. *Biology and Philosophy* 14(2): 167–180.

Brandon, Robert N., and Janis Antonovics. 1996. The coevolution of organism and environment. In R. N. Brandon, ed., *Concepts and Methods in Evolutionary Biology*. Cambridge: Cambridge University Press.

Brigandt, Ingo. 2002. Homology and the origin of correspondence. *Biology and Philosophy* 17: 389–407.

———. 2003a. Homology in comparative, molecular and evolutionary biology. *Journal of Experimental Zoology (Molecular and Developmental Evolution)* 299B: 9–17.

———. 2003b. Species pluralism does not imply species eliminativism. *Philosophy of Science* 70(5): 1305–1316.

Brooks, D. R., and D. A. McLennan. 1991. *Phylogeny, Ecology and Behaviour*. Chicago: Chicago University Press.

Burkhardt, Richard W., Jr. 1983. The development of an evolutionary ethology. In D. S. Bendall, ed., *Evolution: From Molecules to Men*. Cambridge: Cambridge University Press.

Buss, D. M. 2000. *The Dangerous Passion: Why Jealousy Is as Essential as Love and Sex*. New York: Simon and Schuster.

Cohen, Jonathan D., and Frank Tong. 2001. The face of controversy. *Science* 293 (September 28): 2405–2407.

Collias, Nicholas E. 1987. The vocal repertoire of the red junglefowl: A spectrographic classification and the code of communication. *The Condor* 89 (510–524).

Cosmides, L., and J. Tooby. 1987. From evolution to behaviour: Evolutionary psychology as the missing link. In J. Dupré, ed., *The Latest on the Best: Essays on Optimality and Evolution*. Cambridge, MA: MIT Press.

———. 1992. Cognitive adaptations for social exchange. In J. H. Barkow, L. Cosmides, and J. Tooby, eds., *The Adapted Mind: Evolutionary Psychology and the Generation of Culture*. Oxford: Oxford University Press.

———. 2000. Evolutionary psychology and the emotions. In M. Lewis and J. M. Haviland-Jones, eds., *Handbook of the Emotions*. New York: Guildford Press.

Darwin, C. 1964. *On The Origin of Species: A Facsimile of the First Edition*. Cambridge, MA: Harvard University Press.

Ekman, Paul. 1972. *Emotions in the Human Face*. New York: Pergamon Press.

Fodor, J. A. 1983. *The Modularity of Mind: An Essay in Faculty Psychology*. Cambridge, MA: Bradford Books/MIT Press.

Franklin, Allan. 1986. *The Neglect of Experiment*. Cambridge: Cambridge University Press.

———. 1990. *Experiment, Right or Wrong*. Cambridge: Cambridge University Press.

Gaulin, Stephen J. C., and Donald H. McBurney. 2001. *Psychology: An Evolutionary Approach*. Upper Saddle River, NJ: Prentice Hall.

Godfrey-Smith, Peter. 1991. Signal, decision, action. *Journal of Philosophy* 88: 709–722.

———. 1996. *Complexity and the Function of Mind in Nature*. Cambridge: Cambridge University Press.

———. 1999. Adaptationism and the power of selection. *Biology and Philosophy* 14(2): 181–194.

———. 2000. Three kinds of adaptation. In S. Orzack and E. Sober, eds., *Optimality and Adaptation*. Cambridge: Cambridge University Press.

Goode, Richard, and Paul E. Griffiths. 1995. The misuse of Sober's selection of/selection for distinction. *Biology and Philosophy* 10: 99–108.

Goodman, N. 1954. *Fact, Fiction and Forecast*. London: Athlone Press, University of London.

Griffiths, Paul E. 1994. Cladistic classification & functional explanation. *Philosophy of Science* 61(2): 206–227.

———. 1996a. Darwinism, process structuralism and natural kinds. *Philosophy of Science* 63 (3, supplement): S1–S9.

———. 1996b. The historical turn in the study of adaptation. *British Journal for the Philosophy of Science* 47(4): 511–532.

———. 1997. *What Emotions Really Are: The Problem of Psychological Categories*. D. Hull, ed., *Conceptual Foundations of Science*. Chicago: University of Chicago Press.

———. 1999. Squaring the circle: Natural kinds with historical essences. In R. A. Wilson, ed., *Species: New Interdisciplinary Essays*. Cambridge, MA: MIT Press.

———. 2001. Emotion and the problem of psychological categories. In A. W. Kazniak, ed., *Emotions, Qualia and Consciousness*. Singapore: World Scientific.

———. 2004. Emotions as natural kinds and normative kinds. *Philosophy of Science* 71(5) (Supplement: Proceedings of the 2002 Biennial Meeting of the Philosophy of Science Association. Part II: Symposia Papers): 901–911.

———. 2006. Function, homology and character individuation. *Philosophy of Science* 73(1): 1–25.

Griffiths, Paul E., and Russell D. Gray. 2001. Darwinism and developmental systems. In S. Oyama, P. E. Griffiths, and R. D. Gray, eds., *Cycles of Contingency: Developmental Systems and Evolution*. Cambridge, MA: MIT Press.

Hacking, Ian. 1981. Do we see through a microscope? *Pacific Philosophical Quarterly* 62: 305–322.

———. 1983. *Representing and Intervening: Introductory Topics in the Philosophy of Natural Science*. Cambridge: Cambridge University Press.

———. [1982] 1991. Experimentation and scientific realism. In R. Boyd, P. Gasper, and J. D. Trout, eds., *The Philosophy of Science*. Cambridge, MA: MIT Press.

Hall, Brian K. 1992. *Evolutionary Developmental Biology*. New York: Chapman and Hall.

———, ed. 1994. *Homology: The Hierarchical Basis of Comparative Biology*. New York: Academic Press.

———, ed. 1999. G. R. Bock and G. Cardew, eds., *Homology*. Vol. 222, *Novartis Foundation Symposia*. Chichester: Wiley.

Hardcastle, Valerie Gray. 1999. Understanding functions: A pragmatic approach. In V. G. Hardcastle, ed., *Where Biology Meets Psychology: Philosophical Essays*. Cambridge, MA: MIT Press.

Harvey, P. H., and M. D. Pagel. 1991. *The Comparative Method in Evolutionary Biology*. Oxford: Oxford University Press.

Ho, Mae-Wan, and Peter Saunders, eds. 1984. *Beyond Neo-Darwinism: An Introduction to the New Evolutionary Paradigm*. Orlando, FL: Academic Press.

Hull, David L. 1987. Genealogical actors in ecological roles. *Biology & Philosophy* 2: 168–184.

Just, Walter, Wolfgang Rau, Walther Vogel, Mikhail Akhverdian, Karl Fredga, Jennifer A. Marshall Graves, and Elena Lyapunova. 1995. Absence of Sry in species of the vole Ellobius. *Nature* 11(2): 117–118.

Laland, K. N., F. J. Odling-Smee, and M. W. Feldman. 1996. The evolutionary consequences of niche-construction: A theoretical investigation using two-locus theory. *Journal of Evolutionary Biology* 9: 293–316.

Laland, Kevin N., F. John Odling-Smee, and Marcus W. Feldman. 2001. Niche construction, ecological inheritance, and cycles of contingency in evolution. In S. Oyama, P. E. Griffiths, and R. D. Gray, eds., *Cycles of Contingency: Developmental Systems and Evolution*. Cambridge, MA: MIT Press.

Laubichler, Manfred D., and Günter P. Wagner. 2001. How molecular is molecular developmental biology? A reply to Alex Rosenberg's reductionism redux: Computing the embryo. *Biology and Philosophy* 16(1): 53–68.

Lauder, G. V. 1990. Functional morphology: Studying functional patterns in an historical context. *Annual Review of Ecology and Systematics* 21: 317–340.

LeDoux, J. 1996. *The Emotional Brain: The Mysterious Underpinnings of Emotional Life.* New York: Simon and Schuster.

Lewontin, Richard C. 1982. Organism & environment. In H. Plotkin, ed., *Learning, Development, Culture.* New York: Wiley.

———. 1983. Gene, organism & environment. In D. S. Bendall, *Evolution: From Molecules to Man.* Cambridge: Cambridge University Press.

Lorenz, Konrad Z. 1966a. *On Aggression.* Trans. M. K. Wilson. New York: Harcourt, Brace and World.

———. 1966b. Evolution of ritualisation in the biological and cultural spheres. *Philosophical Transactions of the Royal Society of London* 251: 273–284.

Love, Alan. 2001. Evolutionary morphology, innovation, and the synthesis of evolution and development. *Biology and Philosophy* 18(2): 309–345.

———. Forthcoming. Morphological and paleontological perspectives for a history of evo-devo. In J. Maienschein and M. D. Laubichler, eds., *From Embryology to Evo-Devo.* Cambridge, MA: MIT Press.

Love, Alan C., and Rudolf A. Raff. 2003. Knowing your ancestors: Themes in the history of evo-devo. *Evolution & Development* 5(4): 327.

Matthen, Mohan. 1998. Biological universals and the nature of fear. *Journal of Philosophy* 85(3): 105–132.

———. 2000. What is a hand? What is a mind? *Revue Internationale de Philosophie* 214: 653–672.

Mount, A. B. 1964. The interdependence of the eucalpyts and forest fires in southern Australia. *Australian Forestry* 28: 166–172.

Neander, Karen. 2002. Types of traits: Function, structure and homology in the classification of traits. In A. Ariew, R. Cummins, and M. Perlman, eds., *Functions: New Essays in the Philosophy of Biology and Psychology.* Oxford: Oxford University Press.

Odling-Smee, F. John. 1988. Niche-constructing phenotypes. In H. C. Plotkin, ed., *The Role of Behavior in Evolution.* Cambridge, MA: MIT Press.

Odling-Smee, F. John, Kevin N. Laland, and Marcus W. Feldman. 2003. *Niche Construction: The Neglected Process in Evolution.* Vol. 37, *Monographs in Population Biology.* Princeton, NJ: Princeton University Press.

O'Hara, R. J. 1988. Homage to Clio, or towards an historical philosophy for evolutionary biology. *Systematic Zoology* 37(2): 142–155.

Öhman, A. 1993. Stimulus prepotency and fear: Data and theory. In N. Birbaumer and A. Öhman, eds., *The Organization of Emotion: Cognitive, Clinical and Psychological Perspectives*. Toronto: Hogrefe.

Owen, Richard. 1843. *Lectures on the Comparative Anatomy and Physiology of the Vertebrate Animals, Delivered at the Royal College of Surgeons, in 1843*. London: Longman, Brown, Green and Longmans.

Raff, Rudolf A. 1996. *The Shape of Life: Genes, Development and the Evolution of Animal Form*. Chicago: University of Chicago Press.

Raff, Rudolph A., and Alan C. Love. 2004. Kowalevsky, comparative evolutionary embryology, and the intellectual lineage of evo-devo. *Journal of Experimental Zoology (Mol Dev Evol)* 302B: 19–34.

Remane, A. 1952. *Die Grundlagen des Natürlichen Systems, der vergleichenden Anatomie und der Phylogenetik*. Königsstein: Otto Koeltz.

Schlosser, Gerhard, and Denis Thieffry. 2000. Modularity in development and evolution. *BioEssays* 22: 1043–1045.

Schlosser, Gerhard, and Günter P. Wagner, eds. 2004. *Modularity in Development and Evolution*. Chicago: University of Chicago Press.

Shallice, T. 1988. *From Neuropsychology to Mental Structure*. Cambridge: Cambridge University Press.

Simon, H. A. 1969. *The Sciences of the Artificial*. Cambridge, MA: MIT Press.

Sterelny, Kim. 2000. Development, evolution, and adaptation. *Philosophy of Science* 67 (supplement): S369–S387.

Sterelny, Kim, and Paul E. Griffiths. 1999. *Sex and Death: An Introduction to the Philosophy of Biology*. Chicago: University of Chicago Press.

Stotz, K., and P. E. Griffiths. 2002. Dancing in the dark: Evolutionary psychology and the problem of design. In F. Rauscher and S. Scher, eds., *Evolutionary Psychology: Alternative Approaches*. Dordrecht, the Netherlands: Kluwer.

Tooby, John, and Leda Cosmides. 1990. On the universality of human nature and the uniqueness of the individual: The role of genetics and adaption. *Journal of Personality* 58(1) (March 1990): 17–67.

———. 1992. The psychological foundations of culture. In J. H. Barkow, L. Cosmides, and J. Tooby, eds., *The Adapted Mind: Evolutionary Psychology and the Generation of Culture*. Oxford: Oxford University Press.

Wagner, Günter P. 1989. The biological homology concept. *Annual Review of Ecology & Systematics* 20: 51–69.

———. 1994. Homology and the mechanisms of development. In B. K. Hall, ed., *Homology: The Hierarchical Basis of Comparative Biology*. New York: Academic Press.

———. 1996. Homologues, natural kinds and the evolution of modularity. *American Zoologist* 36: 36–43.

———. 1999. A research programme for testing the biological homology concept. In B. K. Hall, ed., *Homology*. Chichester: Wiley.

———, ed. 2001. *The Character Concept in Evolutionary Biology*. San Diego: Academic Press.

Wilson, E. O. 1978. *On Human Nature*. Cambridge, MA: Harvard University Press.

Winther, Rasmus G. 2001. Varieties of modules: Kinds, levels, origins, and behaviors. *Journal of Experimental Zoology (Mol Dev Evol)* 291: 116–129.

7 Reproducing Entrenchments to Scaffold Culture: The Central Role of Development in Cultural Evolution

William C. Wimsatt and James R. Griesemer

Introduction

A perspective on cultural evolution that utilizes human cognitive development throughout the life cycle is critical to understanding how we learn, comprehend, use, and modify complex cultural structures. Evolutionary developmental biology, or "evo-devo" as it is now known to practitioners, has emerged within the last two decades to cast new light on evolutionary problems in half a dozen disciplines. We argue that to understand cultural evolution, we must take cultural development seriously. This requires a sociological view of culture in which cultural inheritance is understood to depend on varying patterns and modes of cultural development. "Thin" views of culture such as memetics and dual-inheritance models black-box human ontogeny and the developmental acquisition of cultural traits and describe cultural inheritance as only a mapping relation between "parents" and cultural "offspring." "Thick" views of culture take the phenomena of interest to be limited to human language users and assume such a rich social milieu in which human cognition and communication occur that it seems impossible for any other species to have culture. This of course also makes it totally mysterious how we could have evolved culture. Most of these writers reject any attempt to give an evolutionary account of the origins of culture or the nature of cultural change.

In this chapter, we argue that an account of units of cultural reproduction can open the black box of development left closed by replicator-based accounts of inheritance (whether cultural memes or biological genes). We utilize generalizable features of developmental processes, in particular, different modes of generative entrenchment, to support a "medium viscosity" theory of culture and cultural evolution. Applying an evolutionary perspective to a developmentally sensitive view of culture requires, in addition, a variational or populational perspective on units of cultural reproduction.

In other words, there must be *countable* units that develop and vary and that form populations of potential interacting reproducers in order for a (Darwinian) evolutionary theory to apply.

Existing theories make little significant use of developmental resources or explain how we acquire and use them (but see Caporael 1997; Ingold 2000; Bateson 2001). But this alone would not suffice: a focus on individual organisms or people, however they develop, still suggests an exclusively "bottom-up" approach to culture and cultural change. The approach must work both from "bottom-up" psychological and from "top-down" sociological and anthropological perspectives because culture is nothing if not shared. Complex cultural structures are not just givens of social organization, however. Nor are they usefully understood as the aggregation over time of a large collection of memes: we need to know why memelike things are attractive to individuals with particular histories, in particular niches or local contexts, given the rich structures already present. The evolution of such structures, how they become structurelike, their social and cultural differentiation, their transgenerational maintenance through recruitment and dissemination processes, and their elaboration through the directive activity of group organization on various scales all require attention to processes of individual human development and of entrenchment in enculturation.

An "evo-devo" theory of culture must not result in a new variety of methodological individualism: human learning and development are irreducibly social (Brewer 1997, 2004). The socializing trajectories of individual learning and development are unintelligible without the structured shaping, guidance, and reinforcement that social institutions and organizations provide. These are not just environments for learning, but interactive products of culture—both produced by and feeding back to form enculturated individuals. Adequate theory requires an integral account of both levels and their reciprocal interaction. Perspectives derived from our respective research programs on reproduction and generative entrenchment yield insights into these processes.

Our new grip on the nature of cultural inheritance emerges through reexamination of an apparent deep difference between biological and cultural inheritance: that the former invariably uses and indeed requires material overlap, while the latter is commonly taken not to (Griesemer 2000a). A process involves material overlap if later stages of the process include material parts that had belonged to earlier stages, for example, when biological offspring include material parts of their biological parents. The contrast is suggested by comparing the semiconservative replication of DNA, a process of direct material transfer of one of two nucleotide strands, with the disembodied transmission of information—for example, when someone reads a copy of this chapter in the library or online. The contrast is actually not quite so simple, and we would argue that more revealing instances of cultural transmission occur when offspring include persistent niche components engineered by their biological or cultural "parents," or when organizations at later stages include individual members, equipment, or physical plant persisting from earlier ones. Developing arguments to see why gives handles on other important properties of cultural inheritance and evolution, including the facts that it can take place at a much faster rate and can lead to the adoption of major "deep" cultural changes whose biological analogues would almost always be profoundly lethal mutations. So how does culture get away with such major changes, and do so, at least sometimes, at such a precipitous rate? These are crucial differences we seek to explain. They relate to the multiplicity of transmission channels, to easing and directing the production of variation, and to how various organizational structures can modulate escape from entrenchment.

Structurally, a scaffold is a supporting framework, often temporary relative to the time scale of what it supports. Scaffolds support materials, tools, agents, and processes, as when builders use scaffolding to support workers and their materials while erecting or repairing a building. Developmental psychologists extend the notion to functional processes critical in child development such as "attachment." When a child "attaches" to mother (placing trust in mother to be a safe guide), mother can then facilitate skill acquisition on the part of the child by creating a safe or otherwise supporting, for example, simplifying, context[1] in which the child can learn and practice a potentially threatening but advantageous activity, like walking, thus blocking or lessening negative fitness consequences of risky behavior. Skill acquisition is a key form of learning relevant to culture because it is *generative*, though it is not the only cultural process that is both crucial and generative. Acquisition of authority is another key generative form of cultural development not explored in this essay.[2] Social and cultural structures and activities that play a role in generating learning trajectories are both known as scaffolding—scaffolding, respectively as state and process (Bickhard 1992). According to Bickhard (1992, 35), "Functionally, scaffolding is precisely the creation of...bracketed trajectories of potential development through artificially created nearby points of stability." In addition to allowing "the child to accomplish something that he or she could not otherwise accomplish alone—it also allows the child to develop further *competencies* through being provided with such bracketings of 'normal' selection pressures. It is this further variation and selection development, made possible by the context-dependent successes, that makes scaffolding a critical aspect of the development of less context-dependent abilities."

Thus, two key features of functional developmental scaffolding are: (1) the lowering of "fitness barriers" to developmental performances or achievements (Bickhard's "artificially created nearby points of stability"), which (2) make accessible new competencies (capacities, skills or authority) that become the "self-scaffolding" of later developmental performances. Looking ahead, a third is generative entrenchment and its consequences. The dependencies created by increasing use of the new competencies make them essential, stabilize many of their features, and engender further elaboration of scaffolding on top of them.

In the United States, differential access to quality early education among blacks and whites and attempts to control broader aspects of the supporting (or not-so-supporting) environments has motivated attempted interventions from Head Start programs on. Scaffolding is not always so direct, nor is it confined to early development. In the remainder of this section, we describe several cases of scaffolding phenomena further analyzed in succeeding sections. In the first, we look at the advanced training of MDs in two empirical studies documenting the different degrees and kinds of scaffolding support received by men and women and its consequences for their career choices and commitments. In the second, we review the impact of the advent of the tutorial system at Cambridge University on the development of mathematical physics over a 180-year span. Both of these studies track individuals' trajectories through institutions, with the first focusing on individual development and the second also looking at the cumulative impact on those institutions. In the third, we consider the particular role of dwellings in culture, and then consider the spread of "kit" houses in America over thirty years in the early twentieth century. Here, we look at the institutional framework supporting this change and features that made this innovation attractive to individuals.

First Pass: Scaffolding the Development of People and Institutions

Scaffolding Cognitive and Professional Development in Advanced Medical Training
B. Wimsatt (1997) documents rich processes of cognitive and professional development along ten dimensions of encouragement and support among eighty residents and fellows of both sexes in advanced medical training. Her data illustrate the importance of scaffolding by mentors and significant

others in helping to acquire skills crucial to advanced cognitive development and the professional development continuous with it. Male and female medical residents and fellows recalled experiences during their training that led to career decisions they made. Men more than women, and increasingly at successive stages of training, chose "encouragement" by mentors as important reasons for their choices. Women tended to downplay their need for faculty encouragement, citing their independence, while for men, it was considered vitally important. However, several women (and no men) volunteered that they depended heavily on support from fathers or spouses.

On closer inspection, Wimsatt found that modes or dimensions of encouragement are (statistically) differentiated by sex. Men described a process in which their interests and cognitive development were shaped in a coordinated (Caporael 2003) interaction with mentors. Further, encouragement can run the gamut of support and investment from emotional to cognitive input, to active and extended sponsorship and promotion. Women more often reported receiving the first type—praise, help with specific tasks, or advice on career decisions—while men far more often reported receiving the more valuable recruitment into research projects and professional sponsorship and promotion. Some women relied primarily on physician-spouses or fathers for support. Interviews revealed that while both men and women felt encouraged in comparable numbers, they actually meant different things by it, had different standards for feeling encouraged, and to a large extent, relied on different sources of encouragement.

Along with this she found coordinated differential choices by men and women. The women rarely selected traditionally male-dominated disciplines, higher-level subspecialty training, or academic research, all of which are higher status and more highly rewarded options. Given that subspecialties and research involve further training, these choices marked asymmetries that increased at later professional stages.[3]

For families with children, it might seem plausible to attribute these differences to sex-role differences. But this cause is not determinative. A follow-up study ten years later interviewed ten couples who had children at the time of the first interview (B. Wimsatt 2001). In *all* cases where there were asymmetries in career development versus child care and home responsibilities, the parent who had more career capital at the time of the earlier interview was the one with fuller career commitments and lighter loads at home ten years later. In two out of ten cases, this was the woman, and so it was the male, also educated at the professional level, who was the stay-at-home parent.[4]

Women commonly marry older men, who tend to have more advanced careers. Thus even if both began on identical trajectories at comparable ages, age differences would tend to generate later asymmetries in age-specific career development, with the younger (usually female) partner more commonly scaffolding the development and maintenance of professional competency and performance of the "more committed" partner (usually male) by taking on otherwise necessary tasks and allowing them more time to dedicate to their careers. So pair-bonding plus gender-differential age structure can produce or enhance other scaffolding effects.

Scaffolding the Discipline of Mathematical Physics
Perhaps the most striking extended discussion of scaffolding in science never mentions the term. It is Andrew Warwick's (2003) study of the rise and development of mathematical physics at Cambridge between 1750 and about 1930. This is a period long enough that one can see not only rich interactions of scaffolding and individual development or career trajectories, but evolution and cumulative entrenchment elaborating a whole discipline, precipitated by significant and competitively evolving means of scaffolding. Mathematics and mathematically rich sciences provide particularly good opportunities to observe entrenchment because of the strong sequential dependence of later methods and results on earlier methods and results. And the content of mathematics and the mathematically rich sciences often utilize a series of successively richer models, each elaborating, but depending on and arising from the former, which they are often expected to regenerate as special limiting cases.

The case of mathematical physics in this period at Cambridge is a particularly rich surprise for those used to focusing just with an "internalist" lens: a change in the examination system from oral disputations to written exams precipitated a series of changes resulting in many of the modes of study and practice known to physics students around the world today, as well as an enormous amplification in the content they had to study. Originally motivated by a desire for increasing objectivity in the examinations, the move to written exams changed the kinds of questions that could be asked from more philosophical and conceptual discussions of texts like Newton's *Principia* to requests to produce proofs and derivations of increasing complexity. Oral effectiveness disappeared as a measure of performance and excellence to be replaced by the ability to reproduce and then increasingly to *produce* multiple new problem solutions on paper in very limited time. Technical competence for the first time became the primary measure of achievement. Those excelled who could construct arduous derivations quickly and had the stamina to withstand the stress of written examinations lasting *many days*. Overall rankings of students became possible, facilitated by the relative ease of generating quantitative measures for performance in mathematics, which also thereby came to assume greater importance in the curriculum. Increasing pressure to perform well amplified and enormously elaborated the practice of studying with outside tutors. Since these tutors were paid and evaluated by how well their students did on the written exams, they experimented with various practices to improve their effectiveness that have become common today.

Strategies invented within the next 100 years included constant practice with solving mathematical problems of diverse kinds. These were also arranged by difficulty, like the "A, B, and C problems" in the back of the math book, so familiar to modern physics students. The A problems served to exercise basic competency in the topic, the B questions to give more challenging, creative, but still standard extensions of it, and the C questions a challenge for the very brightest students, often drawn from the instructor's latest research or even unsolved problems. All of the following techniques emerged in this period: face-to-face teaching with interactive questioning rather than passive attendance at lectures, guided problem solving, regular examinations, textbooks, coaching manuscripts, graded exercises giving feedback on performance, grouping students by ability so that group members can all go at the same pace and learn from each other, and extended private study (Warwick 2003, 41). Because (rag) paper was then expensive (cheap wood-pulp paper came in the mid-nineteenth century), its use for taking notes and doing problem exercises might at first have seemed a wanton waste. The practice of taking notes in classes was employed originally by one tutor to increase the ability of his students to write quickly and for extended periods of time in preparation for the exams, but this exercise in physical fitness quickly took on its own cognitive functions, and spread—cost be damned.

Fitness for the exams and for the amplifying study regimes led to exercise breaks that themselves amplified competitively, and athletic performance became a correlative, and then an independent value. Performance here, but even more on the exams, became increasingly meritorious and the center of ceremony. Although this self-generated competition began in the mathematical sciences, it came to be imitated by the other disciplines. Tutors were outrunning the Cambridge lecturers in competence. They came to have increasing roles in setting questions for the exams and some moved on to become lecturers themselves. Cambridge, embarrassed that the expertise was coming from outside, moved to internalize the tutorial system in written examinations and extensive interactions with tutors, which subsequently spread laterally to other disciplines until it became characteristic of higher education at Cambridge and Oxford. This transformation accompanied changing (and secularized) criteria for lecturers: from "exemplary people taking holy orders and giving example" (c. 1750) to "technical masters capable of teaching cutting edge methods and doing new research" (100 years later). As Cambridge products of the system themselves became tutors and lecturers, many more graduates went out as teachers in the better secondary schools, taking their ideas of the proper teaching and content of the mathematical curriculum with them, thus escalating the number and level of preparation of mathematically inclined students entering universities.

Training for the exams involved studying questions from past exams, entrenching older topics in some places and generating a constant pressure to add newer ones across the board. As these accumulated, they came to be published—initially informally along with notes for their solution by tutors, and later by Cambridge itself. This became an accepted venue of professional publication—for *both* the questions and their answers! As the top graduates of the system became examiners in turn, the cumulative body of exams tended increasingly to include the research questions of the tutors and lecturers who taught them and their extent became so broad that one could not expect to cover them all in courses and training leading up to the exams, so diverse monographs and courses specializing in different topics in mathematical physics emerged and spread.[5] Globally, this marked a transition from studying the paradigmatic text as exemplar (in England, Newton's *Principia*) to studying textbooks and an expanding literature on problems and using methods never envisioned in the *Principia*.

The growth in specialization in all of the sciences—which has often and long been remarked on—took off increasingly, beginning in the nineteenth century. But that it should have been so strongly mediated in mathematics by a change in the structure of an exam, followed by the creative escalation of teaching methods by tutors outside of the official Cambridge educational system, is a striking demonstration of the power and multifarious dimensions of scaffolding in the elaborative evolution of the content and methods of a discipline. A similar story could be told for the earlier rise to hegemony of the medical profession in the United States through the first third of the twentieth century, where the Flexner plan for "scientific medicine" became a tool not only for the generation of increasingly scientifically informed and technologically dependent medical practice, but also for controlling access to medical education, and for disenfranchising other groups such as midwives and homeopaths, who also had prima facie calls in the "healing professions" (Abbott 1988).

These examples indicate the rich diversity of modes of scaffolding: attachment, guidance with complex tasks, ordered design and structuring of curricula and practice, multiple possible dimensions of acquisition of values, encouragement, help and promotion in professional development, and labor and emotional support of networked individuals, in part channeled and differentially conferred by sex, age, kind of relationship, and specialized competence. They also indicate a rich diversity of uses of scaffolding, which are not limited to communicating knowledge, practices, and values. The medical case particularly, but also the course of Cambridge mathematics, is full of examples of the emergence of organizations and of sources of power, as well as the shifting roles of education, institutions, incentives, standardization, and regulation. These penetrate deeply into the Gordian knot of culture, and leave many traces that we cannot pursue here. Many of these diverse modes and uses are not cognitive in kind, but all can affect cognitive development, and all can be inherited.

Does this also affect the common coinage of folk culture? How could it fail to do so? As we turn below to the Sears catalog and Sears kit houses, we can imagine a rural child watching parents search through a Sears catalog and feeling their values expressed in their choices and discussions of them. At the same time they learn to recognize the images, learn to read the text with the pictures, and learn how to work with and use the things ordered from the catalog. As Sears builds a market for their products, the children and adults build and elaborate a structured world of desires, hopes, plans, and subsequent actions, supported by the contents of the catalog, and embodied in their work and dwellings. We return to consider this case after discussing dwellings in general.

Second Pass: Developing Culture and Culturing Development

Culture and the Anthropology of Dwelling—A Theoretical Excursion
Dwellings have long occupied an important place in anthropology and archaeology. They are relevant for us, too, as prime exemplars and iconic metaphors for scaffolding that assume a central importance in our developmental account of cultural cumulation and change. They are complex expressions of material culture, having diverse forms, functions, styles, and traditions with long and rich histories that are specific to society and place, and at the same time reflect broad physical, biological, and human conditions and constraints. They also raise theoretical problems. For

example, should we view dwellings as products of builders' activities—things people make according to preconceived plans—or as ongoing processes of dwelling, continuous and contiguous with our history as a species? Ingold (2000) argues for the latter view on grounds compatible with our argument that a developmental view of culture is prerequisite to understanding cultural evolution.

Understanding dwellings as cultural artifacts and dwelling as a sociocultural process is problematic because the building-builder metaphor shares the limitations of biological dualisms such as phenotype-genotype and acquired-innate. These are frequently invoked to explain products of development, but have been vigorously attacked by developmentalists interested in the richly interactive, "systemsy" quality of units undergoing processes of repeated assembly with recurrent states—that is, life cycles (Oyama 2001; Caporael 2003). Although there is certainly value in treating human dwellers as agents and their dwellings as patients or products, there is also value in interpreting dwellings and dwellers as parts of a developmental system. Dwellings, or their interactions, are generative as are the dwellers. Dwellings afford opportunities to dwellers for their own personal development that they otherwise might lack: places for the stable, safe, and even leisurely acquisition and practice of skills, beliefs, customs, practices, and desires, in addition to the various modes of social exchange and bonding, and to satisfy the needs for protection from extreme weather, predators, places for food storage and preparation, mating, and rest. Dwellings, that is, scaffold the socialization, enculturation, development, sustenance, and reproduction of dwellers, and so play an important dynamic role as generators of those who in turn "make" dwellings and pass them on to others.

The dwelling perspective poses a challenge to "thin" descriptions of culture and cultural transmission offered by some who wish to reconcile biology and culture. Thin descriptions take culture to be any trait or phenotype transmissible by means other than genetics—for example, the dual-inheritance theories of Cavalli-Sforza and Feldman, Boyd and Richerson, and Odling-Smee and colleagues. As such, all sorts of social organisms, that is, organisms that interact with one another—which is to say, all organisms—are potentially capable of having culture, limited only by contingencies of habitat, biological capacity, and evolutionary history. Thin description thus leads to a wide scope for culture. The line between cultural and noncultural species becomes empirical and calls for theoretical explanation (e.g., Boyd and Richerson 1996). "Thick" descriptions take culture to be a symbolic realm open (so far) only to humans in which having and communicating meanings unrelated to the intrinsic or "natural" properties of things require extremely refined capacities for rational thought and articulate speech, that is, language. Such descriptions typically take the difference between humans and other animals to be categorical, even if the biological roots of culture evolved in a particular primate lineage (or lineages). Thick descriptions lead to a narrow scope for culture and makes the difference between cultural and noncultural species a matter of conceptual distinction or definition. We believe that this hides and refuses to address some of the very most interesting problems.[6]

We agree that there is a major difference in kind between human and animal cultures, but perhaps not one intrinsically greater than the differences before and not long after the advent and cultural developments issuing from the origins of written language.[7] The first attempts to push an analysis of culture further should be to see what aspects of a difference in kind are resolvable as differences in degree with the recognition of thresholds leading to major changes in behavior and emergent structure.

Nevertheless, "thin" models are unsatisfying for the vast majority of aspects of cultural phenomena. Like population genetics, which is based on a distinction between genotype and phenotype, thin models of culture presume that cultural transmission just means direct phenotypic transmission, in contrast to the direct transmission of genes or genotypes and indirect transmission of phenotypes assumed by Weismannism (Griesemer and Wimsatt 1989). Thin models of cultural transmission are thus sometimes described as "Lamarckian." These thin models for culture coexist with population genetics models because each is thought to describe a distinct transmission process, separated by a process of development through which genotypes make phenotypes.[8] In both, it is supposed that we need not take seriously the rich contextual details of either development or culture to gain a deep understanding of these evolutionary processes. They have no purchase on the rich details or even on their very *existence*. In abstracting away from these details, they must fall crucially short of an adequate account of the nature and transmission of culture. We believe that there is a productive middle ground that can do so.

How are we to understand cultural transmission when there is such deep disagreement about the nature of culture? Even among those seeking to reconcile biology and culture, there is disagreement. Some developmental systems theorists deny a distinction between biological and cultural evolution on grounds that there cannot be separate channels of inheritance (Griffiths and Gray 1994, 2001). Ingold (2003) writes sympathetically that "the changing forms and capacities of creatures—whether human or otherwise—are neither given in advance as a phylogenetic endowment nor added on through a parallel process of cultural transmission but emerge through histories of development within environments that are continually being shaped by their activities." Such sentiments seem to suggest "medium viscosity" views lying somewhere between thick and thin descriptions and their corresponding conceptions of culture, but give little help in analyzing their nature. They say more about how distinctions such as "nature-nurture" are inadequate than about workable alternatives. This often leaves the field to the obsolete machinery that they criticize.[9]

The difficulty in distinguishing the contributions of resources to a developmental system subject to an evolutionary process are substantial, but there are structural, functional, and dynamic roles that must be distinguished if we are ever to understand cultural processes of transmission and change. We think that the changing forms and capacities of human dwellers emerge through an interplay of reproduction, scaffolding, and generative entrenchment in groups acting and interacting in space and time. This can be illustrated with an example from the recent history of home building: the advent and spread of "kit" houses by Sears, Roebuck and Company in the early twentieth century.

Case Study: Sears Kit Houses

Sears, Roebuck and Company is widely known as a general merchandiser that grew by selling through mail-order catalogs. In the United States, this mail-order business began in Chicago, the birthplace of Sears, with the innovation of Sears' competitor, Aaron Montgomery Ward ("Montgomery Ward & Co. (Am. co.)," 2004). Ward worked as a clerk and a traveling salesman for the Chicago department store, Marshall Field (founded in 1865). "Department stores" had emerged in the mid-nineteenth century in Europe and spread to the United States.[10] Mail-order catalogs could be sent through the mails at very low cost because special subsidized postal rates (fourth class, for printed matter) were set by the U.S. Constitution to encourage the widespread distribution of newspapers, which the framers felt was essential for a well-informed citizenry (Chandler 2000, 8–9). Low postal rates encouraged increasing distribution of catalogs and the growth of mail-order businesses after the Civil War. Sears began selling watches by mail order in 1866 ("Sears, Roebuck and Company (Am. co.)," 2004). The advent of rural free delivery in 1896 and parcel post in 1913 by the U.S. Postal Service, together with the expansion of the railroads and the highway system ("Sears," 2004), as well as cheap paper and steam-press typesetting, furthered this commercial explosion, bringing the contents of the new department stores to rural America.

On the edge of the antebellum American frontier, Chicago marked a boundary between the well-connected network of efficient Eastern transportation and the frontier economy and transport network of the plains and Western states. Ward recognized that mail order would provide better service (indeed usually the only service) to rural customers without ready means to travel to urban centers to shop in department and other retail stores. Ward's and Sears both began as mail-order-only businesses. In the 1890s, Sears expanded to include a wide range of items when Julius Rosenwald, a clothing manufacturer, purchased Roebuck's interest in the company ("Sears," 2004). The first retail store—building on their monetary success and now familiar name—opened in 1925, after Robert Wood joined the company and took advantage of the increased accessibility of urban shopping with the spread of the automobile. By 1931, retail sales exceeded mail-order sales ("Sears," 2004). Thus, mail order emerged as a means of extending the reach of department stores beyond the "neighborhood" defined by the travel distances of consumers, but it also leveraged a national expansion of successful retail stores. Both extensions took place through the catalog, scaffolded by the various institutional, technological, and organizational innovations of society and government mentioned above. The Sears catalog became an icon of American popular culture: in many rural homes in early twentieth-century America, it was one of the few reading materials besides the Bible. According to Thornton (2002, 6), many rural children learned to read using Sears catalogs.

In 1895, Sears introduced a catalog specializing in building materials such as lumber, hardware, and millwork in addition to the 10,000 items of general merchandise in the regular catalog (Thornton 2002, 2). In 1908, Sears introduced a "Modern Homes" catalog of plans and kits for building whole houses (see figure 7.1). Within a few years, the kits included the framing lumber needed to build each plan, and in 1914, the lumber was offered precut, with each piece labeled to facilitate inventory and assembly (Thornton 2002, 12). With the end of World War I, during which the building of new housing was virtually suspended, increased immigration and returning veterans generated significant demand for housing and created a large pool of available construction labor without the capital to realize their "American dream." In 1911 Sears introduced mortgage financing of their kit houses, but by 1918 the terms had become successively more liberal, with much lower down payments than commonly available from banks at that time. Sears' kit-home business peaked in 1929, with ninety different house designs and sales of $12 million (Thornton 2002, 33). In 1930, sales declined to $10.6 million. In 1934, with the Great

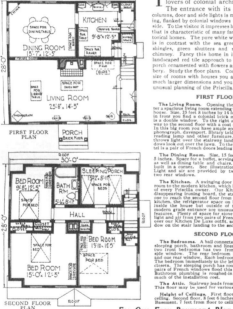

Figure 7.1

Page 34 of the 1926 Sears catalog: The "Priscilla," six rooms and sleeping porch, No. P3229, "Already Cut" and Fitted, Honor Built, $3,198. The house plans collectively were supplemented by forty-one pages detailing house fittings and accessories, construction details for the different grades, the advantages of Sears designs, factory-cut assembly kits, mortgage application, and a good deal of hortatory prose. Note the appeal to "lovers of colonial architecture."

Depression in full swing, Sears foreclosed on $11 million worth of mortgages, bringing their housing bubble to an agonizing end for company and customers alike (Thornton 2002, 29; citing Emmet and Jeuck 1950). Together with Sears' reputation as a purveyor of quality goods and materials, the expanding rail and postal network, the entrenchment of its catalogs, the demand for housing, and the abundance of labor all contributed to the spread of Sears Homes. By the time the company closed the doors on the Modern Homes division in 1940, they had sold around 75,000 house kits and listed between 350 and 450 different designs (Thornton 2002, 1, 5).

These patterns emerged, moreover, in a context of "rationalization" of many segments of American social, economic, and technical life (Gerson 1998).[11] Interest was piqued in a clean, organized, and economical way of life that contrasted with the dusty and cluttered image of Victorian living (Thornton 2002, 7). The rise to prominence of the germ theory of disease in the nineteenth century led to increased interest in cleaning devices, electricity, and changes in domestic patterns, practices, and habits (Thornton 2002, 8; citing Gowans 1986). The emergence of a working middle class without servants also helped generate a market for time- and labor-saving devices and strategies. Sears tapped this market by using their home kits as a marketing tool for the thousands of items in their regular catalog. Though not usually architecturally innovative (Sears tended to try to produce the styles already most popular in an area), they made central heating, indoor plumbing and toilets, electricity, and gas available in houses for middle-class people, and thus often drove styles and standards in their neighborhoods. Sears was also technologically modern and an "early adopter" of the new safer and more economical "platform" construction, when they switched (wholesale, in 1928) from the earlier "balloon construction" design.[12]

Sears also "rationalized" its business practices in this period: standardizing house designs and styles (in contrast to houses individually contracted and crafted by architects), mass-producing parts for houses in its own lumber mills, and precutting lumber on a mass scale to eliminate wood waste and shipping costs. They organized and standardized fittings, furnishings, and financing terms along with systematizing and coordinating their general merchandise with the various home plans, which all contributed to the creation of a market.[13] And despite the common perception that standardization and mass production threaten "craft quality," Sears' rationalizations actually increased their reputation for quality products: by judicious precutting, they were not only able to reduce shipping costs, but also to remove knotholes from lumber, thus elevating their wood product from second- to first-grade (Thornton 2002, 15).

The advantages of "rationalization" to the customer were substantial. Before modern portable electrical tools, precut lumber generated especially significant savings in time and hand labor for the customer, as Sears never tired of saying in advertisements. The coordination problems solved by having a unitary kit should not be underestimated. The "average" housing kit had 30,000 parts and filled two railcars. Even making sure that all of the parts arrived at the building site on time obviated the need to seek them out individually, or to delay while a suddenly recognized necessary part could be found or ordered and shipped. And the downstream advantages for maintenance and repair of having windows, doors, screens, and fittings to standard specifications was enormous (as restorers of older Victorian houses often find out to their dismay).

Sears encouraged customers who sought alternatives to Sears' designs to work with their architects to get a set of plans and kit from Sears—after which Sears could add a new choice to its catalog. Within a design, they advertised that any house could be "mirror inverted," and designs morphed in simpler ways that called for many systematic changes, but did not require major design reconfigurations. (This could include, for example, making the house three feet wider, adding the space to the living room, dining room behind it, and the two bedrooms above it, without disturbing the rest of the design.) One Sears customer cut out a catalog picture of the roof line of one design, taped it to the bottom of another, and mailed it to Sears with a note requesting the hybrid design, which the company agreed to produce (Thornton 2002, 67). Local housing offices in big cities had "showrooms" (sometimes with associated model homes), and places where customers could consult for modifications, or see the combinatorial possibilities for the multiple options at every stage.

To coordinate some systematic changes, Sears offered two grades of construction, "Honor built" and "Standard," with a third for garages and summer cottages, specifying different qualities, thicknesses and kinds of wood, flooring, shingles, windows, furnace and plumbing provisions, trim, and the like (Thornton 2002, 14–15; Sears Roebuck and Co., 1991, 12–13). A customer could order a quality level, and then modulate that design as they chose, avoiding the hundreds of coordinated decisions that such a quality change entailed.

Coming to Sears for a home kit in the virtual sense of shopping their catalog encouraged coming there for other goods, thus coupling the spread of

Sears home styles and the styles of other artifacts. Indeed, this motivated Sears' initial move into the housing market. Sears sold nails, screws, bolts, more complicated fittings, plumbing fixtures for kitchen and bathroom, furnaces, tools, lamps (electricity, gas, kerosene—early electricity and gas distribution was urban only) in different styles and costs, different decorative door, stained glass window, mantelpiece, and bookshelf combinations, even books for your library, and a machine to make concrete blocks (in various decorative styles) for your home's foundation.

Sears also built clusters of homes, and even a large chunk of a town. In 1918, Sears contracted with Standard Oil to build 192 houses in Carlinville, Illinois, and in nearby "Standard City," complete with streets, sewers, and sidewalks (Sears Roebuck and Co., 1991, 2–3; Thornton 2002, 35–67). This was a "company town," but also in some ways a worthy predecessor to post–World War II's paradigmatic suburb, Levittown.

All in all, Sears' activities with kit houses introduced changes and provided new paradigms in lifestyle and made homes affordable for many middle-class Americans. Along many of the older roads of the Midwest, which often paralleled the railroad rights of way, one sees houses (and foundation bricks!) that look like they came right out of the Sears catalogs. Sears brought "Fordism" and the benefits of mass production, scale, standardization, labor savings, and technical innovation to a new sector of the economy at the same time they populated American homes with artifacts from the rest of their catalogs. In this way Sears scaffolded home ownership for consumers, lowering the cost and risk of building a middle-class house outside an urban area, and at the same time extended their catalog sales and projected the same American values that they helped to create in ever-deeper mercantile layers to a widening public.

Perhaps the advantages to consumers of social investment in Sears' "external" scaffolding rather than their own personal investment in direct transmission of goods is best seen in the contrast between pioneers who had to pack up a household in a wagon that could be self-sustaining for months or years as they made their way across the frontier versus the spread of Sears houses after the railroads and other infrastructure were set up as a persistent network of scaffolding that supported rural settlement. As long as the rails are repaired, rights-of-way and law and order maintained, customs, conventions, and institutions sustained for ordering from Sears, then you would not need to move an operating household from the East to the West: just take the train West, stay in a hotel when you got to the provincial town, order your Sears house from the catalog, have it delivered to the railhead, and build it nearby (or contract with a local builder) when it arrived. And the children raised in that and in other houses would see and share the values they would buy in the catalog.

The multilevel richness of this case demands further analysis. We return to it in the section on "cultural reproducers" after we have developed the theoretical apparatus to do so.

A Central Role for Scaffolding

Biologists as well as anthropologists, sociologists, social psychologists, and historians of science have taken a theoretical interest in scaffolding, though not by name. The construction of scaffolding is a special kind of niche construction, but that theory as elaborated by the biologists Odling-Smee, Laland, and Feldman (2003) "black-boxes" ontogeny.[14] Without opening this box to make significant use of development, they take only the first steps (though important ones) toward an adequate theory of human cultural evolution. This beginning does not reach far enough to capture some of the most important parts.[15] Development seems so complex and particular, like the adaptive designs of different phenotypes, that those seeking general theory tend to avoid it. But there *are* abstractable and general features of development, and each of our approaches has sought for and used them.

Consider how extraorganismal cultural resources form repeated assemblies that serve as critical scaffolding for the development and inheritance of culture. "'Repeated assemblies' are recurrent entity-environment relations composed of hierarchically organized, heterogeneous components having differing frequencies and scales of replication" (Caporael 1995, see also her 1997, 2003). Scaffolding leads to self-scaffolding as individuals gain skills they can use to acquire new skills (Bickhard 1992). As skilled individuals contribute resources to the development of behavioral traditions within social groups (Avital and Jablonka 2000), they facilitate the repetition, and thus the accumulation, of the particular configurations of resources that we identify as items of culture, whether artifacts, behaviors, practices, or traditions. It is as if, through a designed structuring of the environment, we are able to sequester the right kind of prebiotic soup (or pizza—see below) to synthesize the kind of artifact we seek to produce, then learn how to package it, and then we are off, producing—If we are wildly successful—a new sector of the economy. Moreover, the order in which resource configurations become repeatedly assembled creates downstream dependencies in the productive role of scaffolding, which entrenches the dependencies in development.

As entity-environment relations (e.g., actor-artifact relations) are repeatedly assembled, so that culture becomes generatively entrenched, it also *thereby* becomes reproductive, forming a level of potential inheritance in its own right. That is, when individuals in a particular social group context exercise their relatively context-dependent abilities, scaffolded "from outside," so as to develop "mature" abilities that are more robust and less context sensitive, they achieve that measure of autonomy and authority needed to serve as agents of scaffolding to others (e.g., biological or cultural kin). The "internalization" of scaffolding—gaining relatively context-independent abilities to perform services for others similar to those experienced in one's own development—is a form of generative entrenchment that then becomes portable and tied to the agent. If it is central enough, the context dependence may become black-boxed and invisible to the sufficiently socialized community that can take it for granted, with its ethnocentrism apparent only to an outsider. The internalized scaffold, like a skeleton, becomes visible to an insider only if something sufficiently untoward and disruptive happens that the fabric unravels and we see again the bones making up the movements we have been following in each of our examples.

Social-group development depends on this internalization of individual ability to perform scaffold services as a means to the maturation of group members and thus to group development. Internalization facilitates social-group reproduction and identification at its own level: those particular features of a social group that distinguish its artifacts, traditions, and practices as its *culture* are propagated to new generations of individuals and groups. Thus we seek also to understand culture in terms of the emergence of new levels of inheritance and the ways the entrenchment of generators at lower levels become the basis for scaffolding development at higher levels.

An important aspect of the emergence of new levels of inheritance is the evolution of developmental mechanisms that depend on the direct transmission of organized packages or propagules at the new level (Griesemer 2000a). Biological transitions to new levels generally achieve this by new organizations of material propagules—for example, genes organized in material chromosomes, chromosomes organized in gametes and cells, cells in bodies, and bodies in social reproductive units (Maynard Smith and Szathmáry 1995). It is an open question to what extent cultural transmission depends on reproduction via organized *cultural* propagules. Many of the products of culture are material artifacts that serve as models for imitation and social learning, and are also transmitted or inherited as property. Parts of the social "infrastructure" that sustain culture are material as well. The theory of the firm in evolutionary economics must treat the firm as a hybrid legal-material object, an institutional/organizational complex that attracts and feeds on investment, uses and transforms material resources, and derives energy (or capital) and evolves through competition, growth, and reproduction. Firms attract trainees and produce "company men" (the sexist language still seems to fit here!). Firms are born, swallowed, sued, and wooed. They "respire," recruiting and relinquishing (or, too often, excreting) workers in the aperiodic seasonality of market ebb and flow. They are agents in our economic and political environment, but they too are scaffolded and enculturated, in ways that sometimes require great care, as multinationals must discover when they buy or bud a firm into a different country.

All of these dimensions, from different national industrial cultures, patent laws, industry-university relations, and the use and production of technology and scientific research are explored in Murmann's (2003) lovely economic and technological history of the growth of the synthetic dye industries in England, Germany, and the United States from 1858 to 1914. Murmann's work illustrates particularly well the importance of a multilevel approach to cultural evolution. The rich interpenetration of organizations, institutions, professions, and networks of people in nationally particular evolving configurations can engender such a context sensitivity of the players at various levels that one longs for entities as well individuated as those in biology. In culture, the evolving system is often more like a messy ecosystem that a species. Such rich embedding makes material and ideational cultures inseparable—not as Siamese twins that can, however laboriously, be separated, but as gently poured and incompletely mixed cream into coffee.[16] Our explorations of the theoretical role scaffolding can play in a developmental model of culture lead us to consider below the role of material organization in cultural evolution.

Models of Culture

A variety of simple models for culture can be used to explore the ways repeated cultural assemblies develop, reproduce, and become generatively entrenched through adaptive evolution. Though they are specified very abstractly here, each tries to capture key features of more complex relationships. For any given case one may need to consider them as alternatives, but in the whole of cultural descent, all of them are represented, and in more complex cases, we may see a tangled web of several of them for different aspects of or actors in the system simultaneously.

Figure 7.2
Biological and cultural Weismannism. In biological Weismannism (left), genotype (G) is the sole or primary developmental determinant of phenotype (P) in a given generation and also the sole or primary hereditary determinant of the genotype in the next biological generation. In cultural Weismannism (right), biological actors (BA) are the sole or primary makers of artifacts (A) and the sole or primary determinant of biological actors in the next generation.

First, if we think of material culture as the set of artifacts produced by actors belonging to a culture, then culture grows and accretes by means of the activity of actors. Particular artifacts may be passed from actor to actor (cultural property inheritance) and even embellished or simplified by descendants (cultural evolution of an attenuated sort). But artifacts (clay pipes, automobiles, Gibson guitars, heirloom watches) do not typically give rise to new artifacts via a material transfer of their parts. The "evolution" of artifacts is the result of descendant actors producing new artifacts that are variations on the themes of the artifacts of their ancestors. Artifact "evolution" is Weismannian, with artifacts as "dead-end" soma (see figure 7.2).

Second, we may instead think of culture not as the artifacts, but as the shared conventional understandings needed to produce artifacts, or the set of behaviors or practices of producing them, in which case artifacts are mere indicators of culture, while culture is activity that can be passed from cultural ancestors to descendants by facilitation of trial-and-error learning or by teaching and social learning. In this way of thinking about culture, artifacts may provide a feedback mechanism for a growing and changing culture insofar as the interaction of actors and artifacts alters the understandings or practices that are transmitted in teaching and learning, but it is the altered actors *producing* artifacts and a population of actors that evolve culturally (see figure 7.3).

Third, under either conception of culture, actors may evolve as a consequence of engineering their environments via the production of artifacts. People who live in dwellings that protect them from the elements may happen or plan to provide a dry place for storing food protected from scavenging animals. They may thereby evolve dietary interests and needs different from those of people whose food is exposed to destruction, loss, and theft. We can track changing artifact types such as dwellings or changing

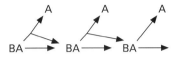

Figure 7.3
Cultural Weismannism with feedback. Biological actors (BA) make artifacts (A) in a given generation and biological actors in the next generation, as in cultural Weismannism (see figure 7.2). But the process of making an artifact in the parent generation affects biological actors in the next generation—for example, by teaching and learning.

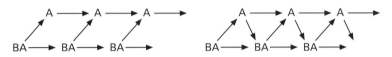

Figure 7.4
Cultural reproduction and coreproduction. In cultural reproduction (left), biological actors (BA) make artifacts (A) and also biological actors of the next generation, but artifacts also contribute to making artifacts of the next generation. In cultural coreproduction (right), in addition to the relations shown for cultural reproduction, the artifacts coproduce with parental biological actors the biological actors of the next generation—for example, when artifacts scaffold the development of the offspring actors.

understandings and practices of actors who produce dwellings, but in either case, the consequences or effects are filtered through the Weismannian channel from actor to actor. This "builder perspective" is Weismannian, focusing on the way actors change as a function of the fitness of their constructed artifacts or cultural "phenotypes" (Ingold 2000, 2001).

Fourth, more radical conceptions of culture (in the sense of departing from conventional biological wisdom that inheritance is always Weismannian) include those in which cultural elements themselves constitute a "channel" of inheritance distinguishable from that of the biological actors who form the generative basis for culture and, still more radical, those in which the generative basis of culture does not lie solely with biological actors. Consider the following two broad classes of such multichannel conceptions (see figure 7.4). In dual-inheritance models, patterns of variation in cultural items have a direct path of influence on the patterns of such items in subsequent generations so that tracking the reproduction of the biological actors alone cannot explain those patterns of variation (Boyd and Richerson 1985; Richerson and Boyd 2004). So-called memetic theory

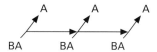

Figure 7.5
Cultural Coreproduction of practices. The practices of biological agents (BA) bringing about artifacts (A) also reproduce such practices in the next generation.

(e.g., Aunger 2000, 2002) would fit here too, though we have deeper problems with it that we will not consider here (Wimsatt 1999, 2002).

Fifth, in the other, more radical departures from Weismannism, there would be *reproductive* elements of the culture besides the biological actors: artifacts or practices that, as it were, give rise to more (and varying) artifacts or practices rather than merely being repeated in subsequent generations due to new acts of making by descendant biological actors.[17] Differently put, in the previous case, if artifacts and practices are tracked separately from biological inheritance, a more adequate prediction of changing patterns of variation can be made than if only biological inheritance is tracked and artifacts and practices are treated as parts of the biological phenotype caused by genes. Models pursuing such an approach aim to overcome the implausible view that genotypes are the sole determinants of cultural phenotypes. In the cultural-reproduction approach, on the other hand, the relation (and mechanism of development) between items of culture across generations would differ from that of the dual-inheritance models (see figure 7.4). Cultural items would directly reproduce, just as biological actors do, contributing parts to offspring items, and our reasons for tracking them would be the same as for biological actors. In this case, the population dynamics of culture would be coevolutionary for a variety of "species" of reproducers interacting in a cultural ecosystem.

Sixth, a variant on these reproduction and coreproduction representations that focuses on the process of producing actor-artifact relations treats the practices linking biological actors and artifacts as the objects reproduced and reproducing rather than their elements (see figure 7.5).

We do not argue in favor of this pure form of cultural reproduction, although we think that to explain cultural evolution a more nuanced account of the cultural development of items typical of human culture (both actors and artifacts) than spartan dual inheritance mapping relations is necessary. It is likely too that all of the alternatives we have described are relatively accurate for different cultural life forms, just as biology has a diversity of reproductive modes (some Weismannian and some not). We do not claim to have given an exhaustive list of the possibilities. Cultural artifacts and practices are more than mere objects of imitation and learning, but probably less than reproducers, so a medium-viscosity theory is required to account for their development and seems the best strategy. Moreover, in addition to feedback effects of the use of artifacts and engagement in social practices, we must consider ways biological actors function as cultural artifacts and practices for other biological actors. That is, the development of a biological actor in a society with culture involves other actors functioning as scaffolding to facilitate the acquisition of skills (and authority; Gerson, forthcoming) on the part of a developing actor. And it is certain that anyone interacting negatively ("impersonally") with a large bureaucracy has experienced the other side of this coin of "person as artifact," in which human actors become so entrained by the policies and stereotypes their roles engender that they seem like, and effectively are, parts of the machinery.

Actor-artifact complexes persisting over extended developmental time have an integrated and constitutive unity that needs to be taken very seriously. Our view is that although this scaffolding function of biological actors and scaffolding uses of artifacts in culture do not render particular artifacts (mathematics exam papers, Sears kit houses) cultural reproducers, they do render cultural systems developmental and biological actors in them cultural reproducers. Because such systems can also become emergent cultural reproducers, the material transfer of "nonbiological" components can take on a reproductive role. This is just how the material transfer of nonreproducing biological components of a cell (such as membrane lipids, cytoplasmic proteins, and DNA strands) render cells reproducers when they transfer developmental capacities via these molecular mechanisms. In other words, *cultural developmental systems become reproductive when components perform generative scaffolding functions internal and vital to the production of cultural items within cultural reproducers*, whether these are artifacts, practices, or biological actors (see figure 7.6).

In our examples, this is probably most obvious in the evolution of mathematical physics. The practices that developed around written exams were passed on and refined, and the cumulative publication of exam questions and answers forced generation of new ones (through desires to avoid repetition). The cumulative use of earlier mathematical techniques and results in new problems gave them a generative role in the production and reproduction of new knowledge.

Finally, the different time scales of production and reproduction for the various interwoven and embedding biological actors, artifacts, and hybrid

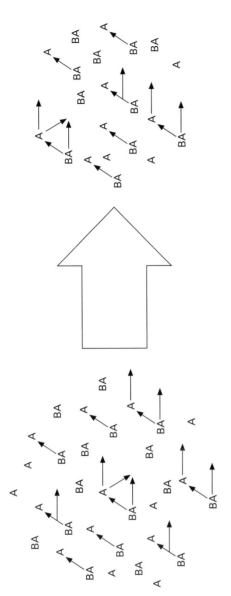

Figure 7.6

Coreproduction of cultural systems. A cultural system consists of a number of biological agents (BA), artifacts (A), and practices, simplified here as a single population rather than a complex network of partially overlapping, structured reference groups and social worlds. The system reproduces (in a very "broadband channel") when material components are transferred in space and time to offspring systems. The developmental mechanisms transferred yield a cultural system-level reproducer when they are generative and provide scaffolding relations within and among elements and substructures (not shown).

cultural systems in our cultural "tangled bank" explain a lot of the ambiguity many feel about whether we should be talking about cultural development, or cultural evolution, or given that ambiguity, about neither. After all, if we cannot pick out the units, and cannot even agree on their life courses enough to say whether they are developing or evolving, why bother? This requires a closer look.

Cultural Inheritance and Cultural Life Cycles

Any Darwinian theory of evolution, whether biological or cultural, must be populational, following Mayr (1964), Boyd and Richerson (1985), and Hull (1988).[18] For cultural evolution, this means modeling the fates of different variants of what are regarded as the "same" cultural entities among individuals or larger social groups (see figure 7.6).

Boyd and Richerson (1985) study the conjoint operation of biological heredity and cultural heredity. Thus their theory is a "dual-inheritance" theory. They argue that earlier attempts to cash out cultural processes and change solely in terms of biological genes and biological fitness will not do. We agree. Recognizing the dual nature that they note for biocultural inheritance gives a good first approximation for the dynamics driving biocultural evolution in cases like those discussed by them and by Durham (1992). But here we diverge.

The nature of embodied reproduction requires us to reconceptualize processes of evolution and inheritance in a developmental mode that reveals significant problems with the replicator concepts of Dawkins (1982) and Hull (1988), and many more problems with their use to conceptualize evolutionary processes (Griesemer 2000a, 2000b, 2000c, 2002). Moreover, generative entrenchment is a phenotypic property with implications for differential evolutionary stability of traits and trait complexes that provides a framework operating directly on the causal structure of development to predict differential rates of evolutionary change without invoking genetic information (Wimsatt 1999, 2001, 2003). Analysis of phenotypic structures in terms of generative entrenchment can thus be used to complement genetic analysis, but can also proceed independently of it (Schank and Wimsatt 1988; Wimsatt and Schank 1989, 2004).

Neither Griesemer's reproducer dynamics and hierarchy of reproducer concepts nor Wimsatt's generative entrenchment require genes or things modeled too closely on them to generate results. This is fortunate. Culture has no genes—at least none determining particular cultural variants or adaptations at any level.[19] And, meme advocates to the contrary, it has no memetics—at least nothing at all similar to genetics, or with its analytic power (Wimsatt 1999; Sperber 2000).

Although we accept Boyd and Richerson's framework involving both genetic and cultural change as having significant heuristic power and a formative influence on problem structure in several respects, we focus here on shorter time scales involving cultural change only.[20] The problems we discuss have not involved significant gene-culture coevolution, though we agree that it occurs and is important on the longer time scales necessary to consider the origins of language, our basic forms of sociality, and the Neolithic origins of agriculture and culture (Boyd and Richerson 1985; Richerson and Boyd 2004). On the time scales that interest us, from minutes to months and from decades to hundreds of years, depending on mode of transmission and scale of analysis, gene-culture coevolution is less important than *gene-culture codevelopment*: the ways culture and genes interact to scaffold individual and group development in the context of social groups and social worlds.

We also agree that biological and cultural inheritance processes must be treated as distinct for purposes of modeling cultural evolution, but this is just a start on what is required for a robust understanding of cultural inheritance. We must consider more than two channels, even on short time scales. To understand the properties of the relevant inheritance processes or the character of cultural innovation, neither biological inheritance nor especially cultural inheritance can be treated as single-channel processes. We return to this issue shortly, but because issues of time scale, stasis, and change come up repeatedly below, we take a short detour to consider them here more generally.

Elements of any complex organization exhibit a spectrum of rates of change. Choosing an object or phenomenon to study determines a time scale on which its behavior is interesting, with trajectories that evolve, develop, decay, oscillate, or simply change. On that time scale, some (commonly larger) things change so much more slowly as to appear constant or nearly so, and are treated as structural, objectified, or background constraints. (Usually smaller) things changing much more rapidly become transients, fluctuations, interruptions, or background noise. We bracket or filter them out to focus on the target entities and relationships between them (Lotka 1925; Wimsatt 1994). When these change at different rates, this difference can dominate the outcome, restricting what can happen (Simon 1962). Thus we cannot eradicate insect or disease pests because their short generation times allow them to simply evolve their escapes, while their large populations quickly give them the mutations to do so. Differences in rapidity of response condition any evolutionary process, and particularly the interactions between biological and cultural change.

Two striking deviations from the general pattern of larger-is-slower and smaller-is-faster are relevant here. The first is cultural evolution itself which, though we tend to think of it as (very) high level, is capable of changing at much faster rates than biological processes commonly do. This claim will be the topic of the next to last section. The second exception is that specific mechanisms can make something much more stable than one might expect for things of that size. Two alternative explanations for this in biology are developmental canalization and generative entrenchment.[21] These (and their cultural analogues) can be important for culture, but entrenchment particularly so, since it plays a role in understanding how scaffolding can support much higher rates of successful cultural invention and change and enormous growth in social complexity.

A particularly curious "time-scale" lacuna in "thin" accounts of cultural evolution marks off our approaches from most others. We both make central uses of development. Very few of the quite diverse theories of cultural evolution do so, though all effectively presuppose it.[22] Cultural evolution today is as devoid of a developmental perspective as population genetics was in 1970. This is a serious mistake. Development figures centrally in understanding the most important differences between biological and cultural inheritance, and consequently the differences between biological and cultural evolution. Let us see why.

Culture involves multiple channels of inheritance. But unlike the genome, which is acquired effectively simultaneously at the formation of the zygote, the sequential acquisition of culture by maturing individuals gives these multiple channels incomparably rich possibilities for interaction and intermodulation. Contrast the processes for cultural acquisition, modification, and transmission in the cultural life cycle of an individual with the much less complex life-cycle diagram for biological inheritance and evolution. We start with the far simpler (while still complex) biological life cycle (see figure 7.7).

Explicit representations of these life cycles in biology texts are usually much simpler than this one, aiming to convey only the features of the simple abstracted model being discussed there. This life cycle is intended to represent all of the major forces acting and information required for a full evolutionary account—forces discussed elsewhere (often in a family of other complementary models) in a more complete treatment. The aim is to bring the description of the biological life cycle as close as possible in its

The Structure of Biological Life Cycles as Used in Population Genetic Models of Evolution

Key: **Objects in bold,** theories or sources of theoretical description (plain), *selection processes in italics.*

stage 1: [DETERMINE A BREEDING POPULATION]

Population(t_0)

=> [demography, mating rules

(*sexual selection*)]

$\Rightarrow$ **mating pairs**

stage 2: [COMPOSE OFFSPRING GENOTYPES]

=> [Mendelian (transmission)/cytological/molecular genetics,

(*molecular, genic, gametic selection*)]

$\Rightarrow$ **offspring genotypes/zygotes**

stage 3: [DEVELOPMENT AND SELECTION → NEXT GENERATION]

=> {(development, physiology, ethology, ecology,

biogeography, geophysics)

$\Rightarrow$ [(**phenotype, environment**) →

(*Darwinian*) *selection*/fitness]}

$\Rightarrow$ **Population(t_1)** [RETURN to stage 1 for NEXT LOOP]

Figure 7.7
This life-cycle diagram (after Wimsatt 1999, figure 1) is more detailed than commonly found in biology texts (to better parallel the complexities of figure 7.8), and includes for each of the usually separated stages, the biological objects present at that stage (**in bold**), the processes of selection acting (*in italics*), and the theories brought to bear to describe and account for the transformations at that stage. It is often simplified into three or even lumped into two boxes, connected by arrows with the author's focal mechanisms in one box, and the rest in another.

The Developmental and Network Structure of Organism-Centered Cultural Life Cycles

Stages 1.1 through 1.n (reflecting age or experience structure): [DETERMINE A BREEDING POPULATION]

Given a population of potential transmitters, determine which ones do so, and to whom* [*iterate n times for reception and transmission periods in age-structured life cycle*]:

COMMENT: This requires knowledge of transformations in 2.1 through 2.n from preceeding iteration

Stages 2.1 through 2.n: [COMPOSE OFFSPRING GENOTYPES] (inseparable from stages 1.1 through 1.n); [*iterate n times for reception periods in life cycle*]:

a. For each individual *i*, and each stage *j*, determine who receives how much from whom, as functions of their prior state *s* and place *p* in the transmission network. [This state includes the structured and coadapted prior assemblage of cultural viruses, acting as filters for new memes, and affecting how they are received (and possibly transformed).]

b. For each individual *i*, determine how the reception of the input vector changes the cognitive/normative/practical Ümwelt (prior state *s*) and social locus (place *p* in the transmission network).

c. For each individual *i*, and each stage *j*, determine whether these changes make *i* a potential transmitter at stage *j*, and of what and to whom. (Consider capability + opportunity + determination?)

Stages 3.1 through 3.n: [DEVELOPMENT AND SELECTION] (Is this separable from stages 1 and 2? No: the composition rules (analogous to Mendelian genetics) but also selection and development are implicitly represented in the transformation rules in **2.1** through **2.n** a–c above, and presupposed in stages 1.1 through 1.n above, *so these processes are almost impossibly confounded*)!

COMMENT: So this is already done in 2.1 through 2.n above. Perhaps clause 2.c belongs here, though some of it with equal justification could be placed at stage 1.

Figure 7.8
An attempt to construct a "life cycle" for enculturating individuals paralleling the biological life cycle of figure 7.7 for a "population genetic"–style model of cultural

relevant complexity to that of the next figure, for cultural evolution. For an example of the simplifications of this diagram in simple population genetic models, the population at t_0 is simply assumed to *be* the mating pairs, and to be in Hardy-Weinberg equilibrium, thus eliminating or trivializing the first stage. Molecular selection (e.g., transposons) and gametic selection are assumed not to occur, simplifying the second stage. The role of the organism in creating its environment is also ignored. The six scientific domains in the flowchart at stage 3 describe processes playing roles in constructing both phenotypes and the selection forces acting on them. But the whole of stage 3 and all of the levels of organic form and behavior is collapsed to assign fitnesses as scalar multipliers for genotype frequencies, ignoring all of these complexities in returning to stage 1. Nonetheless the complexity of this life cycle seems minimal when we frame the dynamics of cultural evolution in the same way (see figure 7.8).

This complex figure of the causal structure of processes of cultural evolution is depicted from the perspective of a population of individual developing and enculturating organisms to maximize the similarity between this representation and the preceeding. However, its complexity is daunting, in effect a reductio ad absurdum showing the futility of trying to treat cultural evolution on the model of population genetics or "memetics." Conceptually, it is useful to provide a matrix on which one can hang all of the different kinds of causal interactions and processes, but unlike the schematic biological life cycle of figure 7.7, it could not readily be turned into a computer program for which one would input data describing the population and expect to get a simulated run of its future cultural trajectory, even for relatively short periods.

One often uses models for biological evolution and inheritance without age structure, but the sequential character and order dependence of cultural acquisition in learning and socialization make this impossible for cultural evolution. This explains the iterated subscripts 1.1 through 1.n,

Figure 7.8 (continued)
evolution (after Wimsatt 1999, figure 2). Here the age structure of the population must be made explicit, and cultural transformations are gerrymandered to fit into boxes corresponding to the three for the population genetic account of the biological life cycle in figure 7.7. But then each of the stages can be seen to presuppose information from the other two in order to determine what happens at that stage. As a result, except for very special and not very useful simplifying assumptions, processes of heredity, development, and selection are almost impossibly confounded.

2.1 through 2.n, and 3.1 through 3.n in the steps of the three *causal stages* of transmission, reception, and generated capacity for transmission, and we must perform an iteration through the *n developmental stages* and for all transmission from actors (i) and reception by actors (j) mediated by the communication network, and all of the selections and transformations performed by each actor from what they receive or take from their environments.

The dependencies that make this so are implicit in the structure of the clauses (a), (b), and (c) of stage 2, which corresponds to the action of Mendelian genetics in assembling new genotypes from the gametes contributed by the two parents. But the two parents with equal contributions of biological inheritance dissolve in a welter of confusion. The cultural information is acquired through the course of development, so we break the acquisition up into stages. (This is intended here as a conceptual or mathematical device to order input, not endorsement of a stage theory like Piaget's.) A given piece of culture, however individuated, may be assembled from a varying number of parents, who may have contributed different amounts. Nor are the amounts additive, but they interact richly as the cultural element takes form, more like the highly nonlinear structures of organic synthesis. But whether it is acquired is a product of both the location of the learner in the social network, and also whether their extant cultural resources lead them to accept or reject it. (This may also depend on who they got it from, as parents quickly learn.) In either case, these will affect how an actor will understand and transform, and apply it, as they take it in or reject it.

In principle, the analysis of the cultural "genome" (if it makes sense to speak of one) seems almost impossibly complex. There is no time at the beginning of the life cycle when we have a completely assembled cultural genome analogous to the zygote for biology, so it is clear that we cannot separate hereditary and developmental "components" for culture. And because the assembly process for the cultural genome involves a major component of cultural selection—the sum of what we think of as individual choice, socialization, and the acquisition of explicit, tacit, and procedural knowledge (Reber 1993) as mediated through cognitive development and education—we cannot separate heredity from selection (Wimsatt 1999). Acquisition of some cultural elements may potentiate, rule out, necessitate, bias, or variously transform subsequently acquired cultural elements. *So heredity, development, and selection are almost impossibly confounded for cultural evolution*, and population genetic–style models will be intractable if we cannot find other sources of constraint to simplify the specification of relevant interactions.[23]

It is obviously possible to abstract from this complexity in special and simplified cases, as biologists do, but to understand the evolutionary dynamics, processes, and character of culture, one needs representation of this causal structure in its full interactive form, or in forms yielding complementary simplifications to study different aspects of the dynamics. *One important consequence of this sequential developmental complexity is that it should be particularly revealing to track individuals in narrative biographical detail.* (This should include both biological individuals and higher-level individuals like organizations, firms, and nations.) Complementing these narrative accounts, we would need more synchronic phenomenological descriptions of characteristic modes of behavior and interaction at cultural strata where they appear, acting as structuring elements in the trajectories of individuals. If these two modes of explanation look familiar, they should. In many respects, we would expect to find the disciplinary structure of the human sciences much as we find them today, so no sociobiology-style "urban renewal" of disciplines is anticipated. We also expect that it would not be particularly revealing to try for a direct theoretical implementation of the structure of this figure in all of its complexity to generate a dynamic simulation: one would have to invent or make up values for too many of the parameters for it to be realistic, and there are too many parameters in it even to figure out what would be interesting starting values, or for us to get significant illumination from the results (Levins 1966). Whether there are interesting hybrids of these two approaches (of significantly reduced complexity) is worth considering further, and there are some encouraging signs.

We live in a world that is almost indefinitely complex in its fine details, and in some of its coarser ones too—a world that has also become a lot more complex by our own hands. But it is also a world rich in pattern, with many sources of predictability, and if its patterns were too complex for us to deal with we would not be here; a kind of biopsychosocial existence proof: we have to get through multiple biological and cultural hoops to survive and reproduce. Furthermore, most of what we acquire in cultural development has to be accessible to us or we could not teach it or acquire it. The *developmental machinery for enculturation has to be mostly accessible to us or else it could not work on us.*[24] Some of this machinery is clearly internal, and cognitive, affective, conative, and normative. But the explosive growth of scaffolding processes—those things that have given us our rich and complex cultures—makes large chunks of it at least partially external.

The environments in which we are raised are not so unpredictable, and we are socialized in regulative and directive ways toward behavior that is convergent in many respects, canceling out much of the variability we face. Individuals make choices, or choices are made for them. None of us has to do everything for ourselves or choose among the whole range of possibilities. Even though in some sense, most of us could learn to do any of an arbitrarily wide range of things sufficiently early, to do almost all of these other things we would have to be differently placed in the socioculturaleconomic network, and for many of them differently placed in space and time. Richard Levins (1968) suggested that organisms evolve in such a way as to minimize the uncertainties in their environments. With some important qualifications (Wimsatt 1994) that also seems an important factor or theme in our biological and cultural evolution (Richerson and Boyd 2004). One of the ways we can enormously reduce the range of possibilities and uncertainty is to have us key quickly into a variety of specific factors characteristic of our culture and our place in it, so that our acquisition and elaboration of skills and roles can proceed in an environment of manageable complexity. The complex interactions in acquiring culture that concern us are mediated through processes of scaffolding and niche construction. They focus the cumulative wealth of culture (or that part of it accessible to us at our location in the linguistic and technosocial network[25]) into nourishing our cognitive, conative, affective, and normative trajectories into tracks that are usually (locally) socially productive and either attractive, or more attractive than the accessible alternatives. This focus bootstraps us to do more than we could otherwise achieve, but also coordinates a tolerable level of social predictability within our particular culture and subcultures. The character of these interactions would be impossible without a temporally extended period of development, a fact that developmental systems proponents of niche construction recognize (Griffiths and Gray 2001, 205), making it absolutely crucial to include development in theories of cultural transmission.

Scaffolding involves a mix of (relatively) static resources and constraints and dynamically interacting processes that together facilitate the acquisition of complex skills, knowledge, and behavioral routines when these interactions are appropriately organized. This kind of interactive structure is the key to new kinds of possibilities for innovation and change in culture. It also generates greater possibilities for "deep" or "revolutionary" modifications and much greater rates of change than are characteristic of biological evolution. So understanding the particular potentialities of cognitive development and individual and social learning requires recognizing a multiplicity of systematically interacting cultural channels.[26]

An earlier version of our view, "hierarchical multilevel Weismannism," observed that cultural inheritance was multichannel, and richer for it (Griesemer and Wimsatt 1989; Callebaut 1993, 425–429). However, none of these earlier accounts goes far enough in recognizing the complexities introduced by the serial acquisition of knowledge and competencies and the social structure of its acquisition, transmission, and distribution through the network of actors sketched in figure 7.8. We propose now to correct these limitations of earlier hierarchical models.[27]

In our earlier work, cultural entities such as scientific theories are treated as abstract (disembodied) generative structures, without scaffolding, and without the natural variability expected in the populations of their advocates and users. Agents who believe the theory are not represented, nor are the support structures for teaching it, learning it, spreading it, modifying it, or applying it. We cannot understand the properties of cultural inheritance without understanding how these abstract disembodied generative structures are embodied in the scientific community, developing, testing, and applying their overlappingly shared theory, while recruiting, training, and embedding new scientists in their community.

Biological and Cultural Reproducers

Inheritance, whether cultural or biological, is a special kind of reproduction process, so we must consider reproduction and its units in order to characterize the special case. Reproduction is the transfer of generative, developmental capacities from parents to offspring. In biological reproduction, the transfer is material: a propagule is organized as a mechanism carrying developmental capacities that begins as material parts of the parent(s) and that subsequently become an offspring. In culture, sequestration and bundling of both material and nonmaterial items can play a similar role, packaging cultural propagules into elements of cultural reproducers. In writing, for example, we bracket off the things we want to say from the rest of the text visually. Bracketing words, for instance, that introduce a new section along with visual section identifiers, sequester the material inscription from the complex, highly reactive context so that the ideas can be organized as the generators that their authors desire them to be. Bracketing words bundle ideas together—makes them an item—that can serve as a robust package or mechanism for generating the intended response in readers, in interaction with the rest of the context, but playing their own identifiable role.

Sequestration and bundling of cultural items into cultural propagules play fundamental roles in the process of cultural reproduction. The organization of these propagules, particularly the relations between material components such as biological actors and other components such as ideas (memes), practices, and behaviors, is key to articulating a view of cultural development. What, indeed, does it mean for an actor, an artifact, and a practice (such as the actor's playing the artifact) to be "bundled" into a propagule?

Genes and Memes

When Dawkins wrote *Memes: The Mew Replicators* (1976, chap. 11), he considered small items of culture: bits of melodies easily hummed, statements on sheets of paper that can be photocopied, visual images repeated on billboards, dogmas and doctrines that can be turned into slogans or sound bites. He was interested in the cultural transmission of such bits on analogy with his ideas about biology: the fundamental units of genetic transmission are small items of biology—bits of DNA (or RNA) that are easily copied, divided, and reaggregated to form the beginnings of new organisms. Dawkins also told stories about how the bits of culture and the bits of biology play similar roles in the life of the individuals that have them: cultural memes and biological genes play the role of "difference makers," things that make a difference in the development of the phenotype of the (biological or cultural) "organism" against the backdrop of other variant, competing difference makers in a process of evolution by natural selection.

For Dawkins, a gene (or meme) is an "active, germ-line replicator." A replicator is anything in the universe of which copies are made; a germ-line replicator can be the ancestor of an indefinitely long lineage of such replicators (however this may be achieved, by whatever mechanism); an active replicator is one that influences the probability it will be copied (Dawkins 1982, 82–83). This analysis works because it depends on taking the context of activity—replicator power—for granted. The context is just there, so all that is needed for a replicator to be active is to make a difference to the probability of being copied. And the same taken-for-granted context provides the machinery of transmission.

Copying only works as the core concept of transmission in sufficiently rich and appropriately structured contexts. DNA is copied only so long as the whole properly organized and functioning transcription-translation mechanism is present to produce replication enzymes (among many others). Our chapter is photocopied only so long as the whole properly organized and functioning photocopier mechanism is present, we insert the manuscript in the right place, and push the right buttons to adjust the parameters of copying to suitable values for the task at hand. If we wish to explain copying, however, we must explain how these mechanisms that form the context and manage the control of copying come to be. That means we must explain their development. Doing that leads to the concept of reproduction by material overlap, not replication-in-the-form-of-copies (Dawkins 1976; cf. Griesemer 2000a, 2000b). One cannot pursue the concept of copying as a means to understand development because it begs the question of where the richness of copying machinery comes from.

Dawkins's ideas about culture and biology are controversial, but attractive. As he is fond of pointing out, the best example of a meme is the idea of a meme itself, which landed him on the cover of *Wired* magazine (Schrage 1995) as well as in the pages of *Nature* (Goodenough and Dawkins 1994). Many dislike the "reductionism" of his idea of the gene—how can such bits be the whole of the fundamental story of living things—can talk of "genes for reading" or "genes for altruism" or "selfish genes" really be explanatory? Many equally dislike the "universal Darwinism" implied by his interpretation of the gene as *the* unit of selection. The controversy is just as heated over his cultural units. Can memetics be a genuine, let alone fundamental, science of culture? We think not (Wimsatt 1999; Sperber 2000).

We will not quarrel directly here with Dawkins's articulation of gene and meme concepts to apply his stories about replication to the processes of biological and cultural transmission. We have each done that on multiple other occasions (Wimsatt 1980, 1999; Griesemer 2005b). We focus on the role of such items in development. The difference-maker approach Dawkins uses makes sense only so long as the rich context of difference making can be taken for granted. Dawkins's "gene for reading" (or reading ability) takes for granted the complex developing organism that learns to read, the developed organisms that teach it, the wealth of books to teach with, and the interactions with the world (including other readers) that give what is read a richer content and a dancing gaggle of linguistically mediated referents. Dawkins's memes—a bird's song, the "St. Jude mind virus," the idea of a celibate priesthood—take for granted the rich communicative, cognitive, social environment of a singing bird or a religious order. How, without such context, could such riches ever be garnered? Dawkins's answer is: by accumulation of the small, favorable bits by natural selection and discarding of the unfavorable ones. But that is only the beginning of an answer because it treats the "richness" of organization in the simplest imaginable way: as a cumulative aggregation of many small items. Neither organisms nor organizations nor cultures are aggregations of items—if they were, they would be far simpler to understand (W. Wimsatt 1997). They are highly organized and their organization matters to the processes of development, transmission, and selection. Dawkins's story does not address the evolution of organization, only evolution by accretion, because he takes the richness of context for granted in order to tell the causal story of genes and memes as a story of difference making. To claim that difference-making genes and memes made the difference, historically, between the rich organizations of developed traits that have been selected is merely to beg the question of the evolution of organization. It is, in effect, to fail to recognize the difference between the unassembled Sears house kit and the assembled one, or worse, between the distributed, uncollected and untransformed raw materials to make the parts, and the finished house. The parts themselves are the product of a redesigned decomposition of the process of house production and assembly, and the social, technological, and cultural infrastructure to accomplish that!

To even attempt to put Dawkins's memes in context requires representation of at least some of the institutional and organizational structure and relations between organizations that may share individuals and interact with artifacts. Think of something like a commercial firm, which recruits new employees and socializes them to the company as well as integrating them into the organization's task and production structure and training them for the particular tasks they will perform. Conversely, institutions and organizations form "reference groups" for the people participating in them. When individuals belong to multiple reference groups, social structures can be complex and nonhierarchical.

Sequestration and Bundling

Gene and meme concepts leave unanalyzed the very conditions of their replication and are consequently incapable of addressing questions of the evolution of organization. A way out of this question begging is to recognize that the problem of the evolution of organization is a problem of the evolution of development and of how the problem of transmission is entangled with it. Dawkins sought untangled accounts of transmission as replication and development as phenotypic difference making (Dawkins 1982, chap. 5). For Dawkins, as for all Weismannists (see Griesemer and Wimsatt 1989), these are simply different properties replicators may have that are aggregated in the fundamental units of selection: active germ-line replicators. In contrast, we see activity and germ-linedness as entwined by the very nature of developmental processes.[28] Understanding how they are entwined suggests a way to describe cultural inheritance more satisfactorily than talk of transmission of cultural "traits" admits.

The difference-maker approach factors out all of the context and content shared by things that make a difference. Following Mill's methods for analyzing causes, the cause *is* the difference that makes a difference. These methods assume that difference making itself is context-free so that differences can be aggregated into total causes. But this is not how biology (or culture) or any complex mechanisms work. Difference making is context-sensitive. A gene difference in a rich environment might lead flies to have white eyes rather than red ones, but in a less rich (or just different) environment it may make no difference at all (e.g., to flies that develop without eyes). In general, the components of complex systems have many moieties and functionalities and so contribute to difference-making in multiple, interacting, context-sensitive ways.

In biology organisms can partially control the context sensitivity of difference making by various means. One of these is *sequestration*. Biochemical substances (and cells) in rich contexts have many reactive possibilities that are (partly) controlled by sequestering them from undesired reactions and, concomitantly, bundling them together with other substances to promote desired reactions. Cells are not mere bags of reactive molecules. They are highly organized, with internal membranes, microtubules, macromolecular binding complexes, and formative devices like the Golgi apparatus, endoplasmic reticulum, and ribosomes. Even the double-helical organization of DNA sequesters the reactive nucleotide surfaces inside the molecule. Single-stranded tRNA molecules fold up into hairpin secondary structures that hide away reactive surfaces while forming others. Ribosomes bundle RNAs and proteins into translation "machines." Controlling reactivity through sequestration has bundling as its flip side.

Not only can bundled items act preferentially and differentially, they can also be held and transmitted together as a package. Packages can carry whole, organized mechanisms from place to place because their organization is not disrupted in transit. Individual bits transmitted separately, like the message packets of e-mail messages transmitted separately over the Internet, must be reassembled *in the correct organization* in order to transmit the structure of a parent organism that survived natural selection's gauntlet. But where does the organization at the receiving end come from? If the receiver is an entity like the sender, then we hit a regress if we appeal to the organization of the sender to explain how the receiver becomes organized in development. Thus, it will not do to explain an offspring's development by appeal to the transmission of bits from the parent unless we have already explained the development of the parent or how organization is acquired from the environment by the receiver and yet is correlated with the organization of the sender. In the difference-maker analysis of selfish replicators, the *organization* carried by reproductive propagules (gametes) is neglected in the account of the transmission of "replicator power" from parent to offspring. At most, on Dawkins's view (1995) the other elements of the propagule serve to "bootstrap" the initial process of development of the offspring, invoking the idea of development but leaving it wholly unexplained and unanalyzed.

Replicators or Reproducers?
Organized contexts in the biological world can persist by being stable on a longer time scale than other entities of interest or by being transferred from place to place and time to time in material propagules that develop a full slate of properties. The former mode is familiar from our usual treatment of the environment as a persistent context. The habitat surrounding the burrow that mother mouse dug before birthing a litter is more or less the same, seasonally adjusted habitat the offspring emerge to discover when they are weaned. Moreover, the sort of burrow and habitat the offspring experience are likely conditioned in an ongoing fashion by those participating in them, including their ancestors. As Ingold (2000, 186) put it,

> For any animal, the environmental conditions of development are liable to be shaped by the activities of predecessors. The beaver, for example, inhabits an environment that has been decisively modified by the labours of its forbears, in building dams and lodges, and will in turn contribute to the fashioning of an environment for its progeny. It is in such a modified environment that the beaver's own bodily orientations and patterns of activity undergo development. The same goes for human beings. Human children, like young of many other species, grow up in environments furnished by the work of previous generations, and as they do so they come literally to carry the forms of their dwellings in their bodies—in specific skills, sensibilities and dispositions.

Odling-Smee, Laland, and Feldman (2003) urge a similar perspective in their work on niche construction. The alternative mode, persistence by material overlap of reproductive propagules and repeated assembly in development, although familiar, needs motivating.

Dawkins and his philosophical defenders (e.g., Sterelny and Kitcher 1988) treat everything beyond a gene as its "environment." This blackboxes development as "beyond" the relevant units of explanation and gets the difference-making analysis going. After a gene is replicated within a cell, it is transferred to a new body—the cellular body of a gamete, say, and then the cellular body of a zygote formed by the fusion of two gametes, for sexual reproduction. The parent cell and its offspring cell (or gamete) are related by "material overlap," physical parts (e.g., the genes, membranes, and cytoplasm) that belonged to the parent at one time are transferred and belong to the offspring at a later time. These "progeneration" processes (see Griesemer 2000a and figure 7.9) involve the transfer of an *organization*, not a mere bag of parts. If the organization transferred is a mechanism for the development of the offspring, then call the process a reproduction process. That is, reproduction is the transfer of organized, material mechanisms of development. If the mechanisms of development transferred are adaptations (evolved because they confer selective benefit on ancestors), then the reproduction process can be called an inheritance process, since the mechanisms of development will cause trait heritability. If, further, the mechanisms of development are "coding" mechanisms, then call the inheritance process a replication process. This gives an alternative to Dawkins's copy-based analysis of replicators.

On the reproducer approach, a gene and its replication must involve a reproduction process. Dawkins mistakenly locates replication of the gene in copying because the transfer of the organized mechanism of development accompanying it and causing reproduction to exhibit a copy relation is taken for granted.[29] In biology, the difference-making power of genes is transferred materially, along with the genes through reproduction. But even the rest of the reproductive propagule is not enough. Some of the power to confer difference-making power lies in the "environment" beyond the propagule.

The genes in the zygote do not have the power to make eyelashes, the zygote must acquire that power through development. Nor do the genes of the zygote have the power to make an offspring *organism*. The developmental units of which they are parts must acquire that power through development. The difference-making language misleads us into locating the causal powers in the wrong place. Development, from the replicators' point of view, is a cascade of sequential acquisitions of replicator powers, culminating in the power to propagate difference-making power in reproduction.[30] Thus, properly speaking, *reproduction is the recursive propagation of the organized capacity to develop and development is the ordered realization of the capacity to reproduce* (Griesemer 2000a, 2000b).

So the sequestered genes do their work only in the protected bundles that hold together and maintain organizations capable of development and transferred in reproduction. But why does it happen that way in biology? Must it? Consider the alternative. If biological reproduction did not

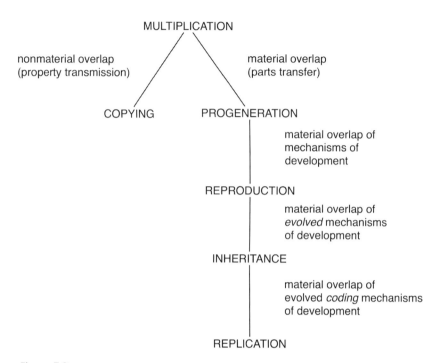

Figure 7.9
Modes of multiplication (after Griesemer 2000a, figure 5, p. S360). Two interpretations of the process of multiplying entities are contrasted. The left branch downward represents Dawkins's copying idea (see text for discussion). The right branch represents the multiplication of entities by a transfer of material parts from parent to offspring (progeneration). When the transferred parts constitute developmental mechanisms, then the progeneration process counts as reproduction. Inheritance and replication are special classes of reproduction process in which the reproduction process is of *evolved* mechanisms of development or evolved *coding* mechanisms, respectively. Other, nondevelopmental progeneration processes are not shown.

propagate organized developmental mechanisms, then the power to confer difference-making power on offspring would have to be found in the "environment." Origin-of-life scenarios often suppose that a "warm little pond," chemical soup, or prebiotic pizza (a structured solid-surface template-catalyst to organize the reactions; Maynard Smith and Szathmáry 1995, 32) provides all the organization needed for a collection of individual bits (diffusing molecules) to become jointly autocatalytic. But in those scenarios, the first hints of life become packaged in membranes that then internalize the environment, creating a self-scaffolding organization that confers the capacity to develop on the next stage.

Presumably, this provides a competitive advantage. Proto–living things that continued to depend on exogenous scaffolding would be at the mercy of environmental instabilities, constrained by the geographic extent of their scaffolding environments. Becoming larger or more complex would raise insuperable problems, for they would simultaneously have to accomplish more complex tasks more reliably, with all of their machinery exposed to whatever else is out there. Proto–living things that internalized the scaffolding of development, for example, by making their own catalysts and relevant catalytic structures, would be more robust to environmental change, better able to propagate their organization to offspring, and able to expand into a broader range of environments with it. Biological reproduction concerns the propagation of developmental packages while Dawkins's replication concerns the propagation of individual bits, taking the package for granted and mistaking the gene for the whole developmental package. The fitness advantage of reproduction over (bare) replication is, presumably, a real biological advantage in the earliest stages of the evolution of life on earth, or else the world would still be populated by naked RNA ribozymes. Even Dawkins recognizes this when he discusses the advantages to genes of cooperating in bundles called "chromosomes." The history of life on earth is not only a history of adaptive accretion, but of evolutionary transition to new levels of developmental organization (Buss 1987; Maynard Smith and Szathmáry 1995; Griesemer 2000c).

Without packaging (sequestration and bundling), *all* of the internal organization achieved in development must be regenerated each generation from exogenous sources. If the environment is reliable in delivering resources and scaffolding to the developing entity, then (presumably) development would go faster for those entities that didn't have to build their own internal scaffolding as part of their developmental process—they would get their developmental organization "for free." But to the extent that the environment is unreliable or ineffective in scaffolding development exogenously, internalizing it in organized living propagules for reproduction would be favored over mere transmission of individual replicators. There must be dynamic trade-offs in how much and what kind of organization to internalize for purposes of successful reproduction. Most multicellular organisms pass through a single-cell "bottleneck" in their life cycle (Grosberg and Strathmann 1998).[31] Within the organism, only cell-level organization seems to be packaged for biological reproduction. The organism-level phenotype is typically, though not always, regenerated through a developmental process rather than propagated directly.

How might these concepts be extended to apply to culture? Are there "cultural reproducers" in addition to biological reproducers, or are there only cultural replicators (memes) lacking development and material transfer across generations? Or does the degree of ignoring context become so extreme for culture (and the standards for recognizing "memes" so unconstrained) that one might say the notion of a "meme" could only make sense if more richly anchored in another way (Wimsatt 1999)? We opt for this course, to anchor memes in cultural reproduction plus the relative stability of parts of the cultural reproducers required by their central role in producing other elements of development—that is, their generative entrenchment. *Generative entrenchment can give deeply entrenched features the stability and generative power to make them in many ways the analogues of genes* (Wimsatt 1981b, 167–171), but also to act as scaffolding.

In this we delight in following in Thomas Hunt Morgan's footsteps. In 1909, Morgan produced a searching critique of "Mendelian factors" for many then plausible reasons, one of which was their disconnection from development, and a second was their lack of independent operational detectability from the ratios they were invoked to explain. This was a deep criticism that held until Morgan himself only two years later was forced (still only partially satisfied) to localize them on chromosomes, through which linkage mapping gave a rich set of alternative handles on Mendel's (now robustly operationalized and materialized) factors. This paper is little known and less appreciated. It would not be reprinted or noted today but for Morgan's famous and productive reversal, leading him to elaborate chromosomal mechanics to reproduce the study of Mendelian heredity as classical genetics.[32] *It seems quite likely to us that one may not be able to say what the "memes" are until one can figure out the hereditary channels for culture, and those can only be determined once one discovers the scope and nature of sequestration and bundling for the relevant cultural processes of development and reproduction one is tracking.*[33]

Cultural Reproducers

Let us return from these general considerations of reproduction to the specific question of cultural reproduction. Consider the Sears kit-house case described above.

The Sears kit-house case is a rich multilevel and multichannel hybrid of persistent, recurrent, and reproductive elements that we think is typical of human culture. We use this case to illustrate developmental problems of cultural evolution because it preserves detailed evidence of modular organization of component cultural artifacts (houses, house parts, furnishings, mechanisms, tools, and plans of construction) that play reproductive roles in the generative process of house consumption, production, and distribution.

The Sears kit-house case comes as close as we think typically plausible to support talk of particular cultural items as reproducers. Clearly, Sears houses are not reproducers: one house does not give rise to another by means of a transfer of material parts from parent house to offspring house. Though there may well be property inheritance of home contents and even of whole houses, it is not typical for house parts to become parts of other houses in ways that play a generative, developmental role in the production of offspring houses. But neither are new houses typically "copies" of old houses in a memetic transfer sense of cultural inheritance. Houses are typically built from common plans rather than as copies of one another, and in the Sears kit-house case, the plans are products purchased through a material, disseminated mail-order catalog. An adequate account lies somewhere in between pure memetic and pure reproductive extremes.

Sears posts a catalog to a would-be consumer, who picks out, purchases, receives, and builds a kit, possibly copying neighbors' choices of plans, "following in the footsteps" of previous consumers, the guidance of editions of the Sears House catalog, and using the rail, road, postal, and linguistic infrastructures external to but shared with predecessors. Perhaps on occasion a child inherits a catalog from a parent or learns to read from a Sears catalog, thus acquiring a taste for Sears houses along with acquiring the material catalog as propagule from the biological/cultural *parent* (linking the child into a genealogy that includes its biological parents and Sears, Roebuck). Maybe she or he builds a Sears house next door, or in the next town down the road, or at the next rail stop. The catalog (and economic infrastructure) can thus be a material component transferred to a new generation of house-dweller-producers via skills, capacities, and dispositions acquired by offspring that have been scaffolded by parents, catalogs, infrastructure, *and houses*. New kit houses are neither copies nor material descendants of old kit houses, but rather new productions that repeat an assembly process (Caporael 2003) and pattern according to the plans and designs contained in a kit, while the acquisition of a kit may involve a material transfer of elements that play a reproductive role in house-building practices.

Of course, *transfer* means something different for actor parts, house-kit parts, and infrastructure parts: material overlap of chromosomes, furniture, and roads is achieved by quite different developmental mechanisms, but they are no less material transfers or developmental mechanisms for all that. Similarly, an organism requires a material transfer of many kinds of parts via various mechanisms. These parts play more generative roles than mere supply of "food" to nourish naked DNA replicators: consuming food and reproducing are different kinds of material transfer processes, even if the food itself is reproductive. Reproduction involves the material transfer of generative developmental organization from parent to offspring—organization that becomes generatively entrenched in the process of repeated assembly by differentially successful development that comes to depend on the scaffolding effects of its presence and activity in the developing organism or cultural system.

There are "persistent" elements in the Sears kit-house story such as the Sears, Roebuck Company, houses that were built in each (cultural and biological) generation, infrastructures of rail, road, and postal delivery, suppliers who provided raw materials and component parts to Sears, language, economic, and political conventions for economic exchange, and so forth. Some of these persist on longer time scales than others and some persist by active maintenance—that is, by persistent efforts of many actors to sustain them in a particular form or condition. All function as economic "externalities" in the sense that they tend to be "there" in the environment of a given actor to be used virtually for free, but they can also function as brakes on innovation.[34] One (or a group) can choose to build one's own persistent elements rather than take advantage, or contribute to the maintenance of externalities, but there is typically an associated cost. Innovation, whether different in kind or only in number, trades off against cost and ease of use.

Persistent elements tend to be artifacts, items produced in the culture that play a role in the continuation and elaboration of the culture but are not reproducers. However, some biological actors such as schoolteachers persist across many student generations and may play the role of persistent artifact (playground ball remover, traffic light, metronome, pencil sharpener) for students, even while functioning as a scaffolding actor in other contexts, as teacher, parent, and so forth. But not all artifacts play cultural roles in the same way. The line between artifact and actor is blurred, just as the line between reproductive and nonreproductive is blurred, when sociocultural contexts are considered that shift focus to the system level.

Some artifacts contribute significant feedback effects to alter the development of the actors who use them while others merely contribute to the trajectory a target developer is already on. Thus, the rail and road systems that make it easier to transport Sears kit houses outside urban areas alter the options for people who desire to leave urban areas. A road connecting rural and urban places, extending amenities of urban life such as shopping in department stores, may increase the demand and need for such extension if the road makes it easier to move out of town.[35] The Sears Modern Homes catalog in turn facilitates the use of roads as a means of leaving town, in turn generating further demand for catalog products, including kit houses. Other artifacts may provide externalities that facilitate some activity of a user, but could be substituted by alternatives for little additional cost and for them there is less advantage to persistence. The point is that even persistent artifacts can play a reproductive role by playing a developmental role, scaffolding the activities of actors (making them easier, cheaper, safer than they otherwise would be) or scaffolding the use of other artifacts (mail-order catalogs, kit houses, automobiles) in ways tending both to enhance the reproduction of biological actors and with them the cultural systems of which they are parts.

Artifacts that provide scaffolding services may be recurrent rather than persistent, especially if it is cheaper to produce new ones than to make longer-lasting durable ones. Toothbrushes and cleaning sponges (or hair and fingernails) as opposed to washing machines and automobiles (or muscles and bones) come to mind. Henry Ford's famous observation that automobile parts still in good shape in the junkyard are overengineered and too durable expresses the value of recurrence: it may be more economical to recurrently produce a cheaper part that lasts only as long as its role in the system has value. Recurrence suggests reproduction, but the question for a theory of cultural evolution is reproduction of what? Biological actors reproduce, and when their offspring make artifacts, these are recurrent *in kind*. For many aspects of culture, however, the recurrence may be the product of reproduction directly, either as the expression of a biological agent's behavior, or as a cultural emergent: when a firm spins off a subsidiary, the recurrence of a managerial or accounting practice might be counted as reproduced, especially if the subsidiary is founded with personnel leaving the parent company as a "social propagule."

Our Sears house story is not fully enough developed to determine whether there is cultural reproduction of the system. It is clear that a number of kit house *companies* started up, but to argue that some of these were reproduced, for example, as spin-offs from other ones, would require company data that we lack. To argue that parts of the system are cultural reproducers is plausible, but we similarly lack data detailed enough to address this question either. The seemingly obvious answer that the Sears catalogs are reproduced is only half right since it is not in virtue of the sort of copy relations a meme theorist would identify, but rather the more subtle role catalogs play in scaffolding cultural development which counts. It seems likely that furnishings would have been handed down from (biological) generation to generation and plausible to think that the deep embedding of Sears catalogs in popular culture might have combined with this artifact inheritance to affect preferences, tastes, capacities, and propensities to purchase a Sears house kit in which to put the inherited furnishings. If so, then there is only a difference of degree between the role of a Sears desk lamp and a chromosome passed from parent to offspring in triggering and shaping the eventually developed home or body of the offspring. Desk lamp and catalog together "instruct" the consumer in dwelling development just as chromosome and metabolism "instruct" the consumer's bodily development.

Mail-order catalogs can be considered material propagules that transfer mechanisms of development for consumers who are parts of a society and culture—cultural reproducers. Consumer culture is scaffolded by the catalog, lowering the cost of consumption by packaging many artifacts, up to the whole house. Culture becomes reproductive as the catalog becomes generatively entrenched as a need to consume artifacts in packages (perhaps because the time available for item-by-item consumption is allocated elsewhere). Packaging facilitates the coordinated transfer of artifacts as well, thus, insofar as catalogs transmit, as cultural propagules, the practices that tend to lead to the reassembly of the packaged artifacts, the system becomes a cultural reproducer. Like biological gametes, mail-order catalogs transfer material parts of the sender to receivers: the catalog is produced by the company and sent to consumers, who then take material ownership of a small piece of the company.

Also like gametes, mail-order catalogs transfer organizations of heterogeneous elements—the arrangement of catalog items into functional groupings, price groupings, customer categories—in short, all the categories of marketing package items in generative configurations, for example, those that can make a house into a livable home. Ward's first catalog was a simple list of items and prices, but they quickly developed much more sophisticated structures.[36] Moreover, mail-order customers did not have to travel to a distant urban center to take advantage of this marketing organization in the form of the store layout and merchandise displays, they could instead browse the catalog. Unlike gametes, catalogs transfer only avatars rather than actual items for sale (pictures, drawings, plans, or references that are material enough—marks on paper—but highly transportable). Again like gametes, what is transferred is not the end product, but crucial parts of a mechanism for its generation. Gametes transfer developmental mechanisms (genes, proteins, membranes) that must act to coordinate development in situ. Catalogs transfer representations and configurations of items that must be realized in situ by cultural developments including: browsing the catalog, choosing items to fit into an existing cultural organization (typically, the home), ordering, waiting, receiving, and placing the items into the home-dwelling context of use. Paralleling the roles of the cell in "gene action" (Moss 2002) we have the operations of the post office, roads and rails, the internal, instructed "metabolism" of the banking industry, and Sears in the transfer of funds and transformation of the order into a deliverable product.

Cultural reproduction by means of catalog propagules can extend beyond the direct company-to-customer path as well. As we pointed out above, the catalogs may also becomes tools of instruction in reading for children, which may vicariously convey a preference for, say, Sears' or Ward's organization of goods. Although one may interpret the transmission of Sears (or Ward's) culture as cultural replication of items or preferences for items along Dawkins's lines, the case of the Sears kit home, with its bundle of 30,000 pieces represented in a complex package in the home building catalog is harder to interpret this way because the key feature—the organization of items into developmental packages—is left out of the account. Thus, a proper account of cultural evolution, by our lights, would include an account of the development and propagation of variant cultural *packages* in populations delimited by the institutional, organizational, and other social forms comprised of the conventions, artifacts, and actors who jointly conspire to produce "culture."

Developmental Scaffolding and the Process of Culture

As a first approximation to a characterization of culture, we consider H. S. Becker's sociological view of culture as shared conventional understandings manifest in act and artifact that people use to guide the organization of collective action (Becker 1986).[37] One step toward a view of culture we can use for our multilevel evolutionary purposes is to track artifacts through cultural *development*, which Becker does only implicitly.[38] Another step is to generalize from people to actors, where an actor is a part of a social system that can have "understanding." This second step puts us onto terrain of interest to cultural evolutionists: we should not take it as given that culture applies only to people. Other animals might have culture, as might other social entities, such as corporations, political parties, scientific research groups, disciplines, and nations.[39] Culture is not universal and is most plausibly a degree property depending on (evolved) capacities for representation and understanding. Becker's view together with our additional steps toward a view of culture helps move the analysis of cultural evolution toward the medium viscosity theory we prefer and away from "thick" views of culture that, by definition, rule out human culture as an evolved product from animal ancestors and "thin" views that make culture ubiquitous as mere nongenetic trait transmission.

Three Kinds of Scaffolding

With these steps in mind, consider three modes of developmental processes for culture in terms of three kinds of scaffolding that facilitate them: artifact, infrastructure, and developmental agent scaffolding. All kinds of scaffolding are relational: scaffolds are convenient, "rationalized," safe developmental contexts for particular things or kinds of things and acts of using them, and may serve only in some circumstances or for some actors and not for others. Scaffolds may differ in their generality—in the range of kinds of things they can act to scaffold, or agents or circumstances under which they can do so, but being very general in application or conditions of application does not make scaffolds any less essentially relational.

Artifact Scaffolding Artifacts can scaffold acts when they make acts possible, feasible, or easier than they otherwise would have been. The painter's scaffold makes it easier for painters to reach the upper floors to do their work than without the scaffold. The *use* of the artifact by the actors in their work is what makes it a scaffold.

Artifacts can become so central to the nature of acts that they are in part constitutive of them—for example, "cross the t" as a naval maneuver for warships, "take off," "peel off," and "strafe," for aircraft, "back up" for computers, "emulate" for virtual machines, and "pass" as something one does in a passing zone on a highway. Human agents participate in each of these actions but the participation of the artifacts is so central to the acts that we treat them as if *they* are the agents. One can then ask in what sense the artifacts are still scaffolding the acts, although at least these acts would have been impossible without multiple layers of scaffolding. Note that constitutive involvement is a strong indicator of generative entrenchment. This theme is further developed from another direction below.

In all these cases, the collective that acts involves a human actor and a non-human one. Their interaction is symmetrical: the human is scaffold to the nonhuman as well, making the actions of the artifact possible, feasible, or easier than they otherwise would have been. The painters' scaffold is built by the painters in the course of the project, used, and then taken down, to be reassembled on another site. The artifacts we describe as scaffolds have "life cycles" just as the humans do. The scaffold can be part of the developmental system *of the humans* if the scaffolding interaction meets the right evolutionary conditions—namely, that such interactions in the genealogical past played a role in the survival and reproduction of the humans' ancestors (Griffiths and Gray 1994, 2001). The symmetry of the interaction implies that the human can also be part of the developmental system *of the scaffold* if the scaffolding interaction meets the right evolutionary conditions—as with Warwick's (2003) account of the development of mathematical physics at Cambridge. However, this is not likely: painters' scaffolds are not typically genealogy-producing reproducers,[40] hence not candidate units of evolution (Griesemer 2000a). Thus, the contributions of symmetrical scaffolding interactions to the two developing systems—human and nonhuman—may, and commonly will, have asymmetrical evolutionary consequences. The problem of cultural evolution, then, includes interpreting the ways in which scaffolding interactions among different sorts of elements of a society contribute to the development of cultural systems or processes whose reproduction may or may not be distinct from those of its participants.

The key to understanding scaffolding as a cultural process is its *developmental* aspect: an artifact plays a developmental role by facilitating collective action and thus contributes to cultural aspects of development in an organism, institution, or organization.[41] A minimal evolutionary concept of development is acquisition of the capacity to reproduce (Griesemer 2000a, 2000b, 2000c). If cultural artifacts (like repainted or repaired buildings) contribute to a cultural system's capacity to "reproduce" (by serving as models of building, paint jobs, successful maintenance projects, or whatever), there is a temptation to see the building as a reproducer, just as we tend to see a gene as a replicator. This would require an expanded view of what counts as the relevant developmental system: a viewer of the building, say, interacts with the building, becoming part of the building's developmental system, and then the viewer becomes a propagule of the building, going off to form a new building (or paint job) in the context of the viewer's developmental system.[42] Without the viewer-*in-context* counting as a material propagule-*in-context* of the building-*in-context*, we ought not to consider the building as a reproducer, though perhaps systems of building that include both buildings and viewers as parts might be. Such a view of cultural reproduction might seem extreme although empirically it might turn out to be the case, and even not so infrequently.[43]

Cultural development, then, is only acquisition of a capacity of a cultural system or process. It need not concern the reproductive propagation of each of its components, nor is this the case in biology: multicellular organisms, which develop by means of the reproduction of their cells, do not require that *all* of their cells reproduce. Even so, the multiple overlapping and interpenetrating hereditary lineages and channels of cultural evolution make the neatly hierarchical standard metazoan model generally impossible for cultural systems. But that is part mythology: a closer look at our life cycles as interwoven with our carried communities of internal and external parasites and endosymbionts raises more general questions about the accuracy of that model anywhere. Indeed, the failings of facile analogies between biological and cultural reproduction often trace to lack of subtlety in the treatment of the developmental aspect of reproduction processes on both the biological and the cultural sides of the analogy. So, artifacts that serve as scaffolds are symmetrically also targets of scaffolding interactions between actors and artifacts.[44]

Infrastructure Scaffolding Some artifacts or artifact types have the character of infrastructure: objects that persist on a much longer time scale than typical artifact scaffolding interactions and which are commonly regarded as parts of the environment or "niche" in which an action takes place. This may be planned from the beginning, or they may acquire an infrastructural role through changes elsewhere in the environment or in the things they scaffold which coadapt to them. Then they become infrastructural basically by becoming more deeply entrenched. A painters' scaffold is put up alongside a building and used to scaffold the repainting while the painters are on the job. It is taken down soon after. It would be "in the way" if it were left up. Road or rail systems, communications satellites, the internet, postal delivery, plumbing, meeting halls, and the like are designed to last and serve repeatedly over many scaffolding interactions.

The persistence of such a system encourages adaptation of other elements to it, which increases its entrenchment. You want your pipes to last for many water-carryings from source to kitchen sink. Home water pipes scaffold your use of water in the home by making it much easier to carry and use—controlling pouring, heating, and filtering water. Piping will get connected not only to the usual minimum array of sinks, toilets, tubs, water heater, and furnace, but in some homes, as input to or output from a sump pump, humidifier, dehumidifier, air-conditioner, fire-retardant sprinkler system, outside shower, swimming pool, and outlets for gardening and lawn sprinkling systems, each with their dependent control systems. Or the water supply may be sufficiently limiting that swimming pool and outside watering are never considered, and one installs a "graywater system" (with quite different piping systems and tankage) to reuse water from sinks and showers for sanitation. Connection to city systems or to well and septic field, and mode of pressure generation in the system also affects design, and all of these generate needs for a host of valves, bypasses, vents, modulators and accessories to "the plumbing."[45]

Why does it make sense to treat scaffolding interactions as contributing to *development?* Some have similar difficulty seeing why the organism-environment interactions that construct niches should count toward the development of a system of interactions rather than merely as changes in the niche, that is, environmental context of development and selection for the replicators in them (e.g., Dawkins 2004). Thus Sterelny (2001) distinguishes ecological engineering from ecological inheritance and expresses skepticism of the empirical scope of the latter (Griesemer et al. 2005 critique Sterenly's criteria). Developmental systems theory aims to address this difficulty by identifying developmental systems (or processes) as the loci of action, so that the environmental "side" of the interaction is interpreted as part of the system that develops. We pursue a different course: we are interested in those cases and conditions under which parts of the environment or niche function in an organism-like way, as reproducers with generative entrenchment relations to the rest of the developmental systems in which they act. Put differently, we are interested in the conditions under which culture develops because these will *delimit emergent cultural reproducers that can propagate the conditions for culture*—conditions in which cultural development fosters cultural inheritance. In turn, genuine cultural inheritance is a kind of process in which cultural variants make more cultural variants so that offspring exhibit fitness differences in a population of cultural reproducers. A key feature of our approach is the role of developmental propagules in reproduction. These are items that can be counted, can form populations, and can share infrastructure (niche components) at the same time that their interactions with shared items of infrastructure are genuinely their own.

Scaffolding interactions with infrastructure are much the same as artifact scaffolding in other respects, although the dynamics of systems of infrastructure scaffolding are bound to differ from more temporary, smaller-scale artifact scaffolding. At first it looks simple: perhaps the up-front social costs and marginal utility of erecting an infrastructure are greater than with more ephemeral artifacts (or even longer-lasting artifacts that, due to smaller scale or fragility, have to be "brought out" for each use and thus can only intermittently function as scaffolds). And no doubt the relation of propagules to the rest of a developing system changes the dynamic properties of the system as a whole (Griesemer 2000a, 2000c). One important way is when reproductive processes entrench the factors producing them. In becoming deeply entrenched, elements become more like infrastructure as they become more polyfunctional: as infrastructural elements are co-opted for more functions because they are there, they become more deeply entrenched, and in consequence, more strongly conserved, causing a mutual positive feedback of amplifying stability relations. Thus one can build infrastructure to facilitate the amplification of structures and processes utilizing them, but one can also simply use what is there to such an extent that it *becomes* infrastructural. And *generative* entrenchment requires reproduction, so the circle is closed.

Perhaps the most important mode of infrastructural scaffolding are forms without which culture and society would not be here at all. Going backwards in time: written language, settlements and agriculture, and animal husbandry and trade practices (developing into economic systems) were major infrastructural innovations central to all that followed. Spoken language with oral traditions and tools use antedate all of these by many tens to hundreds of thousand years. All are generatively entrenched so deeply as to be virtually constitutive of all of our forms of life, limiting the kinds of presence-and-absence comparisons we would like to have to assess their effects. Their origins raise entirely different kinds of methodological problems that they share with archaeology and paleontology—the paucity and gappiness of the direct historical record.

By contrast, the *current* differentiation, cross-pollination, and mixing of language varieties, and use, often of multiple languages in different social strata and crosscutting contexts within the same society, indicate rich areas of current study in cultural evolution. Language change as studied in interaction with its "ecology" (in the broadened sense of Mufwene 2001) already reflects the populational approach urged here. And it is full of phenomena and "model organisms" (see Griesemer and Wimsatt 1989) having all of the features appropriate for rich evolutionary study (Mufwene 2002, 2005), and a phylogeny increasingly richly correlated with archaeological remains and the genetic traces of human migrations (Cavalli-Sforza and Cavalli-Sforza 1995). Language practice, and the expansion and contraction of language varieties interact with other economic and social practices in rich ways that only begin to reflect the deep entrenchments of spoken language in our cognition and our culture in the costs of illiteracy, innumeracy, and rarer deeper linguistic deficits. The very plasticity and robustness of language acquisition indicates its central evolutionary importance for *Homo sapiens*.

Developmental Agent Scaffolding Artifacts and infrastructure function as scaffolding when actors use them to facilitate collective actions in a particular developmental way. The difference between use and scaffolding lies in the developmental function of the latter. We can view artifact and infrastructure as scaffolding when we can identify a developmental function, such as facilitating acquisition of a capacity that the target or an actor in the collective did not have before (or would not otherwise). Intuitively, if the builders' scaffold is used to wash the windows of a bank, that's maintenance (both of light transmission capacity of the windows, and of the reputation of the bank). If it is used to put a clock on the front of the bank building, then it is being used for developmental purposes: to give the bank's façade a function or capacity it didn't have: telling passersby the time, and conveying the image of the bank as civic institution in a new way.

Many of the interesting developmental uses of scaffolding are the development of skills (or authority) in agents, for it is only through these, developed to allow effective organization to perform the immensely complex tasks we find in a highly differentiated society, that culture can be maintained and elaborated. Skill development can be scaffolded by artifacts or infrastructure, as when a person learns to drive a car on an off-road course where mistakes are less likely to lead to injury, or when a gymnast uses an elastic training belt to learn a dangerous move on the high bar. But scaffolding skills in agents where the scaffold is (or includes) another agent are particularly interesting: the scaffold is or involves another person, social group, or organization, often in spatial and temporally organized dynamical arrangements with artifacts. These kinds of interactions can entrench modes of interaction, and build larger social and organizational structures, from office hours through lab meetings and lecture classes to websites, professional meetings, journals, and international societies. These may themselves become elements and contexts in still larger articulations and differentiations of agents, procedures and artifacts.

Consider, for example, the scaffolding interactions between two agents discussed by developmental psychologists (Bickhard 1992). A child is scaffolded by its father when the child commits to trust the father to facilitate a risky operation like crossing the street. The child puts its hand in the father's and father teaches the child to cross the busy street safely. By towing the child across in rhythm with the passing traffic and by calling the child's attention to the relevant cues such as the color of the traffic light, position of the crosswalk, and so forth, the child develops skill in crossing the street in the safe context of the scaffolding interaction. Many such scaffolding interactions are required before the child can exercise the requisite skill on its own, without the guiding visible hand of the scaffolding agent. (Notice also the abundant infrastructure scaffolding—the father scaffolds his *interaction* with the child by use of the traffic light, crosswalk, and other infrastructure to render the interaction with the child in crossing the street safer than it otherwise would be.) The cases elaborated earlier all begin in processes like these child/parent interactions. Barbara Wimsatt's elaboration of advanced medical training, which differs by sex and its (career) developmental consequences, Andrew Warwick's discussion of the elaboration of mathematical physics at Cambridge over 150 years, and the developmental co-option of infrastructural resources and transformation of house designs and financial instruments to deliver over 100,000 kit houses[46] to an expanding U.S. market after World War I all build on the stability inducing trust and adaptive patterns they engender.

But how are these larger structures built, and maintained—and in a way consistent with the flow of agents and resources through them, and creative new alignments tried, maintained, and elaborated on the fly?

Generative Entrenchment vs. Revolutionary Cultural Change: Can Scaffolding Resolve This Paradox?

Recognizing Dependencies: Stasis in the Midst of Change

Evolution is a process of change, but also of stasis: metazoan body plans date to the Cambrian (Raff 1996; W. Arthur 1997); while cell membranes, the genetic code, and primary metabolism are roughly as old as life itself (Morowitz 1992). For evolution to be cumulative, recognizably historical, or even possible, some things must be preserved to build on. The cumulative character of evolutionary change has been debated, but once we remove a naive progressivism and admit the normal range of exceptions characteristic of patterns in biology or the human sciences, this is unproblematic. Evolution is historical by definition, but must it be *recognizably* historical—must organisms and other evolving things wear their ancestry "on their faces"? Indeed they must. Evolution saves things in spades, on all space and time scales. Evolution is not just any change and stasis, but particular patterns of change and stasis, patterns that tend to preserve ancestry. Wimsatt (2001) argues that all nontrivial evolutionary processes involve entities with development, and with their different parts showing differential generative entrenchment. From these it follows that they will also show life *cycles* (since they will be constrained to start each generation in very similar places by entrenchment acting more strongly earlier in development). They will also show the kind of historicity to allow reconstruction of phylogenies and to be recognizably historical as above (Wimsatt 2001; a point first made by Riedl 1978). These entities will also be reproducers, in Griesemer's (2000a, 2000b, 2000c) sense, from which other properties discussed in the preceding sections follow. But why should such features be so pervasive and so important?

Consider an observation of commonsense engineering. Rebuilding foundations after we have already constructed an edifice on them is demanding and dangerous work. It is demanding: unless we do it just right, we will bring the house down, and not be able to restore it on the new foundations. It is dangerous: the probability of doing so is very great, and there are rarely strong guarantees that we are doing it right. We are tempted to just "make the best of it," doing what we can to "patch" problems at less fundamental levels, leaving deeper modifications alone. This is sensible advice, given the difficulty of the task. One can't make just any change one likes with impunity. These are the phenomena and the patterns to be explained and exploited.

This bias against doing foundational work is a very general phenomenon. Scientists rarely do foundational work before their house threatens to come down about their ears. (Philosophers relish it, but like to work on others' disciplines!) If we cannot avoid them, actual deep revisions are preceded by all sorts of vicarious activity. We would prefer to redesign or revise a *plan*, convince ourselves that it will work, rebuild only after we are satisfied with the revisions, checking during reconstruction for problems that inevitably come up, and move in only after the rebuilding is done—complete with local patches and in-course corrections. Planning and the semipublic exchanges between concerned parties accompanying it are two major modes of social feedback. We thereby draw on the expertise of larger groups of people in structured ways to avoid major failures. Construction projects that forgo this, executed by theoreticians, or architects who do not visit the site are the butt of jokes[47]—folk wisdom about the difficulties of foundational revisions, or in getting from new foundations to the finished product. (And why amateur builders who do not know all of the steps and pitfalls may welcome kit houses!)

Big scientific revolutions are relatively rare for just that reason—*the more fundamental the change, the less likely it will work, and the broader its effects; so the more work it will make for others, who therefore resist it actively*. The last two facts are organizational, institutional, social, and social psychological, but the first two are not. They are broad, robust, deeply rooted, structural, and causal features of our world—unavoidable features of both material and abstract generative structures. It is this robustness that can allow dynamic patterns derived from generative entrenchment not only to apply in both biological and cultural realms, but to weave patterns that tie them irrevocably together. The first two features are central to the power of generative entrenchment in biology, but all four are relevant throughout the human sciences and the domain of our products, including how we conduct our investigations in all of the sciences.

Generative Entrenchment as a Cause of Stasis in Evolutionary Processes

To know how to proceed, we must analyze the dependencies in evolved systems. How many things fail or are changed if we change a given element in a system? How many things jump if we pull on it? How much of the porch collapses if you pull out a corner column, a center truss, or a floorboard? Will it affect the rest of the house? The reach of the damage or the spread of the change is a measure of the generative entrenchment of the element which was changed. Its generative entrenchment is the range of things it contributes in the normal operation or functioning of the system. And, to a first approximation, the degree of generative entrenchment (or GE) of an item is a predictor of the amount of change (and potential damage) arising if it is changed. Since the probability that a change will be damaging goes up rapidly with the size of the change and number of things affected, the generative entrenchment of an item is a good predictor of its evolutionary conservatism. We return to this below.[48]

Of course we want to know not only how much will fail, but what will be affected, and how. This is the key to diagnosing what went wrong in a failure after the fact, and to figure out how to design a way of repairing or improving a system if we want to change it. This last task is difficult, and so generally we try to fix things by changing as little as possible. This has broader consequences: *it is why evolution proceeds mostly via a sequence of layered kluges which give local fixes*. Nature is full of larger contingencies as well as smaller ones, but the contingencies that drive evolution usually start quite small and can grow in their effects over time. Local kluges become deeper fixes and ultimately foundational architectures only by persisting through successive generations over evolutionary time, increasingly embedded in and adapted to by a growing network of local fixes.

As their reach spreads, they become, in "evolutionary" time, necessary preconditions or framework design principles for almost everything else in that organism and major dimensions of the niche it comes to occupy, and inescapable for its expanding descendants—a "natural" mode of ex post facto rationalization. These come to define many dimensions of the organismal forms of life, the conditions that are taken to be constitutive in natural and human design and "optimization" problems. Things do not have to persist, but if they do, their resistance to being eliminated tends to increase.[49] Natural processes make foundations as they proceed, but not by "wedging" them in at the ground. Things which are there already tend to *become* more foundational through use, and then are used still more as they become increasingly foundational.

House construction has an apparently "natural order" because of the dependency relations of some elements relative to others. (One proceeds roughly from bottom to top, and frame to finish, now in "ontogenetic" or "developmental" time.) Gravity is one major constraint, and not covering up what you have to work on is a second. Early elements scaffold later ones. These affect both design of the house and the order of work. Protecting workers, work and tools from major disasters, and from the elements is a third, but one that must sometimes force delicate workarounds in the first two. These do not determine outcomes in all details, but they broadly constrain the results, since proceeding in any other way is inevitably more work, more expense, and involves unmaking or redoing things that have been done already. (Contractors can sometimes change the order in which different crews come in, but sometimes must simply stop work until the necessary crew, materials, tools, or weather is available.[50]) Organisms, theories, technologies, and complex functional structures—biological, mechanical, conceptual, or normative—also have orders imposed by relations of dependency and interdependency.

Moreover, we must often dwell in a house while it is being rebuilt.[51] (Here we mean to include metaphorical "houses" across the disciplines!) But these "running repairs" only work because the conceptual organization of science, of engineering practice, of much of technology, and of social organizations and institutions, is usually robust, modular, and local, each of which reduces the reach and magnitude of generative entrenchment. Shaking (local) foundations usually does not bring the house down, and we still have a place to stand (on neighboring timbers) while we do it. This is not an accident, but a deep principle of design.[52] This is the key for how scaffolding can facilitate making deep modifications. With multiple information channels, we have new possibilities for their adaptive interaction, and one or more of them might be able to serve as scaffolding to modify the character of another channel, or its content, or how that content is to be responded to. But these modes of interaction are not interactions among abstract information flows. They are modes of interaction between embodied channels in "carpentered" environments, whose information content can affect the actions and values of agents.

This suggests an argument for why biological and cultural transmission is multichannel: if in general the information flow must be scaffolded because the information carriers have to *develop* in an environment in order to assimilate, use, and carry it, then the scaffolding must propagate (or persist) alongside the information "in" the channel. This might be the case if the information is sufficiently complex to be a generator of a cell, an organism, a theory, a cultural tradition—a generator of any biological or cultural item richer than a gene or a meme. (And these of course are generated easily only within their larger support structures.)

Some have argued that there cannot be multiple channels on grounds that they cannot be statistically independent and therefore do not constitute separate channels.[53] We view matters differently. First, *channel* is ambiguous: it can either mean "channel conditions" as a statistically independent component of carrier plus signal or it can mean a physically separate means of transmission as with two wires. The separateness of channels is a claim about their physical (spatiotemporal) separation, not about the statistical correlation of their states. Thus, channels can be separate, but nonindependent. So symmetry arguments between channel *conditions* and signal are insufficient grounds to reject the multiplicity of separate channels (see Griesemer et al. 2005). Moreover, a single channel can be made to carry multiple signals, whose states are statistically independent of one another, so the single set of channel *conditions* can be switched symmetrically with each of them and still retain statistical independence of all of them.[54] Thus, symmetry of signal and channel conditions is not sufficient grounds to reject the multiplicity of either separate channels or of independent signals, which is all we need to support the idea of a multiplicity of physically separate inheritance processes. The distinction between physical separation and statistical dependency is crucial for the analysis of developmental systems.

Processes of "generative entrenchment" are important generators of complex organization across many disciplines. Generative entrenchment (or GE) causally links development and evolution, and is particularly useful in characterizing systems with both dynamic adaptive and constraining processes—cognitive, cultural, social, or biological. The core insights derive from processes that produce different rates of evolution. Differential generative entrenchment of parts or processes produce and experience differential rates of evolutionary change. This differential conservation in the evolution of the system may also lead to secondary responses, such as reducing entrenchment through sequestration or packaging (differentiating modularity), the generation of parallel redundant systems, or the generation of distributed redundancy across multiple diverse systems ("functional multiplexing"). This can in principle produce changes leading to the multiplication of hereditary channels, as well as the elaboration of adaptive order. These responses may in turn make it easier to modify the generatively entrenched element. Or we may find the production of new systems or defenses to regulate and protect the generatively entrenched elements from being modified (e.g., canalization in biological development, or deciding that a relationship should be treated as a conceptual or definitional truth or as a matter of faith, rather than as a (modifiable) empirical one in a scientific or philosophical argument). These additional relationships suggest new ways to recognize, apply, and strengthen an evolutionary perspective in some areas that have resisted it (Wimsatt 1999, 2001). We turn to this now.

Generative Entrenchment in the Human Sciences and the Study of Cultures
Evolutionary fixity through entrenchment affects other domains where the operative mechanisms and forces are quite different than they are in biology. Unlike most moves to biologize the human sciences, reproduction and entrenchment do not pose reductionist threats, and are quite compatible with accepting descriptions as thick as you like, as long as it is recognized that they are analyzable and explicable in terms of accounts at their own level which nonetheless relate to a broader evolutionary context of reproducers and generative entrenchment, and articulate with the microprocesses necessary to realize them. (They are thus of only "medium viscosity.") Allowing for generative entrenchment and reproduction commonly involve reorganization of factors already recognized to see their effects

systematically in a new light, rather than denying their efficacy. (If anything, this reconceptualization should do the reverse!)

These relationships indicating entrenchment include the serious consequences of early deprivations in linguistic or other cognitive, emotional, or social development, and the fact that early training, and interwoven sources of scaffolding are crucially important for high competence in sequentially dependent and elaborated skills like playing the piano or becoming a gymnast or mathematician—that is for the rich manifestations of culture we embed in others and ourselves. It involves recognition that we build larger competencies by packaging, articulating, and comodulating parts into larger skills, and ways of interacting, an assembly process partially recognized by Simon (1962). What Simon did not appreciate sufficiently was that the stability of the subassemblies was often generated by "black boxing" and entrenching them (Latour 1987)—by contextual, rather than intrinsic (or "top-down" rather than "bottom-up") relationships.[55]

But the impact of generative entrenchment goes beyond the individual. We are highly social creatures. Our culture generates, accumulates, and maintains motivations and scaffolding for us to acquire and integrate diverse skills, empowering our complex societies and cultures. Societies are structured by overlapping entrenchments, modulated by various peculiarly cognitive or social adaptations to produce surprising freedoms of change, but also frustrating anchors against it.[56] Generative entrenchment is a simultaneously conservative and creative mechanism in each of these domains. It provides a systematic account of the development and evolution of richer larger-scale structures that is missing in competing accounts of cultural evolution (Wimsatt 1999, 2001).

Characteristic common transitions of particular elements occur in different domains of human culture. New qualitative features of human minds, cultures and societies emerge along these trajectories. (One must also discuss their integration in larger structures, a nearly universal feature generating the transitions listed below.) Increasing entrenchment can lead contingent associations or practices to take on the character of meaning relations and conventions. Movement along trajectories by elements through time accompanies increased fixity, resistance to change, generative, and normative roles. The second-order transformations of defensive or canalizing mechanisms arise from use of this increasing fixity in these larger structures, making categorical transformations where the underlying basis is merely a difference in degree. Metasimilarities across trajectories are products of role changes for elements with the growth in fixity, generative role, and adaptive complexity of the resulting interpenetrating and increasingly cogenerative structures as they become larger systems:

1. Some actions become individual habits; become social commonplaces, become practices; become conventions; become institutionalized.
2. In science, some empirical patterns become regularities; become generalizations, become laws; become framework principles; become quasi-definitional truths. (Scientific techniques and technologies are not forgotten here; they would be covered under (1) and (3) respectively.)
3. Some behaviors, forms, and mechanisms become common, adapted to, built upon, generatively productive, increasingly "innate," "natural," and "essential" characteristics of the species.[57]
4. Technologies emerge as kluged experiments, become exploratory tools for early adopters, acquire "killer apps," develop competitive markets, radiate third-party suppliers calling for standardization, rationalization, legal regulation and governmental control. In the process they have changed while transforming us and our environment: thus track the mutually elaborating automobile, petroleum industry, highways and metastasizing infrastructures, suburbs and suburban value reorientations, loss of neighborhoods and public transport, with multidimensional impacts on our foreign policy (B. Arthur 1994; Basalla 1987).

These kinds of patterns involving simultaneous increases in generative entrenchment and stability with accompanying changes in the qualitative status of increasingly entrenched elements give powerful handles in the analysis of diverse systems and in the emergence of new kinds of properties with the growth of these complex systems. But the patterns are not as simple as depicted above. And that is fortunate, because the systems are not either.

Exceptions Can Probe the Rule
Now to return to the challenge with which we began. The deeper elements of a generatively entrenched structure should be very stable because of their deep entrenchment. But exceptions to this pattern occur, and sometimes systematically. These provide ways of strengthening and enriching the theory (Wimsatt 1987) if either they can be incorporated in it in a natural fashion, or can be used to delimit its boundaries more precisely. (Most productive changes will do both simultaneously.) The aim should not be to pose reproduction plus generative entrenchment as a universal two-factor explanatory solvent, or superglue, for they are not. The aim is to show

how these important factors interact with others—just as natural selection as an account of evolution is strengthened by recognizing genetic drift, population structure, and the claims of genetic linkage.

The most important exceptions occur when apparently deeply entrenched things seem to be too easily changeable. Major but evolutionarily recent early developmental changes are known in various organisms (Raff 1996). We must understand how this is possible. Analysis of these exceptions reveal other already recognized and important features of biological organization like regulation and canalization, redundancy, and modularity whose interactions with generative entrenchment explain these anomalies and still other phenomena (Schank and Wimsatt 1988, 2004; Wimsatt and Schank 1988, 2004).

Cultural change intuitively occurs so much faster than biological evolution, and with enough deep changes, especially noticeable in science and technology, that revolutions are not uncommon. But how is this possible? This is the target question of this chapter, and forces us to reach beyond reproduction and entrenchment. We should expect these contrasts to reveal important differences between evolutionary processes in biology and culture.[58] (It is reassuring to note that these deeper changes, though much more common for culture than for biology are still much rarer than changes of smaller effect in each of their respective domains—as expected with GE. This fact also suggests a search for systematic differences between biology and culture that modulate the impact of generative entrenchment without destroying it.)

Scaffolding for Speed, Accumulation, and Depth in Cultural Change

Now we must confront the deepest paradox of cultural evolution. While "higher-level" processes generally proceed more slowly than lower-level ones, it is a truism that cultural evolution can proceed enormously more rapidly than biological evolution, and often apparently through major almost explosive innovations and adaptive radiations.[59] How?

The first-order answer is to point out that culture is commonly "horizontally" transmitted. Like a disease, it can thus spread much more rapidly from host to host than a host population could spread or expand through its own ("vertical") biological reproduction. Successive waves of the plague spread from Asia throughout Europe in less than a decade in the fourteenth century, and roughly once a generation thereafter, as new susceptibles increased in frequency enough to sustain epidemic waves. Similarly, Islam spread around much of the Mediterranean in little more than a generation. But not everything was as fast: Christianity spread from the eastern Mediterranean throughout the reaches of the Roman Empire in about 400 years (Stark 1996). By contrast, the biological spread of agriculturalists due to their increased reproductive rate after the first green revolution as they replaced hunter-gatherers—an expansion at blinding speed for such a process—was only about ten miles per generation (Cavalli-Sforza and Cavalli-Sforza 1995).[60]

This point is useful: pointing to the speed of horizontal transmission allow us to think about *maximum* horizontal transmission rates, such as for the spread of news (or rumors) deemed important. And this clearly has increased by many orders of magnitude through technology, as the speeds of material migration and informational transmission have increased. For material transmission, we have walking, running (the original marathon, which served to bring news of an invasion), the mounted dispatch rider and pony express, carrier pigeon, rail, highway, and airplane, with parallel (though slower) expansions for marine transport. For information, we have drum and semaphore, the telegraph, telephone, trans-Atlantic cable, radio, television, and satellite internet transmissions developing through technologies and supporting material networks. It is this that gives the world instant access to election results, the 9/11 disaster, and declarations of war. At a more mundane and metabolic level, it has created a "world market" with activities running "24/7," smearing "market closings" into nonevents. This is the currency of the rampant meme, though (as we now see) the meme advocates have ignored the fact that their spread has increasingly required enormous interpenetrating and multilevel scaffolding. We can see this through another approach to the question.

In another important way, appeals to maximum horizontal transmission rate beg the question. Why didn't Christianity spread as fast as the fastest runner to Rome? It worked to warn Athens and to recruit Sparta against the coming Persian hordes. Why did it take until Augustinian's rise to power 300 years later to really cement the spread of Christianity, and what led Augustinian to declare it the state religion, which was the pivotal act which did so (Stark 1996)? Obviously a lot else was going on, and this had to do with spreading and transforming a culture and a lot of social scaffolding. The significance of the coming Persian army and the scaffolding to mobilize the Greek city-states was already there, in treaties and common perception of a common danger. Or to put it in another way which covers both cases—what makes an individual ready to accept and adopt, or even be capable of understanding a meme? Until we can understand this (and something of the special social and cultural structures through which memes are likely to spread) we have not gotten an adequate account of cultural transmission or evolution, and even this (because of its focus on individuals) is likely to be at best necessary but not sufficient.

Now let's ask the question again: given that culture is acquired developmentally, what kinds of things give cultural evolution the possibility of faster change, accumulation, and the acceptance of deeper modifications than is common for biological evolution? These would all contribute to the rapid hegemony culture has acquired in our relatively recent evolution.

Consider First What Speeds it Up

In his long and pivotally influential essay-review of work on "Evolutionary Epistemology," Donald Campbell (1974) recommends that we recognize a series of ten systematic perceptual, cognitive, and social adaptations as "vicarious selectors." These included distance perception, trial-and-error learning, thought experiments, language, and science. They had in common two things. First, although ultimately products of natural selection (and whatever other vicarious selectors were there already) they could act more rapidly and cheaply than natural selection. Second, like sexual selection, they could sometimes transcend it and drive the results in a different direction than natural selection alone would take them. So how could vicarious selection processes be "faster and cheaper," and sometimes take processes in a different direction? The following provide a large part of the answer.

First we consider things that facilitate the structured generation of large amounts of variation at low cost. One may here consider both the cost of generating variation, and the cost of generating *relevant* variation. The latter (items 2 through 5 below) points to kinds of "directiveness" of variation, and thus to ways in which cultural evolution is "Lamarckian" that biological evolution is not:

0. One could begin with the intuition that it is far cheaper to generate psychological and cultural variation than genetic variation. This is probably true, but it is unclear (due to metrical and conceptual problems) how to measure it. All of the points below address this issue, but it is not clear whether they exhaust it.

1. Once something is tried and does not work, we do not need to repeat it, unlike the situation with mutations where useless and even deleterious mutations may recur repeatedly. (Actually, our efforts are not without wasteful repetition either, but we clearly do better than biological mutagenesis here.)

2. We can localize faults in structures, deciding what is at fault, and direct efforts to producing and trying variations at that point. This important point is developed systematically for theory testing in Glymour 1980, but the point is more general as doctors, auto mechanics, software designers,

and a host of others connected with complex technology must know. More can be said about it, but not here.

3. Cost-benefit selection: with the initial genesis of an idea and the formation of a zygote, each undergoes a development, which may abort or be aborted at any stage. Since continued development involves continued commitment of resources (and opportunity costs) any process which can secure accurate assessments of costs and benefits of continuance and act accordingly will in effect contribute to the efficiency of generating variation. If this induces biases in what survives, it is a directive process. It may also be parsed as developmental (or "internal") selection, but via a higher-order control process.

4. We can learn from past trials, and adjust the "direction" of new variations to improve performance in the light of the effects of past ones.

5. Perhaps most importantly we have structured sets of alternatives for generating simple or compound modifications, and structured ways of searching them (often in material propagules designed for this purpose). Not only do we have catalogs of various kinds,[61] but sections, indices, and (in electronically realized versions) a variety of rich "search tools." In the richest forms we may formalize and define a "problem space" in which states, starting points, desired endpoints, and allowable moves and their effects are all represented (e.g., Simon 1973). This is closely related to the kinds of combinatorial variation generators characteristic of codes, languages, and other symbolic forms. (This is "higher" culture than the preceding, clearly constitutes rich scaffolding, and this loose category embraces several strata of divergent sophistication and complexity.)

6. Coordinated with the search tools to find alternatives, when they require material embodiments (i.e., books for ideas as well as nails for houses) there must be ways of producing and distributing them. When these are material objects (like screws, fittings, and other machine parts), coordination problems for mass manufacture with interchangeable parts can make these alternatives and their production and distribution infrastructural in character. This feature spans so many levels of varying complexity that it must just remain as a marker for a much larger subject. See also item 10 below.

Next, we consider aspects of scaffolding that tend to entrench variation and lead to accumulation.

Elements that build culture

7. scaffolding becomes more structure like (less modifiable, and acting asymmetrically as elements that other things are adapted to) as it becomes entrenched.

8. Scaffolding can persist in a maintenance mode, without apparent development but entrenched for the preservation of social institutions or for the efficiency that comes with black-boxing. (Development reappears once we realize that individuals are recruited, trained, and scaffolded by society to do this maintenance.) Things that become sufficiently entrenched as to become requirements for a wider variety of other things will accumulate secondary structures and processes for their maintenance.

9. Scaffolding interacts with black-boxing: acquire skills in the use of an artifact or procedure without knowing in detail how it works, and how to build or maintain it, thus black-boxing knowledge so that one can use it without understanding it.[62] One can present scientific laws in engineering texts for use in "cookbook" fashion, without knowing how they are or were derived, and one doesn't need to know how to design or build or fix a car in order to drive it. The evolution of artifacts so that they can be used with less maintenance and knowledge of how they work allows one to do things with fewer cognitive resources, expending them elsewhere. It also allows (and requires) the differentiation of tasks and work in a society. In this way more complex cultural elements can be assembled, but their functioning presupposes the maintenance of the scaffolding that permitted that ascent, and components of the culture become both more distributed and differentiated. Elements of the culture become more separable, but at the same time more interdependent. (You buy the car from a firm that has built, marketed and distributed it (now usually subcontracting large chunks of this effort), and take it to someone else to fix and maintain it. The evolution of owners' manuals has increasingly downplayed the information necessary to fix it yourself as the frequency of breakdowns for which you must do so have declined, while the tools and knowledge required have increased beyond reach.)

10. Entrenchments grow particularly quickly and are particularly far reaching for elements that become parts of a standing infrastructure. Water, gas, and power distribution all acquired this status soon after the beginning of the twentieth century. Telephone, wireless communication standards (for radio and later for television), and now Internet protocols place strong constraints on the form of compatible devices, and propagation of standards. These all become infrastructural by becoming broadly polyfunctional, used for a wide variety of devices and to serve an even larger variety of ends, and their presence stimulates a new variety of devices designed to utilize them, as their use and the standards which must be met to use them grow in breadth.

Finally, we consider those features that make cultural evolution unusually tolerant of deep changes, and thus make cultures foci for rich and often revolutionary innovation.

Features Allowing Deeper Changes More Easily

11. Probably the most important feature here is vicariousness: *To consider a change is not to adopt it.* This may range from the "throwaway" options rejected without serious consideration through those requiring more research and detailed planning, up to the production and testing of expensive prototypes and "demonstration projects." It may not cost much more to develop and explore quite large changes than to explore smaller ones if one is not adopting them, and "exploring" risks much less than "adopting." In biology, variations are always real: one cannot "consider" genotypes without adopting them.[63] (An organism may be just a trial at the population level, but *it* does not get another try).

12. One can apply "problem decomposition" to the task of changing a system. Although fixing a component subsystem may impact and require changes spread through the whole system, one can treat the problem of fixing it in isolation, make plausible improvements, and then figure out how the other components must change. This can be repeated for other relevant subsystems. This will not always work, since problems are not always "nearly decomposable" any more than systems are (Simon 1962; Wimsatt 1974), and some modification trajectories may be dead ends. But they often will yield to this approach, and when this is employed at the design stage one can explore and modify deleterious effects of different "macromutations" which would be inaccessible without "vicarious" exploration (item 11 above). This procedure allows construction of solutions that would require a number of modifications to be sufficiently viable, and thus be too improbable to expect in biological evolution.

13. Scaffolding can become disentrenched without major loss of organization, or locally of organizational support[64] only when replaced "in parallel" using production of material culture, nonmaterial culture, or both via an appropriate substitution. In this it is similar to an evolutionary process for which new generated redundancies (from tandem duplications at the genetic level all the way up) are the primary way of reducing generative entrenchment. *For culture, the increase in numbers of parallel channels or means of solution, and consequent disentrenchment of any one of them is crucial.* This parallel redundancy is introduced by markets and populational heterogeneity, with learning and borrowing from your competitor generating an expanding redundancy of alternative solutions.[65]

14. There is a natural redundancy for culture. Changing technology—material or ideational—is never instantaneous, so there is always a population of various older machines, procedures, and users. This does several things: Early adopters will have to invest more in learning the new technology (usually in an environment where it is less reliable, less powerful, and less easy to manage), but then they will know unusually well how to help those following them. Those who know the old way as well as the new can "translate" for an extended period of time, and the older technology provides a "backup" when the newer technology fails, which it will do relatively frequently at first. This coexistence of two systems through the transition period can midwife quite momentous changes, which may be relatively minor at first but open entirely new pathways for explosive transformation. (Consider the transformation from penned manuscripts and records to typewriters, to word-processors to desktop computers.) Cooperation within organizations at this level—simple assistance to others in using the new technology—is expected work performance, so even though there are potential "free rider" issues, we are strongly socialized to help, and experience strong disapproval or worse if we refuse. Parallels for biology require kin or group selection, which some writers are unduly suspicious of, but conditions for this kind of interaction for cultural traits are so common as to be endemic.

15. *Scaffolding can change very rapidly if managed by specialists who know what is necessary to make it transparent to others*, often by "black-boxing" it, and using standards to preserve key elements (thus presuming 9 above). If one can find "functional equivalents" for these elements, they too can be replaced, and many new technologies do exactly that. Terms like "backwards compatibility" or "platform independence" in computer software represent exactly this kind of strategy, and permit higher levels of organization to screen off higher levels of function from lower-level details, making deeper changes possible. This is "supervenience" by design (Wimsatt 2001).

Missing from this list are things particular to culture that aid in constructing new variations, and things that aid in evaluating alternatives. Heuristic procedures of model-building (see, e.g., Wimsatt 1987, 2007), or heuristics for making decisions on the basis of very limited information (Gigerenzer et al. 1999) fit here, but exploring them would go beyond the reasonable scope of this chapter.

Almost all of these have in common the use of scaffolding channels, and almost inevitably, the simultaneous use of multiple partially independent channels (independent enough to adapt to different environments and niches, independent enough to support disagreement and competition, independent enough to interact symbiotically as differentiated complementary elements).

So we see that in humans, generalized notions of reproduction and entrenchment describe processes that interact to produce rich multilayered scaffolding, differentiating multiple channels of cultural inheritance. The existence of multiple channels can lead some of them to differentiate the combinatorial and other features (those listed as 1–15 above, and possibly others) to come to be an informational channel, allowing rapid and massive increases in the speed of evolutionary processes, and possible sources of reflexivity emerging out of close-linked control processes.

Conclusion

We have argued for a middle viscosity account of the nature of culture. Starting from Griesemer's (2000a) contrast between biological transmission embodying material overlap, and cultural transmission apparently without it, we have attempted to show (1) that cultural transmission is more embodied than it seems at first, and that the ways in which it is embodied, scaffolded, and in consequence, embedded, increases both (2) the frequency and (3) the robustness (hence the accuracy) of successful transmission. Furthermore, scaffolding makes possible the robust transmission of much more (4) complicated and (5) context-sensitive things.[66] This complexity of cultural transmission is just another expression of the complex context-sensitivity of culture, widely noted by anthropologists, and essentially ignored by meme theorists. This complex context sensitivity has frustrated the best attempts so far from the so-called hard sciences to analyze intentionality: it was the "brick wall" of the "frame problem" which bedeviled attempts at Artificial Intelligence two decades ago. This problem is the Gordian Knot of culture: things not having it will not be counted as cultural,[67] and the reticulate complexity of things having it seem to defy further analysis in any other terms. We accept the first, but demur from the second.

The Emergence of Information Channels?
The ways in which cultural transmission is freed from the constraints of material overlap and direct implementation characteristic of biology allow modification of transmitted elements in ways that are decoupled from immediate (and usually negative) consequences. The fact that this modifiability is itself structured by entrenched functional scaffolding allows selective sequential vicarious filtering for efficient removal of many possible

dysfunctional tries. By trying only things that have gone through our richly scaffolded cognitive filters, in which a sequence of successively structured contexts commonly narrow and focus the search immensely, we lower enormously the fitness cost of learning. Functional scaffolding allows trying more alternatives in increasingly refined ways, so that many more successful deep modifications can be produced, transmitted, and successfully integrated. And this ultimately explains the unparalleled lability as well as the cumulative complexity and richness of culture, and at the same time the enormous degree of intracultural differentiation. This all smacks faintly of "Lamarckian inheritance," but we have gone far beyond that to a much richer vision.

We propose that whenever there appears to be a relatively "free floating" cultural information channel in which any one of an array of messages can be transmitted, either in a preset or easily modifiable form, there is a material propagule or set of propagules (analogous to a Sears catalog), a scaffolding structure or set of structures, or both which permits this freedom, by giving the resources to choose alternatives, and make modifications, often from a specified and sometimes combinatorial set and transmit the resulting messages or consider other possibilities, all without implementing or acting upon their contents. But in addition, there must be a consilience between transmitter and receiver for a message to be comprehensible to the receiver. Scaffolding is required to calibrate the receiver to the source, and in a world without preestablished harmony, successful calibration enabling transmission of messages from a large combinatorial array would require feedback-laden sequential training.

Consider the circumstances under which an information-like channel will appear to "bootstrap itself" to increased reliability and capacity. Without propagules or scaffolding, there is no reason to expect the right sort of consilience to work between transmitter and receiver for cultural systems that can transmit a large (combinatorial) variety of alternative "messages."[68] Scaffolding is required in effect to coordinate the receiver to the source, but it seems unlikely that there could be successful coordination without feedback-laden sequential training of the sort that constitutes the imposition or emergence of conventions and standards, or at least routines, of operation. Differently put, without scaffolding, there is no reason to expect the coordinated, concerted conduct of multiple actors in activities that would constitute a social world (Gerson 1998, chap. 2). The cost of conduct would in general be too high.

Cultural channels must be quasi-independent of scaffolding to be able to transmit any one of the large variety of messages that leads us to call them channels. But this is only for some purposes: treating cultural channels as independent of scaffolding vastly underestimates the cost of communication. The transmission of cultural elements without material overlap via propagules is likely to be supported by scaffolding with substantial and variegated forms of material overlap: people who have trained together or one of whom has trained the other, who may be working on the same project, using material parts made to the same specifications, from the same manufacturer, working from designs with a long history of successive refinement to work well in applications suitably like this one. An important source of the irreducibly social nature of culture is the functional role that people, as well as artifacts and social conventions, play as scaffolding for others.

When collaborating by e-mail to write this chapter, we do not have to detail the books and training and examples we share, each other's beliefs, writings, competencies, and interests we know, all contexts fixed and shaped through interactions we have had over the last quarter century. This means that we can often use telegraphic shorthands for our ideas, and do not need to spend long hours elaborating contexts to disambiguate them. But even when we do, the corrections are more readily localized and can often be in shorthand. Thus shared experience functions as a (small-scale) social convention that scaffolds our interactions, and we *seem* to transmit more than our words can say. This even applies in such "objective" and "public" areas as mathematics, a field in which philosophers have been even less inclined to notice material propagules and scaffolding than other areas of science. We saw in Warwick's study of the rise of mathematical physics at Cambridge how much common training methods, content, and style were critical to communicating and being able to read textbooks results or do problems constructed at Cambridge—things that did not transport to other supposedly comparably advanced students at the University of London, much less Paris, Göteborg, or the United States.

And below our level of intentional communication, our technology manages multiple layers of black-boxed transformations: the fact that we use the same kinds of computer and same (or laboriously-made-partially-compatible) software, that our e-mail messages are broken into packets, labeled, and sent independently through different routes to be reassembled on our respective mail-servers, and that manuscripts on our hard drives are broken into packets, stored on noncontiguous sectors, and subvisibly assembled in the right order (almost) every time we call them up. That too is scaffolding, and thankfully, almost always invisible to us!

The functional freedom of the channel when the scaffolded developmental context can be taken for granted permits easy and controlled variation (directed, intelligent, or planned), including freedom to vary relatively deep generative elements of the transmitted message. This happens far more commonly for culture than for biology, but deeper changes are always relatively rarer in either domain than shallower ones. There are various reasons why changes in deep generative elements are not almost always immediately lethal, as would be commonly so in biology: there are vicarious ways of evaluating the effects of the modification without implementing it, and the fact that implementing the message is a separable act that can follow upon evaluation allows us to avoid many certain failures. These modifications are normally in the thing transmitted, and separable from the scaffolding partially responsible for its synthesis and manipulation. That is, the things characteristic of the functioning of the scaffolding are not under modification (at least not on the same time scale), and it is one of the reasons why we call this an information channel that the message can be changed without changing the machinery for transmitting and interpreting it.[69] This has two consequences: first, the "fitness" of a cultural element (material, behavioral, or ideational) is assumed (to a first approximation) not to affect its possibility of transmission and interpretation (or more generally, how it is acted upon by the scaffolding) and (2) implementing or acting upon the message (activity which realizes the fitness of having received the message) is a separable action from its transmission.

But even this cultural "genotype/phenotype" distinction is still too simple, for it suggests that there is a supporting and never modified structure, transmitting a symbolically encoded message (regularized so that it can be systematically produced and modified), and that there is a rigid separation of domains, of scaffolding with material overlap and no modifications during the operation, divorced from "informational" domains without material overlap and subject to modification.[70] It might seem as if scaffolding provides a shelter surrounding a region of free experiment with ideas. But what is instead true is that while the whole structure (or at least a large chunk of it) must be in place for anything to happen, different parts of the scaffolding differentially support different parts or aspects of the conceptual structure being worked on. So in cultural change, we are in a situation of rebuilding Neurath's boat, but never one of building or rebuilding it from scratch. Indeed there is so much elaborated and contextually focused scaffolding that most of the changes are minimal, quasi-automatic, almost immediately assimilated and invisible semistandard patches. So we have a multiply embedded and entrenched sociotechnocultural scaffolding interwoven and interpenetrating with a multiply entrenched and embedded artifactual-ideational structure with which we are experimenting. *Scaffold-*

ing is an interactive, not a static structure, and a dynamic account utilizing reproduction and generative entrenchment has taken us through it part of the way toward building a successful account of culture and cultural change.

Acknowledgments

We thank the Konrad Lorenz Institute and its Scientific Director, Werner Callebaut, for making it possible to work together, overlooking the Danube for two crucial weeks in summer 2003. Mark Bickhard's fortuitous presence boosted our account of cultural scaffolding to a richer one, aided by discussions later that summer with Linnda Caporeal at ISHPSSB. The SOIT seminar at Chicago chose to read Warwick at a fortuitous time to provide one crucial example while Barbara Wimsatt provided another and rich input and feedback on scaffolding. Lisa Lloyd provided helpful comments on the manuscript, as did Emily Schultz who gave important perspectives on cultural anthropology. We have both been scaffolded by Elihu Gerson for a quarter century, who taught us both more sociology than we know, whose gentle comments often ring in our ears, and from whom we have both learned above all not to "delete the work" in our accounts of science. We have been major scaffolding to each other for even longer, and it is a delight to collaborate again after 18 years in writing another paper on cultural evolution. We thank Roger Sansom and Robert Brandon for patience that we know must be finite (on theoretical grounds) but so far has seen no bounds. Griesemer's research is supported by the Dean of the Division of Social Sciences, College of Letters and Science, UC Davis. Wimsatt is supported by research funds from the Humanities Division and the Big Problems program at Chicago.

Notes

1. "Baby talk," explicit teaching using ordered or artificially simplified tasks, and so on.

2. Elihu Gerson, personal communication, January 29, 2006.

3. To evaluate differential loss of men and women on different career tracks toward high standing positions in the profession, one must disentangle the increasing proportions of women entering medical school in later cohorts (ratios went from about 10 percent in 1970 to nearly 50 percent by 1990), differential losses of men and women from the various specialty and subspecialty tracks, and differential continuing career commitments afterward by men and by women. All three appear to be happening, but Wimsatt's 1997 study and 2001 follow-up (which gives hard to find longitudinal data) point respectively to the latter two as real phenomena.

4. Most but not all twenty of the professionals were MDs. When only one was an MD, that partner's career always took primacy, suggesting that financial considerations were important. However, in an earlier pilot study with some couples having an MD and a university faculty member, primacy often went to the academic career at least up through the awarding of tenure, though there too, the higher income of the MD was crucial in allowing hiring of extensive support for child care, cooking, and housekeeping.

5. The extent to which this accumulating knowledge was embedded in practice and in the broader structure of education is revealed in the difficulty found in exporting it to other contexts, not only in France, Germany, and America, but even to other universities in England (Warwick discusses the University of London). This is especially striking since mathematics is taken to be the most context-independent of subject matters. But Cambridge monographs presupposed the background of experience, high technical competence, and detailed mathematical methods taught at Cambridge. Exercises "left to the (Cambridge) reader" might remain unintelligible to a mathematics student elsewhere.

6. There's a tendency among cultural anthropologists to assume any attempt to build mathematical models of cultural inheritance or evolution will be thin necessarily. We use the term *thin*, not to subject all mathematical approaches such as dual inheritance models or niche construction to criticism from the thick pole of the spectrum, but to characterize models of cultural evolution in terms of the degree to which they involve rich accounts of development or of culture. Boyd and Richerson's framework presupposes developmental processes to a significant extent and includes multiple modes of social interaction, but we place our concerns on a continuum between theirs and those of cultural anthropologists of the "thick" description variety. Thus, we seek to reject a categorical distinction between "thick" and "thin," not to reject mathematical models. We endorse rich characterizations of development and enculturation in the expectation that future progress will naturally include mathematical models to address them within a populational and evolutionary perspective. But for generalizable theory we must commonly (though not exclusively) seek robust qualitative predictions over detailed quantitative ones (Levins 1966; Wimsatt 1981b). But we also think it essential to mine detailed case studies for intuitions about what matters, and particular interactions not covered in general theory. Most "thick" description advocates seek particularistic qualitative accounts and resist attempts at generalization.

7. Goody (1977), Conference on "Margins of Writing, Origins of Cultures" at the Oriental Institute, University of Chicago, February 25 27, 2005, and Saunders (2006).

8. Brumann (2002, 509) argues that part of the purpose of a thick concept of culture was to establish a division of labor between sociology and anthropology and traces this to Talcott Parsons. An analogous division of labor between genetics and embryology led to the theoretical inadequacies that evo-devo is attempting to repair.

9. To clarify, we explore cultural evolution here mainly in terms of examples and problems arising in the rich, "thick" case of human culture, but we are not satisfied either to assume culture applies broadly, if thinly, to a wide variety of social organisms or to define culture so as to apply only to humans. Medium-viscosity theories like ours aim to leave it an open question how widely the concept of culture applies (without the easy reductionism of "thin" models). We acknowledge the "thickness" of the concept, but at the same time insist that "thick" does not mean "unanalyzable." A plausible candidate for "medium viscosity" descriptions of culture would be Gaddis's (2002) discussions of the "fractal" character (the relevance of detail on multiple spatial and temporal scales) of historical descriptions, explanations, and analysis. (These would be produced by successively nested entrenchments of kluged contingencies of broadening scope with a "random fractal" order like those used to generate realistic looking river deltas and mountain ranges (see, e.g., generation by "successive random additions" of perturbations of successively smaller scale in Turcotte 1997, 153). This need not be either hierarchical or regular, unlike the more widely known self-similar fractals that have a snowflake-like homogeneity. (Thanks to Emily Schultz for noticing this ambiguity.) Gaddis urges that we draw on the new historical, chaotic, and fractal physical sciences for methodological intersections with the human sciences. Something that interfaces with rich mathematical theory from another science while not sacrificing the demand for cultural richness should qualify as "medium" in our terminology. One striking characteristic of human intentional action is that it can causally couple these diverse spatial and temporal scales frequently enough to matter. Similar interactive multilevel structures are also common in biology (Wimsatt 1994). The operation of generative entrenchment over time predicts such a macroscopic fractal order of nested contingencies on all scales in biology, and if we are right, in human history (Wimsatt 2001).

10. See S. Schoenherr, "Evolution of the Shopping Center," http://history.sandiego.edu/gen/soc/shoppingcenter4.html.

11. Rationalization "in the economist's sense of doing more with the same resources, or the same work with less resources," see Gerson, forthcoming.

12. In balloon construction, single vertical framing members went from basement to roof, leaving vertical air shafts inside walls. In platform construction, each floor was built separately and sealed off from the floors below. The earlier design presented a fire hazard, since flames could propagate unimpeded (and hidden) from the basement to the attic up these shafts. Flames could rise in a platform building only by burning through floors (or spiraling up stairways). For similar reasons platform homes were more economical to heat (Thornton 2002, 16–18).

13. Sears was one of at least six companies to produce mail-order houses in the early decades of the twentieth century. The others include Sears' chief competitor in the housing arena and the first to offer kit houses, Alladin; their chief competitor in catalog sales, Montgomery Ward; and also the Bennett, Loizeaux, and Gordon–Van

Tine lumber companies (Thornton 2002, 20; this invaluable source mentions all but Bennett and Loizeaux).

14. Odling-Smee, Laland, and Feldman (2003, 192) state that "Niche construction is an ontogenetic process" but this does not make their account a theory of development. Development is treated in their theory as a black box that "allows individual organisms the opportunity to gain sufficient energy and material resources from their environments to survive and reproduce. It thereby both contributes to the building of the next generation of a population of organisms in the conventional manner and causes changes in the niche-constructing organisms' own selective environments" (pp. 192–193). In this minimal use, niche construction interacts with ontogeny too nonspecifically to derive a theoretical structure for its character or interactions.

15. We accept their perspective, but our elaboration of a particular kind of niche construction leads in such very different directions with different tools and explanatory foci as to be almost orthogonal to their approach. We believe that our accounts are broadly consilient, and expect that the whole story for culture will require both. See also Gray 2001 and Griffiths and Gray (2001) for efforts to integrate niche construction into developmental systems theory.

16. Perhaps the fact that we can sometimes successfully separate Siamesed twins but cannot do so for the coffee with cream is misleading: the biologist would point to the much richer and more multidimensional forms of interconnectedness possible for the twins. We suggest that the same is true for material and ideational culture: in some ways it is so easy to separate them that we fail to recognize the violence done to both by the result.

17. There is yet another course than considered here. This characterization of Weismannism assumes that inheritance is only in the biological germ line, not in either the biological or cultural soma or anything else we are forced to recognize. But the multilevel hierarchical Weismannism discussed in Griesemer and Wimsatt (figure 1), and in Callebaut 1993, 427–429, assumes the possibility of other hereditary channels with other (entrenched, and thus stable) germ line like lineages with their own somas, thus recognizing a genotype-phenotype-like relation for cultural objects. This is a useful perspective for some purposes, though it is probably best seen as a predecessor of the development here in terms of reproducers and entrenchment. The use of reproducers gives more powerful tools for determining when and how we should recognize temporal patterned series as autonomous or semiautonomous lineages.

18. No modern biologist would do otherwise, and some theorists closer to the biology (e.g., Boyd and Richerson 1985) just develop a populational theory without special note. Philosophers sometimes reify theories as idealized abstract types (appropriate for some analytic purposes) but then act as if they were just a single lineage of improving theories. Hull (1988) demonstrates that this will not do any more for culture than it will for biology. We argue below that the nature of cultural inheritance gives an additional role for populations as units.

19. When comparing the intelligence of New Guinean hunter-gatherers and modern Europeans, Jared Diamond (1997) suggests that the primary action of biological selection on human populations since the dawn of agriculture and the origin of growing permanent settlements was immunological: resistance to the diseases we have acquired mostly from our domesticated animals and nurtured in the close quarters and larger population sizes, which generate epidemics. Durham's (1992) treatment of malaria and sickle-cell anemia in West Africa shows rich interactions of historical migrations, rainfall and malaria incidence, linguistic groups and other cultural patterns, agricultural practices and different primary crops in different regions, which paradigmatically illustrate what "thicker description" approaches in biocultural evolution can do. The few documented counterexamples to this "disease thesis" (for example, a gene conferring extended lactose tolerance in Northern European animal herders) illustrate how little genes affect or constrain culture directly. Yoghurts, kefirs, and cheeses all provide culturally mediated ways of reducing the lactose in dairy products widely used (and likely originating) elsewhere, and are not less successful in Northern Europe where the gene giving extended lactose tolerance occurs at a higher frequency. Avital and Jablonka (2000) make similar arguments showing that genes-only explanations of culture fail to account for many details of behavior and behavioral transmission in birds and mammals—details that are better explained by social learning and direct transmission of "behavioral traditions."

20. More exactly, we suppose a time scale long enough for substantial cumulative cultural change but short enough that there has been no significant genetic change affecting culturally relevant traits. Except perhaps for genetic loci affecting disease resistance, this seems a good approximation for cultural changes over the last ten to twenty biological generations, and thus for the time since the Copernican Revolution.

21. The semiconservative replication mode of DNA is sufficiently important to make redundancy a third type. The unreasonably high stability of the genes was a driver of important creativity in attempts to understand them in the first half of the twentieth century.

22. Linnda Caporael's (1997) repeated assembly in core configurations model is mindful of development, is group/social focused, and is both top-down and bottom-up in approach.

23. For culture, it is as though we would have to solve all of the complexities of developmental genetics and transmission genetics together before we could get usable information from either. But transmission genetic techniques and their molecular genetic descendants were essential to locating and isolating the genes used in developmental genetic analyses. Analogous problems may exist for determining what can be a meme.

24. This argument applies only for those elements of the process that are conscious, so a potentially important exception here is the role of the acquisition of tacit knowl-

edge of all kinds. Reber (1993) argues that mechanisms for acquiring it are phylogenetically deep, likely pervasive in other processes, and that both the knowledge and how we acquire it are not readily open to conscious inspection, so its extent and importance are likely underestimated. His account of tacit knowledge draws heavily on generative entrenchment to make predictions about its character and mode of action (chapter 3), so we may be able to analyze it with our tools, even if we can't make it conscious!

25. This is, roughly, our social network amplified by our technology, in which we have many new "one-way" connections like movies, radio, and television.

26. Recent work in biology shows that genetic inheritance is multichannel as well. Epigenetic inheritance processes help regulate gene expression but operate, and are transmitted, in parallel to the transmission of DNA. Genetic inheritance cannot function without them because it requires the regulated gene expression of development in the parent to occur (Jablonka and Lamb 1995). In general, it is hard to see how complex, hierarchically organized life could have evolved without the emergence of multiple inheritance processes (Maynard Smith and Szathmáry 1995; Jablonka and Szathmáry 1995; Griesemer 2000c).

27. In "Hierarchical multilevel Weismannism," cultural transmission involved multiple channels, and roughly, any aspect of phenotype could be taken as an element to be transmitted. But speaking of a channel involved transmission of a smaller generative set that (in context of other operating channels) could be used to generate the rest, as an axiom set could be used to generate the theorems that follow from them. Then the generative set became "heterocatalytic" and genelike (Wimsatt 1981b), and the whole adaptive structure resulting from their elaboration became phenotype-like. But this picture did not sufficiently recognize the richness of interaction of these different phenotypes.

28. Activity and germ-linedness are not necessarily coupled. There can be active replicators that are not germ-line (in somatic cells) and germ-line replicators that are not active (germ-line "genomic parasites"). But these separations are themselves products of the evolution of developmental organization and must be explained, not assumed in definitions, especially since the separation of germ and soma has a *very* patchy phylogenetic distribution (Buss 1987).

29. This might stem from the view that since all features of the cell trace back to genetic activity, at one time or another genes determine everything. But this does not give us genetic determinism. On Laplace's classical definition, in a deterministic system, from a totally specified state of the world at a single moment one could predict the state of the world arbitrarily far forward or retrodict arbitrarily far backward in time. But to make the parallel argument work, at any given time, we would have to include operating cells and relevant larger multicellular, environmental and social contexts to predict the behavior of the genes. So the imagined genetic determinism does not satisfy Laplace's definition.

30. We do not accept Dawkins's view that genes are the sole or even the fundamental difference makers in development, but we grant him that assumption for the sake of formulating the view of reproduction. Griesemer (2000a, 2000b, 2000c, 2002, 2003) develops the argument without granting that assumption.

31. One might stabilize more of the environment if one went down to a smaller propagule of cells rather than all the way to a single cell. But Grosberg and Strathmann argue that going to more exposed single cells better escapes transmission of mutations and parasites which might survive better in one or another cell of a semi-protected propagule to destroy descendants of the whole propagule later. See Griesemer et al. 2005 for further discussion.

32. Indeed, thirty-some years ago, it was easy for otherwise penetrating students of the history of genetics to have underestimated the contemporary power of Morgan's criticisms in this paper. John Moore introduced the reprinting of Morgan's paper in 1972 with the comment (p. 123) that "This article is not especially important for any intrinsic merit—the arguments are not very convincing—but in a few years its author's name was to replace that of Mendel as the principal formulator of genetic theory." Historical scholarship on Morgan's developmental views and the appreciation by geneticists and evolutionists alike of the importance of development have changed the landscape dramatically since 1972.

33. On the mutual dependencies of embryological and genetic explanations, the notion of tracking processes of development and transmission, and their changing historical relationships, see Griesemer, forthcoming.

34. Elihu Gerson, personal communication, January 29, 2006.

35. Once upon a time, department stores were icons of urban rather than suburban living.

36. Modern catalogs themselves have complex organizational scaffolding aids to help the customer find what they want, and often to discover what they need. The latter requires an explicit educational function, often supplemented by technical experts available by telephone or e-mail. The Cole-Parmer scientific catalog serves to equip laboratories, and to aid in the construction of often one-of-a-kind research equipment. It thus seeks to offer a general problem-solving toolbox for configuring laboratories, but also aids for how to use it. Even sophisticated scientists need this kind of help. Grants require itemized budgets well before the laboratory is configured. And paralleling the offering of Kit houses by Sears, catalogs offer both individual and bundled systems, solving problems of compatibility of which the customer may not even be aware. The Cole-Parmer 2003–2004 catalog is 2,096 pages plus 300 special pages including 100 pages of indexes. It includes (1) specialized supplies, (2) whole systems as well as modular parts for building various (often prototyped) systems, (3) multiple elaborate indices, (4) color coded keys, (5) other finding aids, (6) four pages of introduction to how to use the catalog, (7) A 100-page

supplementary catalog of new products released within the last year. The information load is daunting. Catalogs and yellow-page sections of the telephone directory are now increasingly offered on CDs, or directly on the internet. This both reduces production and mailing costs and allows much more powerful electronic search functions, but assumes that the potential customer has an appropriately configured computer. Not sure? The friendly internet offers to download on command the software it determines you need to read the latest state-of-the-art catalogs. That is distributed intelligence!

37. Becker contrasts his sociologist's notion of culture as a resource for people in groups to use "in order to coordinate their activities" (1986, 13) with that of anthropologists who take culture to be a system of independently existing patterns that make societies possible. Extending Becker to our multilevel context, it's social groups all the way down: society—that is, groups whose members' activities are coordinated—is a precondition for evolution at any level. But we also accept the anthropologist's insight that culture is emergent, even if not independent of social order. (We thank Elihu Gerson for referring us to Becker's essay.)

38. Becker's wonderful and lucid story of his experiences as a "Saturday night musician" is a case in point. It would be hard to be a pickup piano player if pianos were not present in clubs (Gerson, personal communication, January 29, 2006). But in the hands of some theorists, this is the immediate prelude to claim that (therefore! sic!) we do not need explicit representation of artifacts in the theory, which should be formulated in terms of ideas, conventions, or practices (e.g., Boyd and Richerson 1985; Richerson and Boyd 2004). We disagree fundamentally: for a theory of cultural evolution, the pianos must receive representation on equal footing with piano players, and with their conjoint activities in social contexts, in order to analyze the generative and reproductive aspects of developing cultural systems.

39. One might object that the move to superindividuals is to change the category of the question, but we think not, or at least not automatically. We certainly speak of the culture of a firm, and allow for it to change. And issues about the phylogenetic origination of culture (for the first time, in the evolutionary history of Homo) are different than questions about the differentiation of cultures (which many of the latter represent), but only in the same way as the (mesoevolutionary) origins of language differs from the microevolution and speciation of languages.

40. Of course, the "same" scaffolds may act or be capable of acting as reproducers in some contexts, and not in others, just as with organic lineages.

41. Compare this to the role of enzymes in a cell: enzymes make reactions more likely (and thus more rapid) than they otherwise would be without them. This analysis distinguishes the enzyme from the reaction system it belongs to and assigns the action or role to the enzyme, even though the action could be said to be of the collective. The enzyme that flexes so that a substrate is bound is like the proverbial "arm raising" in action theory: the arm behaves and the person acts. Similarly, we say that

the system, rather than its parts, develop. As Levins (1974) notes, this changes the dynamics, and effective organization of the cell, by creating near-decomposability (Simon 1962) and connectivity along new lines. Usually, enzymatic activity in metabolism is described as contributing to self-maintenance or growth. But in functional terms, self-maintenance and growth serve development and development serves reproduction (Griesemer 2000a, 2000b, 2000c). Reproduction is of a complex organized system of parts, so "action" is something assignable only to systems or processes.

42. This kind of perspective is not inconceivable: economist Kenneth Boulding, an early advocate of cultural evolution, used to delight in remarking to staid audiences of the late 1950s that "a car is just an organism with an exceedingly complicated sex-life."

43. For example, if humans regularly built new automobiles by scavenging some key parts of old ones and adding in new parts to supplement the old, then there would be relations of material overlap between old and new such that automobiles would have genealogies. Suppose further that the new parts are produced by scaffolding operations of the human builders in which templates are constructed from damaged old parts and modified to facilitate production of new parts that fit together with the scavenged old parts. Then the materially overlapping parts would be playing a developmental role in helping to confer reproductive capacity on the offspring automobile. But just this happens in economies like that of Cuba (and rural America today—see Harper 1987)! In the "hot rod" communities of the 1950s and 1960s, this proceeded so far as to generate a new manufacturing framework. There were preferred cars to modify (e.g., 1930s Ford V-8's, and late 1950s Chevys), and engines and carburetors to switch. As the movement took hold, "third-party" suppliers emerged to provide parts that initial one-of-a-kind trials had shown to be promising and imitation had generated a demand and an opportunity. (These ranged from mountings to allow the substitution of specific nonstandard components up to whole bodies and different engines in various tunings). These manufacturers scaffolded the individual hot rods thus produced, and the hot rodders that produced them. Manufacturers even emerged which would take cars delivered directly from the original manufacturer and modify them according to various standardized performance and accessory specifications, to be delivered to the customer who had ordered the final product "as new," with warranty in place. The Ford "Cobra" was one such well-known example—so well known that it (and its niche congeners) were reviewed in sports car magazines, listing specifications, performance (against more commercially standard models), price, and manufacturer (with address). This, like racing, was encouraged by manufacturers like Ford and GM to enhance their image with customers who would favor their standard production numbers.

44. Hutchins's Navigation team from chapter 1 of *Cognition in the Wild* fits this framework like a hand in a glove, though it is richer still, and also involves developmental agent scaffolding.

45. Hughes (1987) describes in detail the development, entrenchment, and shaping of societies in the evolution of large technological systems, such as electrical grids. The latter were modeled on gas distribution, which was in turn modeled on water distribution (Basalla 1987).

46. This seems a safe estimate, given Sears' 75,000 and the involvement of two other large competitors (Aladdin and Ward's) and at least three smaller ones.

47. Cornell's Student Union, Willard Straight Hall, is the subject of one such revealing but affectionate story—retold in Wimsatt 2007, chapter 12.

48. We have kept the link between generative entrenchment and its measurement deliberately loose. The appropriate operational measure could vary with the mode of organization of the system, the representation of it we are using, and the tolerances we should use for registering a change as significant. The range of things affected if we change an element in a system may depend on how it is changed, and by how much, and under what circumstances (conditions on other parts of the system and its environment), and of course on what it is connected to. Then too, the reach of the damage arising when a component fails may propagate across normally separate functional systems in unanticipated ways, and commonly does so in catastrophic failures. A DC-10 crash at takeoff at Chicago's O'Hare Airport twenty-five years ago resulted from a failure in a bolt in an engine mount on the left engine. The engine ripped out of the wing, disabling hydraulic lines to flaps and ailerons on that side, and with this loss of control the plane crash was inevitable. (The DC-10 can fly with one engine out, even if lost on takeoff, but not without controls.) Controls were subsequently redesigned with additional redundant hydraulic lines. But no one would suppose it to be part of the function of the bolt to maintain their integrity. So it is better to speak of a family of measures of generative entrenchment, and how sensitive conclusions are to the measures chosen for the purpose at hand. Schank and Wimsatt (1988) compared different measures of generative entrenchment in their simulations of (idealized) evolving gene control networks, and found that their (particular) conclusions were robust across very different measures.

49. Wimsatt and Schank 1988; Wimsatt 2001. This famously does not happen for species or higher taxa, but only for aspects of their design architectures. See Van Valen's (1973) "Red Queen" hypothesis: lineages do not get any better at surviving. Their environments are constantly degrading because all of their competitors are escalating too and "You have to run as fast as you can just to stay in the same place."

50. These are "physical" conditions. In any complex construction project a series of legally mandated inspections must be performed at key stages of the construction. Thus water, gas, and electrical inspections must come after framing, but before the wallboard, and for good functional reasons.

51. The stereotypical house falls short of the richer dynamical reality. Organisms are constantly being rebuilt, even as they develop and evolve. The same could be said for

our organizations and institutions, and on the dwelling perspective, for our houses, as occupied and maintained. (To recognize this is to recognize that the relevant unit is larger than the physical house.) Such dynamic organization requires even more of the kinds of organizational forms of this and the following note.

52. Such modular structures in cultural systems show crucial elements of biological design that have been ignored by philosophers—design for robustness and redundancy, and the ability to handle imprecision and local contradictions, which let systems deal with failure and stress, and age gracefully. Philosophers' buildings and machines would be like the famous "One Hoss Shay" of Oliver Wendell Holmes's 1858 poem—a structure that ran perfectly for 100 years without scratch, wear, or aging, and then collapsed into a pile of dust—presumably at the appearance of the first contradiction. These are all connected. See Wimsatt 1981a, reprinted in Wimsatt 2007.

53. Griffiths and Knight (1998) offer a "parity argument" to suggest that channel and informational signal coming from a source are symmetrical: they can each be viewed as serving the function of the other, so neither can be privileged as "the information" (see also Griffiths 2001; Griffiths and Gray 2001). Moreover, they claim, the relevant causal relation to explain the transmission of information from source to receiver over a channel is "systematic dependence" of the state of the receiver on the state of the source—that is, they offer a statistical criterion for causal connection to bolster their parity argument. This assumption clearly needs to be reexamined, but the issues are more complex than discussed here. (Thus, does prior communication of the structures necessary to decode signal from channel violate their "systematic dependence" assumption for all subsequent messages? Does the knowledge of how it will work by those who set up the channel violate this?)

54. Consider a set of detectors at a receiver that are tuned to different characteristic frequencies in the Fourier decomposition of a single complex wave carrying information from a source. There is no physical separation among the components of the wave until the interaction at the detectors, which are parts of the receiver, (where they will produce physically separable outputs) and not parts of the channel conditions. FM transmission involves similar principles to combine signal with carrier wave, transmitting them together as a modulated carrier signal, and then "demodulating" at the receiver to separate signal from carrier. If one could decide in advance what the bandwidth of each channel was, there is no reason why such a design could not be used for multiple channels sent at once as part of the same (very wide bandwidth) transmission signal. (Thus if one wanted to combine five voice channels assigning 20 KHz to each, adding 100 KHz of bandwidth to the carrier frequency, and making the modulating frequencies 0–20 KHz for the first signal, 20–40 KHz for the second, and so on, their composition should allow a unique demodulation of the five signals at the other end.)

55. One can defend Simon's analysis at this point by pointing out that he was talking about forming structures through aggregation, not by development, and that his

justly famous argument required at least some intrinsic stability of the subassemblies. The real processes that we consider require invocation of both aggregation and articulation operations, and differentiation and developmental ones.

56. This simultaneous presence and interaction of substantial resistances or insensitivities along some dimensions and remarkable highly context-specific sensitivities in others is a deep conceptual requirement (though not yet sufficient) to speak of cognitive agency.

57. We have used quotes here because we would actually reject these terms. But generative entrenchment can explain the *phenomena* which innateness is invoked to explain (and much more) without the antidevelopmentalist biases and paradoxes produced by the latter (Wimsatt 2003).

58. This contrast is easily misunderstood. Most cultural mechanisms permitting deep changes have biological analogues. But new possibilities with cognition and culture significantly amplify their effects.

59. There is no common metric for comparing their speeds, since it is not clear what we would compare with what—no "cultural code" whose sequence could be compared for rates of substitution with the genetic sequence. But no one would accept an analysis that had the opposite result. So we can try to get more clarity by analyzing the mechanisms involved, which must be part of any such solution. Both biological and cultural evolution actually proceed at a wide range of rates. Boyd and Richerson (1985) note that early methods of tool making ("point technologies") persisted unchanged for hundreds of thousands of years. And the transmission of oral tradition is practiced in such a way as to resist change, and does so for spans covering centuries—easily long enough to encompass the industrial revolution. On the other side, significant bacterial evolution can happen in a few days. So maximum (or minimum) speed alone is not the issue.

60. This case is interesting because it shows that culture need not be transmitted horizontally. Agricultural practices were apparently transmitted from parents to offspring, but as cultural traits, not as biological ones, even though having the cultural trait clearly impacted biological fitness through the increases in numbers of children the practices engendered, and the interference competition agriculturalists generated for hunter-gatherers by clearing forests that housed most of their game species and diverse edible plants.

61. Herkimer 1952 is a rare find of this type, self-baptized an "Engineer's Illustrated Thesaurus." It is a deliberate attempt to give engineers a wide variety of designs of given categorical types to solve kinds of problems. It contains 8,000 figures—drawings of mechanisms—representing kinds of solutions (with different side constraints) that an engineer designing an artifact can look to for ideas. (Hughes (1989) argues that patent office volumes were frequent resources in the workshops of American inventors of the late nineteenth and early twentieth centuries.) The "The-

saurus" idea deliberately exploits the concept of functional (near-) equivalents. Herkimer says that such collections of commented drawings are not new, but that the particular way he has organized them (first by function, and then subdivided by broad mechanical type), is new and especially useful.

62. Latour (1987) points out how new innovations become "black-boxed" as they become entrenched as part of the procedural "furniture."

63. There is a partial biological analogy here in unexpressed variations that may accumulate and be expressed in other circumstances via means such as heat-shock proteins, DNA-methylation, and pseudogenes, but these we defer for another time.

64. Loss of scaffolding does happen and must be kluged on the fly, usually with substantial effort and more modest standards for success. Schultz (2005) discusses Zentella's (1997) story of Puerto Rican transnational migrants as they navigate New York while learning ESL and maintaining contact back home. Globalizing forces can even disenfranchise "migrants" remaining at home, revealing the rich scaffolding we count on, and once again confirming the maxim that we learn how functional systems work by seeing how they break down. These cases with adaptation on smaller scales to failures of social support are just as important for understanding culture as the romanticized "progressive" changes of science and technology.

65. New technologies commonly show a rapid growth of alternatives, with a subsequent reduction of variation as some of the variants become dominant. This "bottom heavy clade" is common for new adaptive radiations. Thus, just a generation after the invention of the internal combustion engine, then still in its early adolescence, Page's (1918, 30–32) book on aviation engines has a three-page (partially overlapping) classification of 66 engine types, including a diagrammed double-row *rotary* engine (not found in a total review of aircraft in the first world war; it presumably existed in prototype form only), and a new phylum—a six-cycle engine (like discovering *cyanobacteria*). Both are nonexistent, indeed unknown today, as are many of these types. Only a few have (much transformed) modern descendants.

66. Clarification: Take item (2) in this list of conclusions as frequency of successful transmission of a message somewhat like that intended, (3) as a measure of the dispersion of successful messages from the exactly correct message, supplemented by the ease of recovery of essentially the whole message from partially degraded instances of transmission, (4) as true because of the possibility of transmitting accurately a number of message parts (possibly from different sources) that have to be assembled correctly to produce the complex whole, and (5) that in the context of scaffolding skills and learned prior knowledge, message parts are successfully disambiguated in ways that are computationally local though possibly semantically global, showing context dependence in the first and context relevance in the second.

67. Simpler things, like sticks and stones used incidentally in a game, that would not count as cultural may be counted as such within the context of a cultural system, just

as viruses can count as living in some extended sense in terms if their relations with the living things that they parasitize and transform.

68. Indeed the only potential causes left of consilient senders and receivers are traditional information flows over channels and correlations induced by development in a common persistent environment. Neither of these is a plausible explanation for the organization of receivers of any organizational complexity: both are too low bandwidth to generate culture of any but low viscosity—that is, the "thin" culture of transmitted memes such as "preferences." Haugeland's (1998) essay "Mind Embodied and Embedded" makes extensive use of the distinction between high and low bandwidth in distinguishing between information channels and embodiment, suggesting a deeper resonance between our views.

69. So the genetic code machinery is not typically modified by changing the message. Essentially this was first noted by H. J. Muller in 1922, when he argued that transmission genetics must use distinct causal pathways from gene expression because mutations that caused diverse and serious disruptions of phenotypic function were all equally heritable. The operation of the telegraph and telephone are similarly independent of or invariant over the contents they transmit. Though each of them could be modified by the appropriate message sent and acted on, they commonly are not, and their proper function depends on this fact. This fact was relied upon by biochemists who exploited rare mutations that *did* change the genetic code machinery to work out the code itself (see Schulman and Abelson 1988).

70. This idealization ignores embodiment and social embedding of the performing structures and processes. There are some purposes and perspectives for which this is a useful idealization, and this disembodied compartmentalization fits some systems better than others.

References

Abbott, A. 1988. *The System of Professions: An Essay on the Division of Expert Labor*. Chicago: University of Chicago Press.

———. 2001. *Chaos of Disciplines*. Chicago: University of Chicago Press.

Aladdin Company. 1995. *Aladdin "Built in a Day" House Catalog*, 1917. New York: Dover Publications.

Arthur, W. Brian. 1994. *Increasing Returns and Path-Dependence in the Economy*. Ann Arbor: University of Michigan Press.

Arthur, W. 1997. *The Origin of Animal Body Plans*. New York: Cambridge University Press.

Aunger, R., ed. 2000. *Darwinizing Culture: The Status of Memetics as a Science*. Oxford: Oxford University Press.

Aunger, R. 2002. *The Electric Meme: A New Theory of How We Think*. New York: Free Press.

Avital, E., and E. Jablonka. 2000. *Animal Traditions: Behavioural Inheritance in Evolution*. Cambridge: Cambridge University Press.

Basalla, G. 1987. *The Evolution of Technology*. New York: Cambridge University Press.

Bateson, P. 2001. Behavioral development and Darwinian evolution. In S. Oyama, P. Griffiths, and R. Gray, eds., *Cycles of Contingency: Developmental Systems and Evolution*, 149–166. Cambridge, MA: MIT Press.

Becker, H. S. 1986. Culture: A sociological view. In *Doing Things Together: Selected Papers*, 11–24. Evanston, IL: Northwestern University Press. (Reprinted from *Yale Review* 71 (summer 1982): 513–528.)

Bennett Lumber Co. 1920/1993. *Bennett's Small House Catalog*, 1920. (Reprinted New York: Dover, 1993.)

Bickhard, M. H. 1992. Scaffolding and self-scaffolding: Central aspects of development. In L. T. Winegar and J. Valsiner, eds., *Children's Development within Social Contexts: Research and Methodology*, 33–52. Mahwah, NJ: Erlbaum.

Boyd, R., and P. J. Richerson. 1985. *Culture and the Evolutionary Process*. Chicago: University of Chicago Press.

———. 1996. Why culture is common but cultural evolution is rare. *Proceedings of the British Academy* 88: 77–93. (Reprinted in R. Boyd and P. J. Richerson, *The Origin and Evolution of Cultures*, 52–65. Oxford: Oxford University Press, 2005.)

Brewer, M. B. 1997. On the social origins of human nature. In C. McGarty and S. A. Haslam, eds., *The Message of Social Psychology*, 54–62. Oxford: Blackwell.

———. 2004. Taking the social origins of human nature seriously: Toward a more imperialist social psychology. *Personality and Social Psychology Review* 8: 107–113. doi:10.1207/s15327957pspr0802_3.

Brumann, C. 2002. On culture and symbols. *Current Anthropology* (43)3: 509–510.

Buss, L. W. 1987. *The Evolution of Individuality*. Princeton, NJ: Princeton University Press.

Callebaut, W. 1993. *Taking the Naturalistic Turn, or How Real Philosophy of Science is Done*, 425–429. Chicago: University of Chicago Press.

Campbell, D. T. 1974. Evolutionary epistemology. In P. A. Schilpp, ed., *The Philosophy of Karl Popper*, 413–463. LaSalle, IL: Open Court.

Caporael, Linnda R. 1995. Sociality: Coordinating bodies, minds and groups. *Psycoloquy* 6(01) Group Selection (1). http://psycprints.ecs.soton.ac.uk/archive/00000448/.

———. 1997. The evolution of truly social cognition: The core configuration model. *Personality and Social Psychology Review* 1: 276–298.

———. 2003. Repeated assembly. In S. Schur and F. Rauscher, eds., *Alternative Approaches to Evolutionary Psychology*. Dordrecht: Kluwer.

Cavalli-Sforza, L. L., and F. Cavalli-Sforza. 1995. *The Great Human Diasporas*. Reading MA: Addison-Wesley.

Chandler, A. D., Jr. 2000. The Information Age in historical perspective. In A. D. Chandler Jr. and J. W. Cortada, eds., *A Nation Transformed by Information*, 3–37. Oxford: Oxford University Press.

Clarke, A. 1997. *Being There: Putting Brain, Body and the World Together Again*. Cambridge, MA: MIT Press.

Cole-Parmer. 2003–2004. *General Catalog* (2003). Cole-Parmer Instrument Company, 625 East Bunker Court, Vernon Hills, IL., 60061.

Dawkins, R. 1976. *The Selfish Gene*. New York: Oxford University Press.

———. 1982. *The Extended Phenotype: The Gene as the Unit of Selection*. New York: Oxford University Press.

———. 1995. *River out of Eden*. New York: Basic Books.

———. 2004. Extended phenotype—but not too extended. A reply to Laland, Turner and Jablonka. *Biology and Philosophy* 19: 377–396.

Diamond, Jared. 1997. *Guns, Germs, and Steel: The Fates of Human Societies*. New York: Norton.

Durham, W. 1992. *Coevolution: Genes, Culture, and Human Diversity*. Stanford, CA: Stanford University Press.

Emmet, B., and J. E. Jeuck. 1950. *Catalogs and Counters: A History of Sears, Roebuck and Company*. Chicago: University of Chicago Press.

Gaddis, J. L. 2002. *The Landscape of History*. New Haven, CT: Yale University Press.

Gerson, Elihu M. 1998. *The American System of Research: Evolutionary Biology, 1890–1950*. PhD dissertation, Department of Sociology, University of Chicago.

———. Forthcoming. Reach, bracket, and the limits of rationalized coordination: Some challenges for CSCW. In Mark S. Ackerman, Christine Halverson, Tomas Erickson, and Wendy A. Kellogg, eds., *Resources, Co Evolution, and Artifacts: Theory in CSCW*. New York: Springer-Verlag.

Gigerenzer, Gerd, Peter M. Todd, and the ABC Research Group. 1999. *Simple Heuristics That Make Us Smart*. New York: Oxford University Press.

Glymour, C. 1980. *Theory and Evidence*. Princeton, NJ: Princeton University Press.

Goodenough, O. R., and R. Dawkins. 1994. The "St Jude" mind virus. *Nature* 371(6492): 23–24.

Goody, J. 1977. *The Domestication of the Savage Mind*. New York: Cambridge University Press.

Gordon-Van Tine Co. 1923/1992. *117 House Designs of the Twenties*. New York: Dover.

Gowans, A. 1986. *The Comfortable House: North American Suburban Architecture*. Cambridge, MA: MIT Press.

Gray, R. D. 2001. Selfish genes or developmental systems? In R. Singh, K. Krimbas, D. Paul, and J. Beatty, eds., *Thinking about Evolution: Historical, Philosophical and Political Perspectives: Festschrift for Richard Lewontin*, 184–207. Cambridge: Cambridge University Press.

Griesemer, James R. 2000a. Development, culture and the units of inheritance. *Philosophy of Science* 67: S348–S368.

———. 2000b. Reproduction and the reduction of genetics. In P. Beurton, R. Falk, and H.-J. Rheinberger, eds., *The Concept of the Gene in Development and Evolution: Historical and Epistemological Perspectives*, 240–285. Cambridge: Cambridge University Press.

———. 2000c. The units of evolutionary transition. *Selection* 1: 67–80.

———. 2002. Limits of reproduction: A reductionistic research strategy in evolutionary biology. In Marc H. V. van Regenmortel and David L. Hull, eds., *Promises and Limits of Reductionism in the Biomedical Sciences*. Chichester: Wiley.

———. 2003. The philosophical significance of Gánti's work. In T. Gánti (J. Griesemer and E. Szathmáry, eds.), *The Principles of Life*, 169–186. New York: Oxford University Press.

———. 2005a. Genetics from an evolutionary process perspective. In Eva M. Neumann-Held and Christoph Rehmann-Sutter, eds., *Genes in Development*, 343–375. Chapel Hill, NC: Duke University Press.

———. 2005b. The informational gene and the substantial body: On the generalization of evolutionary theory by abstraction. In Martin R. Jones and Nancy Cartwright, eds., *Idealization XII: Correcting the Model, Idealization and Abstraction in the Sciences*, 59–115. Poznan Studies in the Philosophy of the Sciences and the Humanities, vol. 86. Amsterdam: Rodopi.

———. 2007. Tracking organic processes: Representations and research styles in classical embryology and genetics. In J. Maienschein and M. Laubichler, eds., *From Embryology to Evo-Devo*, 375–433. Cambridge, MA: MIT Press.

Griesemer, J., M. Haber, G. Yamashita, and L. Gannett. 2005. Critical notice. Cycles of contingency: Developmental systems and evolution. *Biology and Philosophy* 20: 517–544.

Griesemer, James R., and William C. Wimsatt. 1989. Picturing Weismannism: A case study of conceptual evolution. In M. Ruse, ed., *What the Philosophy of Biology Is: Essays for David Hull*, 75–137. Dordrecht: Kluwer Academic Publishers.

Griffiths, P. E. 2001. Genetic information: A metaphor in search of a theory. *Philosophy of Science* 68(3): 394–412.

Griffiths, Paul E., and R. D. Gray. 1994. Developmental systems and evolutionary explanation. *Journal of Philosophy* 91(6): 277–304.

———. 2001. Darwinism and developmental systems. In S. Oyama, P. Griffiths, and R. Gray, eds., *Cycles of Contingency: Developmental Systems and Evolution*, 195–218. Cambridge, MA: MIT Press.

Griffiths, P. E., and R. D. Knight. 1998. What is the developmentalist challenge? *Philosophy of Science* 65(2): 253–258.

Grosberg, Richard K., and Richard R. Strathmann. 1998. One cell, two cell, red cell, blue cell: The persistence of a unicellular stage in multicellular life histories. *Trends in Ecology & Evolution* 13(3): 112–116.

Harper, Douglas. 1987. *Working Knowledge: Skill and Community in a Small Shop*. Chicago: University of Chicago Press.

Haugeland, J. 1998. *Having Thought: Essays in the Metaphysics of Mind*. Cambridge, MA: Harvard University Press.

Herkimer, H. 1952. *Engineer's Illustrated Thesaurus*. New York: Chemical Publishing Co.

Holmes, Oliver Wendell. 1858. The Deacon's masterpiece, or; the Marvelous One-Hoss-Shay. http://www.eldritchpress.org/owh/shay.html.

Hughes, T. P. 1987. The evolution of large technological systems. In W. Bijker, T. Hughes, and T. Pinch, eds., *The Social Construction of Technological Systems*, 51–82. Cambridge, MA: MIT Press.

———. 1989. *American Genesis: A Century of Invention and Technological Enthusiasm, 1870–1970*. Chicago: University of Chicago Press.

Hull, David L. 1988. *Science as a Process*. Chicago: University of Chicago Press.

Ingold, T. 2000. Building, dwelling, living: How animals and people make themselves at home in the world. In *The Perception of the Environment: Essays in Livelihood, Dwelling and Skill*, 172–188. London: Routledge. (Reprinted from Marilyn Strathern, ed., *Shifting Contexts: Transformations in Anthropological Knowledge*, 57–80. London: Routledge.)

———. 2001. From complementarity to obviation: On dissolving the boundaries between social and biological anthropology, archaeology, and psychology. In S. Oyama, P. Griffiths, and R. Gray, eds., *Cycles of Contingency: Developmental Systems and Evolution*, 255–279. Cambridge, MA: MIT Press.

———. 2003. [Untitled comment on "Three Roots of Human Recency" by Robert N. Proctor.] *Current Anthropology* 44(2): 232.

Jablonka, E., and M. Lamb. 1995. *Epigenetic Inheritance and Evolution*. Oxford: Oxford University Press.

Jablonka, E., and E. Szathmáry. 1995. The evolution of information storage and heredity. *Trends in Ecology & Evolution* 10(5): 206–211.

Latour, B. 1987. *Science in Action*. Cambridge, MA: Harvard University Press.

Levins, R. 1966. The strategy of model-building in population biology. *American Scientist* 54: 421–431.

———. 1968. *Evolution in Changing Environments*. Princeton, NJ: Princeton University Press.

———. 1974. The limits of complexity. In H. Pattee, ed., *Hierarchy Theory: The Challenge of Complex Systems*, 109–128. New York: George Braziller.

Loizeaux Lumber Company. 1927/1992. *Classic Houses of the Twenties*. (Originally Loizeaux's Plan Book, #7.) New York: Dover.

Lotka, Alfred J. 1925. *Elements of Physical Biology*. Baltimore: Williams & Wilkins Company.

Margulis, L. 1981. *Symbiosis in Cell Evolution: Life and Its Environment on the Early Earth*. San Francisco: W. H. Freeman.

Maynard Smith, J., and E. Szathmáry. 1995. *The Major Transitions in Evolution*. Oxford: W. H. Freeman Spektrum.

Mayr, E. 1964. Introduction. C. Darwin, *On the Origin of Species*. A facsimile of the first edition, vii–xxvii. Cambridge, MA: Harvard University Press.

Montgomery Ward & Co. (Am. co.). 2004. *Encyclopædia Britannica*. From Encyclopædia Britannica Standard Edition 2004 CD-ROM. Copyright © 1994–2003 Encyclopædia Britannica, Inc., May 30, 2003.

Moore, J. A., ed. 1972. *Readings in Heredity and Development*. New York: Oxford University Press.

Morgan, T. H. 1909. What are "factors" in Mendelian explanations? *Proceedings of the American Breeder's Association* 5: 365–368. (Reprinted in J. A. Moore, ed., *Readings in Heredity and Development*, 123–126. New York: Oxford University Press.)

Morowitz, Harold J. 1992. *Beginnings of Cellular Life: Metabolism Recapitulates Biogenesis*. New Haven: Yale University Press.

Moss, L. 2002. *What Genes Can't Do*. Cambridge, MA: MIT Press.

Mufwene, S. 2001. *The Ecology of Language Evolution*. Cambridge: Cambridge University Press.

———. 2002. Competition and selection in language evolution. *Selection* 3(1): 45–56.

———. 2005. Language evolution: The population genetics way. In Guenther Hauska, ed., *Genes, Language, and Their Evolution*. Universitaetsverlag Regensburg, vol. 29 (fall), 30–52.

Muller, H. J. 1922. Variation due to change in the individual gene. *American Naturalist* 56(642): 32–50.

Murmann, J. P. 2003. *Knowledge and Competitive Advantage: The Coevolution of Firms, Technology, and National Institutions*. Cambridge: Cambridge University Press.

Odling-Smee, F. J., K. N. Laland, and M. Feldman. 2003. *Niche Construction: The Neglected Process in Evolution*. Princeton, NJ: Princeton University Press.

Oyama, S. 2001. Terms in tension: What do you do when all the good words are taken? In S. Oyama, P. Griffiths, and R. Gray, *Cycles of Contingency: Developmental Systems and Evolution*, 177–193. Cambridge, MA: MIT Press.

Oyama, S., P. Griffiths, and R. Gray, eds. 2001. *Cycles of Contingency: Developmental Systems and Evolution*. Cambridge, MA: MIT Press.

Padgett, J. F., and C. K. Ansell. 1993. Robust action and the rise of the medici, 1400–1434, *American Journal of Sociology*, 98 (no. 6, May): 1259–1319.

Page, V. W., Lt. 1918. *Aviation Engines*. New York: N. W. Henley Publishing Co.

Raff, R. 1996. *The Shape of Life*. Chicago: University of Chicago Press.

Reber, A. 1993. *Implicit Learning and Tacit Knowledge: An Essay on the Cognitive Unconscious*. New York: Oxford University Press.

Richerson, P. J., and R. Boyd. 2004. *Not by Genes Alone*. Chicago: University of Chicago Press.

Riedl, Rupert. 1978. *Order in Living Organisms: A Systems Analysis of Evolution*. R. P. S. Jefferies transl. Chichester, New York: Wiley.

Saunders, S. A., ed. 2006. *Margins of Writing, Origins of Cultures, Oriental Institute Seminars, number 2, The Oriental Institute of the University of Chicago*. (downloadable from the Institute website at http://oi.uchicago.edu/pdf/OIS2.pdf)

Schank, J. C., and W. C. Wimsatt. 1988. Generative entrenchment and evolution. In A. Fine and P. K. Machamer, eds., *PSA–1986* 2: 33–60. East Lansing, MI: The Philosophy of Science Association.

Schrage, M. 1995. Revolutionary evolutionist. *Wired* 3(7). http://www.wired.com/wired/archive/3.07/.

Schulman, L., and J. Abelson. 1988. Recent excitement in understanding transfer RNA identity. *Science* 240 (June 17): 1591–1592.

Schultz, E. 2005. Niche construction and the constructing of niches: Missed opportunities in the study of cultural change. Conference paper, The Evolution of Cognition, Duke University, April 23–24, 2005, http://www.duke.edu/philosophy/bio/conference2005.html.

Sears Roebuck and Co. 1991. *Small Houses of the Twenties: The Sears Roebuck 1926 House Catalog*. New York: Dover.

Sears, Roebuck and Company (Am. co.). 2004. *Encyclopædia Britannica*. From Encyclopædia Britannica Standard Edition 2004 CD-ROM. Copyright © 1994–2003 Encyclopædia Britannica, Inc., May 30, 2003.

Simon, H. A. 1962. The architecture of complexity. (Reprinted in H. A. Simon, *The Sciences of the Artificial*, 3rd ed., chap. 7. Cambridge, MA: MIT Press.)

———. 1973. The Structure of Ill-structured problems. *Artificial Intelligence* 4: 181–203.

———. 1996. *The Sciences of the Artificial*. 3rd ed. Cambridge, MA: MIT Press.

Sperber, D. 2000. An objection to the memetic approach to culture. In R. Aunger, ed., *Darwinizing Culture: The Status of Memetics as a Science*, 163–173. Oxford: Oxford University Press.

Star, S. L., and J. R. Griesemer. 1989. Institutional ecology, "translations," and boundary objects: Amateurs and professionals in Berkeley's Museum of Vertebrate Zoology, 1907—1939. *Social Studies of Science* 19: 387–420.

Stark, Rodney. 1996. *The Rise of Christianity*. Princeton, NJ: Princeton University Press.

Sterelny, K., and P. Kitcher. 1988. The return of the gene. *Journal of Philosophy* (85)7: 339–361.

Sterelny, Kim. 2001. Niche construction, developmental systems, and the extended replicator. In S. Oyama, P. Griffiths, and R. Gray, eds., *Cycles of Contingency: Developmental Systems and Evolution*, 333–350. Cambridge: MIT Press.

Stevenson, K. C., and H. W. Jandl. 1986. *Houses by Mail: A Guide to Houses by Sears Roebuck and Co*. New York: Wiley.

Thornton, Rosemary. 2002. *The Houses That Sears Built: Everything You Ever Wanted to Know about Sears Catalog Homes*. Alton, IL: Gentle Beam Publications.

Turcotte, D. L. 1997. *Fractals and Chaos in Geology and Geophysics*, 2nd ed., New York: Cambridge University Press.

Van Valen, Leigh. 1973. A new evolutionary law. *Evolutionary Theory* 1(1): 1–30.

Warwick, Andrew. 2003. *Masters of Theory: Cambridge and the Rise of Mathematical Physics*. Chicago: University of Chicago Press.

Wimsatt, Barbara H. 1997. *Women in Advanced Medical Training: Sponsorship and Achievement*. PhD dissertation, Department of Education, University of Chicago.

———. 2001. Evolution in family structure and the persistence of tradition: A study of work, family and social/emotional support in 10 medical families. Working paper 02/17/2001. New York: Alfred P. Sloan Center on Parents, Children, and Work.

Wimsatt, William C. 1974. Complexity and organization. In K. F. Schaffner and R. S. Cohen, eds., *PSA-1972 (Boston Studies in the Philosophy of Science*, vol. 20), 67–86. Dordrecht: Reidel. (Reprinted in William C. Wimsatt, *Re-Engineering Philosophy for Limited Beings: Piecewise Approximations to Reality*. Cambridge, MA: Harvard University Press, 2007.)

———. 1980. Reductionistic research strategies and their biases in the units of selection controversy. In T. Nickles, ed. *Scientific Discovery, Volume II: Case Studies*, 213–259. Dordrecht: Reidel.

———. 1981a. Robustness, reliability, and overdetermination. In M. Brewer and B. Collins, eds., *Scientific Inquiry and the Social Sciences*, 124–163. San Francisco: Jossey-Bass. (Reprinted in William C. Wimsatt, *Re-Engineering Philosophy for Limited Beings: Piecewise Approximations to Reality*. Cambridge, MA: Harvard University Press, 2007.)

———. 1981b. Units of selection and the structure of the multi-level genome. In P. D. Asquith and R. N. Giere, eds., *PSA-1980*, vol. 2, 122–183. East Lansing, MI: Philosophy of Science Association.

———. 1986. Developmental constraints, generative entrenchment, and the innate-acquired distinction. In P. W. Bechtel, ed., *Integrating Scientific Disciplines*, 185–208. Dordrecht: Martinus-Nijhoff.

———. 1987. False models as means to truer theories. In M. Nitecki and A. Hoffman, eds., *Neutral Models in Biology*, 23–55. London: Oxford University Press. (Reprinted in William C. Wimsatt, *Re-Engineering Philosophy for Limited Beings: Piecewise Approximations to Reality*. Cambridge, MA: Harvard University Press, 2007.)

———. 1994. The ontology of complex systems: Levels, perspectives and causal thickets. In Robert Ware and Mohan Matthen, eds., *Canadian Journal of Philosophy*, suppl. vol. 20: 207–274. (Reprinted in William C. Wimsatt, *Re-Engineering Philosophy for Limited Beings: Piecewise Approximations to Reality.* Cambridge, MA: Harvard University Press, 2007.)

———. 1997. Aggregativity: Reductive heuristics for finding emergence. *Philosophy of Science* 64(4 SUPPS): S372–S384.

———. 1999. Genes, memes and cultural heredity. *Biology & Philosophy* 14(2): 279–302.

———. 2001. Generative entrenchment and the developmental systems approach to evolutionary processes. In S. Oyama, P. Griffiths, and R. Gray, eds., *Cycles of Contingency: Developmental Systems and Evolution*, 219–237. Cambridge, MA: MIT Press.

———. 2002. The selfish meme: Do ideas replicate themselves like genes? (Review of R. Aunger, *The Electric Meme*.), *Natural History* 111 (November): 68–72.

———. 2003. Evolution, entrenchment, and innateness. In *Reductionism and the Growth of Knowledge*, Proceedings of the 1999 Piaget Society Meetings, 53–81. Mahwah, NJ: Erlbaum.

———. 2007. *Re-Engineering Philosophy for Limited Beings: Piecewise Approximations to Reality*. Cambridge, MA: Harvard University Press.

Wimsatt, William C., and J. C. Schank. 1988. Two constraints on the evolution of complex adaptations and the means for their avoidance. In M. Nitecki, ed., *Evolutionary Progress*, 231–273. Chicago: University of Chicago Press.

Wimsatt, W. C., and J. C. Schank. 2004. Generative entrenchment, modularity and evolvability: When genic selection meets the whole organism. In G. Schlosser and G. Wagner, eds., *Modularity in Evolution and Development*, 359–394. Chicago: University of Chicago Press.

Zentella, A. C. 1997. *Growing Up Bilingual*. Malden, MA: Blackwell.

Index

Activity, 33–34, 130–132, 155, 264
 cultural, 247, 273, 300
 gene, 96–98, 102, 109, 110, 124, 135, 151, 265, 306n29, 309n41
 and germlinedness, 306n28
 group, 228
 modified by environment, 266
 variable of, 135–142
Adaptation
 constraints on, 158
 coupling with environmental factors, 133
 and entrenchment, 279, 288
 externalist view, 113
 and homology, 206–208
 lock-and-key model, 195
 and modularity, 202–205, 217
 according to population genetics, 16
 rate, 153, 174
 reinterpretation by organic systems approach, 36, 42, 46, 50
 from the reproducer perspective, 267
 testing, 214
Adaptationism, viii
Amundson, Ron, 4, 28, 58, 158, 210, 212
Antiadaptationism, 209. *See also* Gould; Lewontin
 cultural support, 66n22
 within evo-devo, 191
 explanatory, 206–208

 conflict with internalist/structuralism, 113
 rejection by organic systems approach, 25
 widespread rejection, 27, 62, 62n2
Autocatalytic network, 121–122, 269

Bauplan, vii, 30, 48
Bickhard, Mark, 229, 230
Black-boxing
 cultural scaffolding, 245, 288, 294, 296, 299, 313n62
 development, 27, 227, 244, 266, 304n14
 inheritance, 58
Boundary
 of culture, 239
 of generations, 149, 159n4
 of modules, 175
 of units of selection, 147
Boundary condition, 42, 51, 115–116. *See also* Constraints
 as constraints, 125–128, 156
Boyd, R., 252 253, 528n6, 312n59
Brylski, P., viii
Buss, Leo, 55

Canalization, vii, 30, 102, 290
 in cultural evolution, 254, 287, 288
 significance for the organic systems approach, 36, 46, 57

Capacity variable, 135–143, 151, 152, 153
Caporeal, Linnda R., 37, 244, 305n22
Cause
  proximate, 4, 9, 11, 17, 26, 41, 195, 202, 215
  ultimate, 4, 17, 26, 195
Central dogma, 4, 30, 38. *See also* Gene centricism; Gene selectionism
Change. *See also* Stasis; Generative entrenchment; Variation continuous, 54
  resistance to, 284, 285, 288, 312n59
  revolutionary, 260, 282–284, 295
Chromosome, 15, 219n13, 245, 269–274
Cis-regulatory region, 101, 132, 135, 153
Coevolution, 133, 135
  probability, 143–145, 153–157
Cognitive development, 36, 227, 231, 235, 258, 260
Comparative biology, 32, 35, 61, 204, 212, 216, 218n7, 219n14
Comparative method
  to support adaptation or genealogical hypotheses, 207–209
  Haeckel's, 7–8
  as an investigation of constraints, 115, 154–156
  as a test for a unit of evolution, 155–156
Complexity
  of culture, 257, 259, 293, 297–298
  emergence, 9, 154, 252
  and self-organization, 35
Constraints, 36, 45, 108, 191, 204, 235, 253, 260, 294, 298. *See also* Generative entrenchment
  developmental, viii, x, 3, 31, 52, 53, 108, 113–160, 191
  external, 124–126
  functional, 113–160
  generative, 117–125, 135
  internal, 113, 125, 127
  probability, 143–145
  stability, 117–123
Context-dependence, 94, 230, 245
Context insensitivity, 152, 145, 160. *See also* Robustness
Contingency, 57, 67n29, 68n30, 119, 158, 288
Coupling, 115, 128–134, 137, 140–144, 149, 151, 158n9, 242
Cultural transmission, 229, 236–238, 245, 247, 253–254, 258, 261–266, 270, 274–275
  ideational, 234, 246, 261–263, 293, 296
  symbolic, 236, 293, 300
Cultural virus. *See* Meme
Culture
  animal, 237, 266, 276
  cultural modularity, 229, 242, 271, 286, 287, 288, 290, 311n52
  cumulative nature of, 230–234, 250, 260, 263, 282, 298
  of Darwinism, 29, 66n22
  of evo-devo, 18
  sequestration and bundling, 261, 262, 264–266, 269, 270
  thick versus thin and medium viscosity, description of, 227, 236–238, 242, 250, 276, 287, 302n6, 302n8, 303n9

Darwin, Charles, 5–8, 10, 13–14, 32, 58, 206, 211–212
Dawkins, Richard, vii, xi, 29–30, 34, 41, 43, 173, 206, 207, 212
  on cultural evolution, 252, 262–269, 275, 279
Deep change, 290, 295. *See also* Hopeful monsters

Development
  of competencies, 229–235, 261, 288, 299
  conceptual separation from evolution, vii, 10, 11, 21, 26
  disciplinary, 232–235, 283, 285, 287
  versus evolution (ambiguity in culture), 252, 277
  institutional, 230–232, 246, 264, 277, 281, 294
  professional, 230–235, 282
Developmental systems theory (perspective), ix, 28, 37, 237, 260, 280, 289
Difference-makers, 263–266
Dissociated coevolution, 154–157
Double dissociation test, 197–202
Downstream dependencies. *See* Generative entrenchment
Drift, 52, 55, 59, 126–128, 156, 157, 290

Eco-devo, 44, 189–190
Embedding
  biological, 67n28, 94, 99, 152
  cultural, 246, 250, 261, 274, 285, 288, 297, 300
Emergence, 9, 25, 28, 40–44, 52–56, 196
  Cultural, 235, 237, 245, 250, 273, 279, 289, 297–298
Environment. *See also* Niche
  as a product of cultural evolution, 228, 238, 247
  determining phenotype, 12, 26–30, 45, 46, 95, 110, 265
  and the dialectical account of biology, 36
  within dissociated coevolution, 155
  dynamics and evolvability, 184, 185, 190 (*see also* Adaptation, lock and key model)
  entity-environment relations, 244, 245, 279
  evolving, 180
  as external constituents, 122, 124–125
  and gene selectionism, 266
  imposing functional constraints, 124–125, 133–134
  and novelties, 51
  partial predictability of, 259–260
  persistence of, 266–267, 269
  problems determined by organisms, 203–205 (*See also* Scaffolding)
  robustness to, 93, 269, 270
  and selection, 43, 48, 113, 114
  stability, 259–260
  uncertainty of, 260
Epigenesis, 9, 39, 41
Epigenetics, 12, 33, 40
Epistasis, 36, 138–147, 151, 156
Evolution
  macro, 2, 3, 43
  micro, 2, 3, 101
Evolvability, viii, 31, 32, 36, 46, 153, 173–175, 181–190
Explanation, 36, 38, 39, 64n15, 66n24. *See also* Adaptationism
  causal-mechanistic, 25, 28, 29, 41, 42, 52, 211
  complementary, 26, 127, 157, 159n8, 183
  contingency in evolutionary, 26, 67n29
  developmental evolutionary, 58
  equilibrium, 42
  evolutionary, 62n2
  functional (adaptive, selection, Darwinian), 29, 114, 156, 207, 208, 211, 212
  genetic, 27, 64n15, 195, 305n19
  mechanical of development, 10
  probabilistic, 126, 157
  proximate/ultimate (cognitive), 26, 202, 215
  structural, 114

Fitness, 120, 123, 124, 248. *See also* Adaptation; Evolvability
  of cultural elements, 252, 257, 300, 312n60
  epistasis, 138–147, 151, 152, 157, 159n9
  in explanation, 26
  of genes, vii, 16, 40, 252, 257
  heritable variation in, 43, 113, 125–128, 134
  landscape, 61, 66n22, 102, 104, 106–107, 146
Fontana, W., 36, 55
Frame problem (of artificial intelligence), 297
Function, 96, 123, 210n14, 280. *See also* Coupling
  categorization of organisms and traits, 206–208
  and causal depth, 213–217 (*see also* Adaptation; Adaptationism; Scaffolding)
  context, 94
  delineation, of 189–190
  distributed, 197
  emergence, 43
  epistemic demand, 206–213
  equivalents, 296
  functional biology, 26, 62n3, 65n15
  functional genomics, 145
  functionalism, 113–114, 125
  functional morphology, 131
  functional novelty, 50, 51
  functional organization, 124, 157, 175, 176
  functional redundancy, 123, 144
  maintaining functionality, 124–125, 144
  mapping function, 93
  multiplexing, 287
  polyfunctional, 294
  of scaffolding, 229
  specialization of neural subsystems, 197–202

Gene
  centricism (central dogma), 25, 30, 42, 65
  circuits, 51, 56
  network, 56, 154, 155
  regulation, 35, 51, 57, 94
  selectionism, vii, xi, 34, 35, 40–45, 189, 195–196, 263–265
  selfish, vii, 40, 43, 263 (*see also* Dawkins)
Generative entrenchment (GE), 37, 60, 143, 177–181, 252, 254, 270, 272
  as a cause of stasis, 284–287
  in culture, 227–230, 238, 244–246, 274, 277–284
  cumulative entrenchment, 232
  developmental lock model, 68n34
  of scaffolding, 278, 293–301
  in sciences and study of culture, 287–290
Genetic determinism, 38, 44, 62, 304n29
Genetics
  developmental, viii, 3, 11, 17, 18, 32–34, 94
  molecular, 12, 16, 32, 35
  population, 12, 16, 17, 27, 126, 237, 254, 257
  quantitative, 2, 16, 99, 143
Genotype, 16, 107
Genotype phenotype relationship
  and the builder perspective of culture, 247, 249
  according to developmental genetics, 34
  evolution of, 47–48, 56
  interactionist consensus, 28
  modularity of, 153
  according to organic systems approach, 45–46, 52
  according to population genetics, 2, 15, 26, 27, 29, 30, 62n3
  structural modeling of, 36, 54, 93–111
  and thin models of culture, 237, 300

Index

Germ line, vii, 1, 10, 123, 135, 149–151, 262, 264
Gilbert, Scott, 19, 27, 28, 35, 68n33
Goldschmidt, Richard, 11, 12, 16, 184
Gould, Stephen J., vii, 17, 41, 43–44, 68, 184, 191
Grain problem, 203–205
Griessemer, James, ix, 36, 38, 41, 43, 58, 64, 122
Group
  development, 37, 245, 253, 272, 282
  inheritance, 215
  reproduction, 245, 251
  selection, 43, 189, 212, 296

Haeckel, Ernst, 2, 5, 7, 9–11, 18, 32
Hall, Brian, viii, 3, 26, 65n18, 195, 206
Hierarchical levels, 36
  of constraints, 141
  of coupling, 128, 138
  of evolution, 43, 148
  of gene networks, 55
  of mechanism, 34, 65n19
  in models of culture, 261, 278, 304n17, 306n27
  of modules, 148, 197
  of natural selection, 38, 43, 189
  of organization, 36, 244
  of phenotype (development, organism), 44, 121
Homology
  in evolutionary psychology, 196, 206–217
  used to define novelty, 50
  organizational, 53
  Von Baer's explanation of, 210–211
  Wagner's definition and examples of, 51
Homoplasy (homoplasies), 54, 155–156
Hopeful monsters, 184–186. *See also* Punctualism
Hull, David, 43, 58, 304n18

Induction (in embryology), 133, 197, 217
Information
  accounts of ontogeny, 26–30, 62n3, 64n14
  biological (for identifying homology), 211
  channels, 229, 258, 286, 291, 294, 297–301
  cultural, 258, 286, 291, 307n36, 311n53
  disembodied transmission, 229
  encapsulation in brain modules, 198, 200
  environmental, 68n34
  inherited, 8–15, 122, 270
Infrastructure, 271, 276, 278–281, 294
  material, 246, 271, 272, 278
  social, 281, 282
Ingold, T., 236, 237, 266
Inheritance, 1–5, 11–14, 48, 56, 58, 64, 122, 214, 302n6. *See* Hopeful monsters; Inherency; Lamarckianism; Reproducers; Weismannism
  conceptual separation from development, 10, 17
  cultural, 227–229, 235, 252–261, 265, 279
  of cultural artifacts, 271, 274
  cultural channels of, 253, 287, 297
  definition of (as type of reproduction), 267, 268
  dual inheritance model according to the reproducer perspective, 36, 252, 64n10
  dual inheritance models, 227, 236, 227, 248, 249, 252
  epigenetic, 306n26
  role of scaffolding in cultural inheritance, 244, 245
  selection for accuracy of, 186–189
Inherency, 31, 57, 61, 67n28

Innovation, 25, 31–32, 44, 49–52, 54–57, 239
  in cultural infrastructure, 239, 243, 280
  in culture, 238, 253, 260, 272, 290, 295, 313n62
Input-output-transformation (IOT), 111, 135, 149–152
Integration, 31, 46, 53, 56, 153, 175, 181–186

Kauffman, Stuart, 35, 58, 59
Keller, Evelyn Fox, 36, 61
Kluge, 285, 289, 303n9

Lamarckianism, 12, 17, 26, 36
  of cultural infrastructure, 237, 292, 298
Language
  Evolution, 280, 281
  oral versus written, 280
  origins of, 237, 253
Legacy, 187–191, 192n6, 192n10
Levels
  of inheritance, 245
  multilevel organization, 271, 276, 290–296, 303n9, 308n37 (*see also* Hierarchy)
  organization, 27, 36, 43–44, 51, 55, 56, 196, 245, 246, 251, 270
  selection, 40, 43–44
Lethality 134, 229
Lewontin, Richard, 191
  criticism of genetic determinism, 27–28
  definition of evolution, 43
  on metaphor in science, 37
  on niche construction, 192n11, 218n6
  on limits of population genetics, 63n6
  on quasi-independence, 175, 178, 180, 191n4
Life cycle, 122–128, 132–134, 141, 148–154, 157, 158n3, 236, 270, 283
  cultural, 252–261, 277
  and cultural evolution, 227–278

Lineage. *See also* Legacy
  of cell, 11, 44
  complexity of cultural, 262
  evolvability, 31
  types of, 187–189
  units within, 148–149, 154–158

Maintenance
  of culture, 228, 232, 272, 281–282, 288, 294
  of function, 124–125, 144
  of organism, 121, 133, 267
  of phenotype, 27, 48
  self-maintainance, 54, 121–123, 128, 159n4, 308n41
Material
  overlap in cultural inheritance, 36, 37, 228, 251, 263, 266, 267, 272, 278, 288, 297–300, 309n43
  propagules, 123, 245, 261–262, 266–267, 270–275, 278–280, 293, 298–299
  transfer, 229, 247, 250, 270, 272
Maynard Smith, John, 40
Mayr, Ernst, 7, 26, 62n3
Meme
  as account of cultural evolution, 227, 228, 248, 262–264, 291
  failures of, 252, 252, 257, 262–264, 270–274, 291, 297, 305n23
Mendelianism, 16, 47, 48, 58, 258, 270
Model organism, 12, 16, 18, 154, 155, 281
Modern synthesis, vii–viii, 2, 12, 16, 25–27, 43, 60, 214
Module, modularity. *See also* Quasi-independence
  and adaptivity, 175–184, 202–205, 216–217
  developmental, x, 31, 51, 148–158, 175–184, 216–217
  environmental, 192n11

within evo-devo, viii, 20, 31, 36, 54,
  175–184
evolutionary (modules of evolutionary
  transformation), 115, 148–158
functional, 175–184, 190, 191, 196–
  199, 216–217
of genotype-phenotype map, 152
mental, 153, 199–205, 216–217
virtual functional, 196, 199–201,
  216
Molecular biology, 16–19, 28–30, 36,
  60, 214
Morgan, Thomas Hunt, 12, 14–17, 270,
  307n32
Morphospace, 61, 116, 117
Mosaic evolution, 148, 153
Mutation. *See also* Constraints; Hopeful
  monsters; Punctualism; Variation
difference between biological and
  cultural, 290, 292, 295
genetic, 135–141, 149–153, 159n4
to genetic code, 314n69
gradual, 173–186, 198, 201
according to neo-Darwinism, 55
according to organic systems approach,
  50, 52
quantitative mathematical modeling,
  95–101, 107–110
rate, 113, 127, 290
studies, 16, 33, 34

Natural selection, 101–115, 127–128,
  212, 215, 219n14, 255. *See also*
  Adaptationism; Constraints
on brain structure, 199, 202–206
cultural selection, 258, 292–294
in Darwin's definition of evolution, 6
equations, 126
for heritability, 48
internal selection, 125, 127, 159n7
against lethal variants, 134
and novelty, 51–56
and self-organization, 59

stabilizing selection, 178, 180
tweaking functional joints, 181, 184
Neo-Darwinism. *See also* Gene
  centricism; Gene selectionism
cultural support, 66n22
commitment to externalism/
  selectionism, 57, 62n5, 63n9, 206
commitment to genetic program, 28,
  46, 62n5, 63n9
omissions by, 54, 55, 64n13, 114, 195,
  217n1
Niche, 49, 190. *See also* Niche
  construction
Niche construction, 34, 192n11, 209,
  218n6, 203–204, 209, 302n6, 304n14,
  318n15
by cultural evolution, 229, 244, 260,
  264, 278–280, 285
Nonlinearity, 105
gene interactions, 93–100
interactions within culture, 258
contrast with linearity, 42, 44, 100
phenotypic surfaces, 102–107
and punctuated equilibrium, 54

Ontogeny and phylogeny
conceptual separation of, 2
ontogeny creates phylogeny, 25–26
ontogeny recapitulates phylogeny, 7–8
Organic Systems Approach (OSA), 25,
  44–45
use of equilibrium explanations, 42
explanation of generation, 45–49
explanation of innovation (novelty),
  49–52
explanation of organization, 52–54
compatibility with philosophical
  naturalism, 41
Organization, 25, 27, 34, 36, 45, 52–54,
  56, 57, 64n13
determining constraints, 117–118,
  121–128, 133
homology of, 53

Organization (cont.)
  levels and orders of, 27, 36, 43–44, 51–56, 196, 245, 246, 251, 270
  multilevel organization, 44, 271, 276, 290–296, 303n9, 308n37 (*see also* Hierarchical levels)
  origination of, 25, 55

Packaging
  bracketing, 229, 261
  bundling, 261, 269
  and coordination, 231, 242, 274, 293, 298, 308n37
  and modularity, 287 (*see also* Module, modularity)
  and organization, 288, 244, 245, 261–270, 281
  sequestration, 261, 269, 287,
Persistence
  of agents, 229
  of artifacts, 229, 250, 271, 272, 273, 278, 279, 285
  of infrastructure, 229, 243, 271, 272, 278, 285, 286
  of stable states, 121
Phenotypic evolution, 17, 25–28, 33, 36, 44, 49–57, 205
Phylogeny, 33, 55, 154, 155, 215, 281. *See also* Ontogeny and phylogeny
Plasticity, 36, 45, 46, 65n16, 110, 204, 281
Pleiotropy, 31, 50, 152, 153. *See also* Constraints; Nonlinearity, phenotypic surfaces
Pluralism
  of multilevel selection, 40, 43–44
  of theories, 60–62, 66n24
Population genetics, 12, 16, 17, 27, 63n6, 126
  and thin models of culture, 237, 254, 257
Population size, 119, 126, 127, 128, 147, 254, 305n19, 113

Process-oriented view, 116, 128, 149
Process structuralism, 35, 39
Progeneration, 122, 267, 268
Propensity, 31, 121, 124, 139
Punctualism (Punctuated equilibrium), 54–55. *See also* Hopeful monsters

Quasi-independence. *See also* Semidecomposability; Module, modularity
  of cultural channels, 298
  mutations, 175–186
  units of evolution, 147, 148

Raff, Elizabeth, 7
Raff, Rudy, 7, 19
Recombination, 146, 147
Recurrent artifacts, 271, 273
Redeployment, 51, 155
Reductionism, 65n19, 263, 303n9
Redundancy, 46, 123, 191n2, 287, 290, 295, 296, 311n52
  functional, 123, 144
Reference groups, 251, 264
Regulatory region. *See* Cis-regulatory region
Replication, 41, 121–122, 228, 263–269, 275
Replicators, 43, 262–272, 279, 306n28
Reproducers
  biological, ix, 36, 41, 43, 267
  cultural, 228, 244, 249–252, 261, 271–279, 283, 287, 304n17
  versus replicators, 266–271
Reproduction, 7, 8, 55, 173–174. *See also* Reproducers
  recurrent reproduction of a state, 122–127, 130–133, 139–151, 157, 158n2, 158n3, 236
Resources
  cultural, 244, 246, 258, 246, 258, 282
  developmental, 28, 196, 198, 228, 238, 293

Index

Richardson, Michael K., 18, 59
Richerson, P. J., 252, 253, 528n6, 312n59
Robustness, 32, 93–98, 153, 284, 297, 311n52

Scaffolding
  accumulating adaptation, 244, 292, 293
  artifactual, 276–278
  and chunking or black-boxing, 45, 288, 294, 296, 299, 313n62
  and making deep changes, 290, 295
  definition, 229
  developmental agent, 276, 281–282
  functions, 229, 230
  as generative, 233–236, 250, 251, 272, 284
  infrastructural, 276, 278–281
  internalized, 245, 269, 270
  and niche construction, 244, 260, 266
  parent-offspring, 248, 261, 265–274
  as reproducer, 244, 246, 249, 250, 261–279, 283
  self, 230, 244, 269
  as shaping, 228, 231, 238, 266, 274, 299
  increasing speed of evolutionary change, 290–292, 297
  as supporting differential specialization, 234, 235
Schank, J. C., 177, 178, 180, 181
Selective constraints. *See* Functional constraint
Self-maintenance, 54, 121–123, 128, 159n4, 308n41
Semidecomposability, 198
Sexual reproduction, 123, 146, 147, 187, 188, 267
Signaling cascades, 97–98, 129, 135, 143, 197
Simon, H. A., 288, 311n55
Sober, Elliot, 42, 62n5

Soma
  cultural, 247, 304n17, 306n28
  distinction from germ line, vii, 1, 135, 149–151, 158
Stasis, 154–156, 253
  generative entrenchment as a cause, 284–287
  recognizing, 282–284
Sterelny, Kim, 279

Transcription factor, 129, 132, 135, 139, 140, 153. *See also* Gene circuits; Gene network; Gene regulation

Unity of type, 206–209

Variability, 29, 48, 114, 124, 260, 261. *See also* Variation
Variance/covariance matrix, 109, 143. *See also* Nonlinear phenotypic surfaces
Variation, 16, 27–29, 36, 45–46, 55–56, 65n16, 110–115. *See also* Constraints; Coupling
  between genotypes and phenotypes, 93–110
  heritable, 13, 124–125, 134–135, 139–144, 149–157
  according to the organic systems approach, 49–52
  in population genetics, 2
Von Baer, Karl Ernst
  explanation of homology, 210–211
  laws explained by generative entrenchment, 68n34, 177–178
  laws as influences on Darwin, 6

Waddington, C. H.
  anticipation of genetic integration, 56
  evo-devo renegade, 17, 27
  "Waddingtonian microevolution," 101
Wagner, Günter
  the causal link between molecular and phenotypic evolution, 32–33

Wagner, Günter (cont.)
  development, 65n17
  evolution of modularity, 186, 203
  evolvability, 173
  the nature of evo-devo, 19, 31
  definition and examples of novelty and homology, 51, 211, 219n12
Weismann, vii, 1–5. *See also* Weismannism
  and the disintegration of generation, 9–10
Weissmannism, 26, 38 62n3
  hierarchical multilevel Weismannism, 261, 304n17, 306n27
  models of culture, 237, 247–248
Williams, George C., 40, 41, 57, 62n2
Wilson, E. B., 12–15, 40, 66n24
Wimsatt, William, 42, 60, 68n34, 158n11, 177, 178, 180, 181
Wray, Greg, 18

QH
366.2
.I52
2007